elementary algebra

2e

elementary algebra 2e

Charles P. McKeague

Cuesta College

Academic Press

New York/London/Toronto/Sydney/San Francisco
A Subsidiary of Harcourt Brace Jovanovich, Publishers

OTHER BOOKS BY THE AUTHOR:

BASIC MATHEMATICS (Wadsworth Publishing Company)
584 pages, paperbound, 1981

BEGINNING ALGEBRA (Academic Press)
467 pages, paperbound, 1980

INTERMEDIATE ALGEBRA (Academic Press)
494 pages, hardbound, 1979

INTERMEDIATE ALGEBRA: A Text/Workbook (Academic Press)
572 pages, paperbound, 1981

INTRODUCTORY MATHEMATICS (Wadsworth Publishing Company)
370 pages, hardbound, 1981

Cover photo by Dan Lenore

Academic Press, Inc.
111 Fifth Avenue, New York, New York 10003

United Kingdom Edition published by
Academic Press, Inc. (London) Ltd.
24/28 Oval Road, London NW1 7DX

ISBN: 0-12-484755-2
Library of Congress Catalog Card Number: 80-68938

Printed in the United States of America

Contents

Preface to the Student

I often find my students asking themselves the question "Why can't I understand this stuff the first time?" The answer is, "You're not expected to." Learning a topic in algebra isn't always accomplished the first time around. There are many instances when you will find yourself reading over new material a number of times before you can begin to work problems. That's just the way things are in algebra. If you don't understand a topic the first time you see it, it doesn't mean there is something wrong with you. Understanding algebra takes time. The process of understanding requires reading the book, working problems, and getting your questions answered.

Here are some questions that are often asked by students starting a beginning algebra class.

How much math do I need to know before taking algebra?

You should be able to do the four basic operations (add, subtract, multiply, and divide) with whole numbers, fractions, and decimals. Most important is your ability to work with whole numbers. If you are a bit weak at working with fractions because you haven't worked with them in a while, don't be too concerned; it will come back to you. Also, we will review fractions as we progress through the book. I have had students who eventually did very well in algebra, even though they were initially unsure of themselves when working with fractions.

What is the best way to study?

The best way to study is to study consistently. You must work problems every day. A number of my students spend an hour or so working problems and reading over the new material in the morning, then another hour working problems in the evening. It is often difficult to get into a schedule that is easy to follow if you have been away from school for some time. After a while though, it becomes less difficult, and sometimes almost automatic.

If I understand everything that goes on in class, can I take it easy on my homework?

Not necessarily. There is a big difference between understanding a problem someone else is working and working the same problem yourself. There is no substitute for working problems yourself. The concepts and properties are understandable to you only if you yourself work problems involving them.

If you have decided to be successful in algebra, then the following list will be important to you. The list contains those things that will assist you in being successful.

How to be Successful in Algebra

1. **Attend all class sessions on time.** There is no way to know exactly what goes on in class unless you are there. Missing class and then expecting to find out what went on from someone else is not the same as being there yourself. All you get from someone else is their perspective of what went on. It is not the same experience you would have had if you had been there.

2. **Read the book.** It is best to read the section that will be covered in lecture before you go to class. Reading in advance, even if you do not understand everything you read, is still better than walking into class with no idea of what will be discussed.

3. **Work problems every day.** The key to success in mathematics is working problems. The more problems you work, the better you will become at working problems. The answers to the odd-num-

bered problems are in the back of the book. Be sure to check your answers with the answers in the back of the book. If you have made a mistake, find out what it was.

4. **Do it on your own.** Don't be misled into thinking that someone else's work is your own. Having someone else show you how to work a problem is not the same as working the same problem yourself.

5. **Don't expect to understand every new topic the first time you see it.** Sometimes you will understand everything at once and sometimes you won't. That's just the way things are in mathematics. Expecting to understand each new topic the first time you see it will only lead to disappointment and frustration. Remember, the process of understanding algebra takes time. It requires that you read the book, work problems, and not be afraid to ask questions when you reach a dead end.

6. **Spend as much time as it takes for you to master the material.** There is no set formula for the exact amount of time you need to spend on algebra to master it. You will find out as you go along what is or isn't enough time for you. Even if you end up having to spend 2 or more hours on each section to get comfortable with the material, then that's how much time it takes. Trying to get by with less will not work.

Preface to the Instructor

This second edition of Elementary Algebra retains the same basic format and style of the first edition. It is designed to be used in a lecture situation for students with no prior background in algebra.

The first eight chapters cover material usually taught in an introductory algebra class. Chapter 9 contains additional topics that can be covered in any order. They are included in the book to make it more flexible in meeting the needs of different programs. Each chapter is divided into sections, each of which can be discussed in a 45–50 minute class session.

As in the first edition, the examples in each section preview the problems in the problem sets. The relative level of difficulty increases as the problem set progresses. As an aid in assigning homework, each even-numbered problem is similar to the odd-numbered problem that precedes it. Answers to the odd-numbered problems are provided in the back of the book. This method of organizing the problem sets gives students a chance to check their work on an odd-numbered problem, and then try a similar problem where no answer is provided.

Each chapter ends with a summary that lists the main topics covered in the chapter. Most chapter summaries also contain a section on common mistakes. Also included at the end of each chapter is a sample chapter test to allow the student to determine his or her mastery of the material in the chapter. Full answers to the sample chapter tests are provided.

The following changes have been made in the second edition:

Sequence of Topics The section on solving compound inequalities has been moved from Chapter 9 to the end of Chapter 2. Likewise, the section on graphing linear inequalities in two variables has been taken from Chapter 9 and placed at the end of Chapter 3. The section that covers long division with polynomials has been moved to Section 6.2. At first glance this may seem to be an unusual place to put this material. I think, however, that after reading Sections 6.1 and 6.2 you will find this sequence of topics works very well.

New Topics Three new sections are introduced in this edition: Section 6.7 covers complex fractions, Section 9.5 covers scientific notation, and Section 9.6 covers ratio and proportion. Also, a few problems using the Pythagorean Theorem have been added to Section 5.7 and the problem set that follows.

Earlier Review of Fractions Fractions are covered as they are needed beginning in Chapter 1. Section 1.2 includes multiplication of fractions, as does Section 1.6. Division of fractions is covered in Problem Set 1.7 as a logical consequence of the definition of division for real numbers. Reduction to lowest terms is covered in Section 1.7. Addition and subtraction of fractions with the same denominator is found in Section 4.3. In Chapter 6 all the operations with fractions covered previously are briefly reviewed. Addition of fractions with different denominators is covered in Section 6.4, just before addition and subtraction of rational expressions. Thus, all operations with fractions are reviewed before the corresponding algebraic questions are introduced.

Review Problems Starting with Chapter 2, each problem set ends with a number (6–10) of review problems. Generally, these review problems cover material that will be used in the next section. I find that assigning the review problems helps prepare my students for the next day's lecture, and makes reviewing part of their daily routine. For those classes that will cover all of Chapter 9, each section in Chapter 9 ends with a set of review problems from one of the first eight chapters of the book. Thus, while covering the last chapter, your class will also be reviewing for the final exam.

Chapter Summary Examples The format of the Chapter Summary and Review sections at the end of each chapter has been changed. In most cases there is an example in the margin next to the topic being reviewed. These examples are intended to refresh the students' memory as to the kind of problem that accompanies the topic being reviewed.

Problem Sets Most of the problem sets have been lengthened. Many have new problems toward the end that are more challenging. Problem sets from the first edition that produced somewhat awkward answers have been rewritten.

Answers to Odd-Numbered Problems The answers to the odd-numbered problems in the back of the book have been expanded. For example, many of the answers for word problems now include the equation necessary to solve the problem. These equations are shown more often at the beginning of the book. Also, all the possible solutions for equations that contain radicals or rational expressions are given along with the actual solutions.

I have tried to make this book flexible and easy to use. It has been written to assist both you and your students in the classroom.

Acknowledgments

There are many people to thank for their assistance with this revision. Steve Guty, my editor at Academic Press, was very helpful in guiding and encouraging me throughout the process. Mike O'Connell spent many hours working problems and proofreading the manuscript. The staff at Academic Press did a fine job of implementing the changes I had in mind. The typing was done by two very capable people; Carolyn Williams and Susan Gutierrez. My wife Diane has contributed a number of suggestions on different aspects of this revision, and my children, Patrick and Amy, have been very good collators.

Finally, I want to thank the people who reviewed this second edition for their suggestions and contributions. I was very lucky to have such a fine group of instructors contribute their thoughts.

The Basics

Chapter 1 contains some of the most important material in the book. It also contains some of the easiest material to understand. Be sure that you master it. Your success in the following chapters depends upon how well you understand Chapter 1. Here is a list, in order of importance, of the ideas you must know after having completed Chapter 1.

1.   You *must* know how to add, subtract, multiply, and divide positive and negative numbers. There is no substitute for consistently getting the correct answers.

2.   You *must* understand and recognize when the commutative, associative, and distributive properties are being used. These properties are used continuously throughout the book to create other properties and definitions. They are the fundamental properties upon which our algebraic system is built.

3.   You should know the major classifications of numbers. That is, you should know the difference between whole numbers, integers, rational numbers, and real numbers.

If the material in Chapter 1 seems familiar to you, you may have a tendency to skip over some of it lightly. Don't do it. Look at the above list as you proceed through the chapter and make sure you understand these topics as they come up in the chapter. You will be off to a good start and increase your chance for success with the rest of the material in the book.

1.1
Notation and
Symbols

Since much of what we do in algebra involves comparison of quantities, we will begin by listing some symbols used to compare mathematical quantities. The comparison symbols fall into two major groups, equality symbols and inequality symbols.

We will let the letters a and b stand for (represent) any two mathematical quantities.

Comparison Symbols

Equality:	$a = b$	a is equal to b (a and b represent the same number)
	$a \neq b$	a is not equal to b
Inequality:	$a < b$	a is less than b
	$a \not< b$	a is not less than b
	$a > b$	a is greater than b
	$a \not> b$	a is not greater than b
	$a \geq b$	a is greater than or equal to b
	$a \leq b$	a is less than or equal to b

The symbols for inequality $<$ and $>$ always point to the smaller of the two quantities being compared. $3 < x$ means 3 is smaller than x. $5 > y$ means y is smaller than 5.

Along with symbols for comparison, we have symbols for operations, which we've all seen before. We should be familiar with most of them.

Operation Symbols

Addition:	$a + b$	The *sum* of a and b
Subtraction:	$a - b$	The *difference* of a and b
Multiplication:	$a \cdot b, (a)(b), a(b), (a)b, ab$	The *product* of a and b
Division:	$a \div b, a/b, \dfrac{a}{b}, b\overline{)a}$	The *quotient* of a and b

When we encounter the word *sum*, the implied operation is addition. To find the sum of two numbers, we simply add them. *Difference* implies subtraction, *product* implies multiplication, and *quotient* implies division. Notice also that there is more than one way to write the product or quotient of two numbers.

Grouping Symbols

Parentheses (   ) and brackets [   ] are the symbols used for grouping numbers together. (Occasionally braces {   } are also used for grouping, although they are usually reserved for set notation, as we shall see.)

The following examples illustrate the relationship of the symbols for comparing, operating, and grouping to the English language.

▼   **Examples**

Mathematical expression	*English equivalent*
1. $4 + 1 = 5$	The sum of 4 and 1 is 5.
2. $8 - 1 < 10$	The difference of 8 and 1 is less than 10.
3. $2(3 + 4) = 14$	Twice the sum of 3 and 4 is 14.
4. $3(7 - 5) \neq 10$	Three times the difference of 7 and 5 is *not* equal to 10. ▲

The symbols for comparing, operating, and grouping are to mathematics what punctuation symbols are to English. These symbols are the punctuation symbols for mathematics.

Consider the following sentence:

Paul said John is tall.

It can have two different meanings, depending on how it is punctuated.

1.   "Paul," said John, "is tall."
2.   Paul said, "John is tall."

Without the punctuation we do not know which meaning is intended. It is ambiguous without punctuation.

Let's take a look at a similar situation in mathematics. Consider the following mathematical statement:

$$2 \cdot 7 + 5$$

If we add the 7 and 5 first and then multiply by 2, we get an answer of 24. On the other hand, if we multiply the 2 and the 7 first and then add 5, we are left with 19. We have a problem that seems to have two different answers, depending on whether we add first or multiply first.

We would like to avoid this type of situation. That is, every problem like $2 \cdot 7 + 5$ should have only one answer. Therefore, we have developed the following rule for the order of operation.

Order of Operation

When evaluating a mathematical expression we will perform the operations in the following order:

1. Do what is in the parentheses first, if you can. (In some cases it is not possible to do what is in the parentheses, as is the case when one of the quantities is a number and the other is a variable.)
2. Then perform all multiplications and divisions left to right (in the same direction you read).
3. Perform all additions and subtractions left to right.

▼ **Examples**

5. $2 \cdot 7 + 5 = 14 + 5$ 　　　Multiply $2 \cdot 7$ first (multiplica-
   　　　　　$= 19$ 　　　　　　　tion before addition or sub-
   　　　　　　　　　　　　　　　traction)

6. $5 + 8 \cdot 2 = 5 + 16$ 　　　Multiply $8 \cdot 2$ first (multiplica-
   　　　　　$= 21$ 　　　　　　　tion before addition or
   　　　　　　　　　　　　　　　subtraction)

7. $2 \cdot 7 + 3(6 + 4) = 2 \cdot 7 + 3 \cdot 10$ 　Do what is in the parentheses
   　　　　　　　　　　　　　　　　　　first
   　　　　　　　　　$= 14 + 30$ 　　Multiply left to right
   　　　　　　　　　$= 44$ 　　　　Add

8. $10 + 12 \div 4 + 2 \cdot 3 = 10 + 3 + 6$ 　Multiply and divide left to
   　　　　　　　　　　　　　　　　　right
   　　　　　　　　　　$= 19$ 　　　Add left to right　　　▲

Notice in Example 8 that we divided 12 by 4 and multiplied 2 times 3 before we did any addition. The rule for order of operation indicates that we always multiply and divide before we add when simplifying expressions like the one in Example 8.

Problem Set 1.1

Write an equivalent statement in English. Include the words sum, difference, product, and quotient when possible.

1. $7 + 8 = 15$ 　　　　　　　　2. $6 + 3 = 9$

  3. $7 < 10$

  4. $12 < 15$

  5. $8 + 2 \neq 6$

  6. $10 - 5 \neq 15$

  7. $21 > 20$

  8. $32 > 22$

  9. $x + 1 = 5$

  10. $y + 2 = 10$

  11. $x < y$

  12. $r < s$

  13. $x + 2 = y + 3$

  14. $x - 5 = y + 7$

  15. $x - 1 < 2x$

  16. $x + 3 > 3x$

  17. $2x + 1 = 10$

  18. $3x - 2 = 5$

  19. $4(x + 1) = 6$

  20. $5(x - 3) = 2$

Mark the following statements true or false.

  21. $16 < 17$

  22. $18 < 15$

  23. $10 = 19$

  24. $11 \neq 21$

  25. $3 + 2 < 5$

  26. $5 + 1 > 6$

  27. $11 \not< 10$

  28. $9 \not< 8$

  29. $3 \cdot 6 < 2 \cdot 4 + 1$

  30. $4 \cdot 1 > 3 \cdot 2 - 4$

Use the rule for order of operation to simplify each problem as much as possible.

  31. $2 \cdot 3 + 5$

  32. $8 \cdot 7 + 1$

  33. $5 + 2 \cdot 6$

  34. $8 + 9 \cdot 4$

  35. $3 \cdot 7 + 5 \cdot 2$

  36. $8 \cdot 6 + 4 \cdot 3$

  37. $7 \cdot 4 - 8 \cdot 1$

  38. $6 \cdot 3 - 5 \cdot 2$

  39. $19 - 2 \cdot 3 + 4$

  40. $8 - 1(5) - 2$

  41. $9 + (3 \cdot 2 + 5)$

  42. $14 - (3 \cdot 5 - 2)$

  43. $20 + 2(8 - 5) + 1$

  44. $10 + 3(7 + 1) + 2$

  45. $5 + 2(3 \cdot 4 - 1) + 8$

  46. $11 - 2(5 \cdot 3 - 10) + 2$

  47. $8 + 10 \div 2$

  48. $16 - 8 \div 4$

  49. $3 + 12 \div 3 + 1 \cdot 6$

  50. $18 + 6 \div 2 + 3 \cdot 4$

  51. On Monday Bob buys 10 shares of a certain stock. On Tuesday he buys 4 more shares of the same stock. If the stock splits Wednesday, then he has twice the number of shares he had on Tuesday. Write an expression using parentheses and the numbers 2, 4, and 10, to describe this situation.

  52. Patrick has a collection of 25 baseball cards. He then buys 3 packs of gum, and each pack contains 5 baseball cards. Write an expression using the numbers 25, 3, and 5 to describe this situation.

  53. A gambler begins an evening in Las Vegas with $50. After an hour she has tripled her money. The next hour she loses $14. Write an expression using the numbers 3, 50, and 14 to describe this situation.

  54. A flight from Los Angeles to New York has 128 passengers. The plane stops in Denver, where 50 of the passengers get off and 21 new passengers get on. Write an expression containing the numbers 128, 50, and 21 to describe this situation.

1.2
Real Numbers

In this section we will get an idea of what real numbers are. In order to do this we will draw what is called the *real number line*. We first draw a straight line and label a convenient point on the line with 0. Then we mark off equally spaced distances in both directions from 0. Label the points to the right of 0 with the numbers 1, 2, 3, . . . (the dots mean "and so on"). The points to the left of 0 we label, in order, $-1, -2, -3, \ldots$. Here is what it looks like:

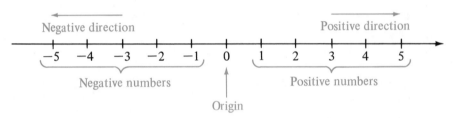

The numbers increase in value going from left to right. If we "move" to the right, we are moving in the positive direction. If we move to the left, we are moving in the negative direction. When comparing two numbers on the number line, the number on the left is always smaller than the number on the right. For instance, -3 is smaller than -1 since it is to the left of -1 on the number line.

Drawing and labeling the number line in this way allows us to associate numbers with points on a line. For every number there is one point on the line, and for every point on the line there is a unique number. There are points associated with the numbers $-3.5, -1\frac{1}{4}, \frac{1}{2}, \frac{3}{4}$, and 2.5.

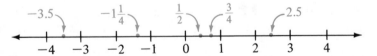

DEFINITION The number associated with a point on the real number line is called the coordinate of that point.

In the above example, the numbers $\frac{1}{2}, \frac{3}{4}, 2.5, -3.5$, and $-1\frac{1}{4}$ are the coordinates of the points they represent. The numbers that can be represented with points on the real number line are called real numbers. Real numbers include whole numbers, fractions, decimals, and other numbers that are not as familiar to us at this moment. We will look at the different classifications of numbers on the line in the last section of this chapter. For the time being, we define real numbers as any numbers that have a place on the real number line.

As we proceed through Chapter 1, from time to time we will review some of the major concepts associated with fractions. The number line can be used to visualize fractions. Recall that for the fraction $\frac{a}{b}$, a is called the numerator and b is called the denominator. The denominator indicates the number of equal parts in the interval from 0 to 1 on the number line. The numerator indicates how many of those parts we have. For the fraction $\frac{3}{4}$, the numerator is 3 and the denominator is 4. If we want to visualize $\frac{3}{4}$ on the number line, we divide the interval from 0 to 1 into 4 equal parts, and then count 3 of them.

Fractions on the Number Line

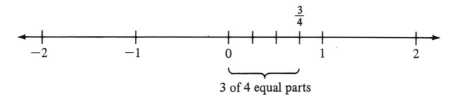

3 of 4 equal parts

Representing numbers on the number line lets us give each number two important properties: a direction from zero and a distance from zero. The direction from zero is represented by the sign in front of the number. (A number without a sign is understood to be positive.) The distance from zero is called the absolute value of the number, as the following definition indicates.

DEFINITION The *absolute value* of a real number is its distance from zero on the number line. If x represents a real number, then the absolute value of x is written $|x|$.

▼ **Examples**

1. $|5| = 5$ The number $+5$ is 5 units from zero.

2. $|-5| = 5$ The number -5 is 5 units from zero.

3. $\left|\frac{1}{2}\right| = \frac{1}{2}$ The number $\frac{1}{2}$ is $\frac{1}{2}$ units from zero.

4. $\left|-\frac{1}{2}\right| = \frac{1}{2}$ The number $-\frac{1}{2}$ is $\frac{1}{2}$ units from zero.

5. $|-3.5| = 3.5$ The number -3.5 is 3.5 units from zero. ▲

The absolute value of a number is *never* negative. It is the distance the number is from zero without regard to which direction it is from zero.

Another important concept associated with numbers on the number line is that of opposites. As will see later on in this chapter, the opposite of a number is also called the additive inverse of the number.

DEFINITION Numbers the same distance from zero but in opposite directions from zero are called *opposites*.

▼ **Examples**

	Number	Opposite	
6.	5	-5	5 and -5 are opposites.
7.	6	-6	6 and -6 are opposites.
8.	-3	3	-3 and 3 are opposites.
9.	$-\frac{1}{4}$	$\frac{1}{4}$	$-\frac{1}{4}$ and $\frac{1}{4}$ are opposites.
10.	-2.3	2.3	-2.3 and 2.3 are opposites. ▲

Each negative number is the opposite of some positive number, and each positive number is the opposite of some negative number. The opposite of a negative number is a positive number; the opposite of a negative is positive. In symbols, if a represents a positive number, then

$$-(-a) = a$$

Opposites always have the same absolute value. When you add any two opposites, the result is always zero.

$$a + (-a) = 0$$

Reciprocals and Multiplication with Fractions

The last concept we want to cover in this section is the concept of reciprocals. Understanding reciprocals requires some knowledge of multiplication with fractions. To multiply two fractions we simply multiply numerators and multiply denominators.

▼ **Example 11** Multiply $\frac{3}{4} \cdot \frac{5}{7}$.

SOLUTION The product of the numerators is 15 and the product of the denominators is 28.

$$\frac{3}{4} \cdot \frac{5}{7} = \frac{3 \cdot 5}{4 \cdot 7} = \frac{15}{28}$$

▼ **Example 12** Multiply $7\left(\dfrac{1}{3}\right)$.

SOLUTION   The number 7 can be thought of as the fraction $\frac{7}{1}$.

$$7\left(\frac{1}{3}\right) = \frac{7}{1}\left(\frac{1}{3}\right) = \frac{7 \cdot 1}{1 \cdot 3} = \frac{7}{3} \qquad \blacktriangle$$

Note: In past math classes you may have written fractions like $\frac{7}{3}$ (improper fractions) as mixed numbers, such as $2\frac{1}{3}$. In algebra, it is usually better to leave them as improper fractions.

We are now ready for the definition of reciprocals.

DEFINITION   Two numbers whose product is 1 are called *reciprocals*.

▼ **Examples**

	Number	Reciprocal	
13.	5	$\frac{1}{5}$	because $5(\frac{1}{5}) = \frac{5}{1}(\frac{1}{5}) = \frac{5}{5} = 1$
14.	2	$\frac{1}{2}$	because $2(\frac{1}{2}) = \frac{2}{1}(\frac{1}{2}) = \frac{2}{2} = 1$
15.	$\frac{1}{3}$	3	because $\frac{1}{3}(3) = \frac{1}{3}(\frac{3}{1}) = \frac{3}{3} = 1$
16.	$\frac{3}{4}$	$\frac{4}{3}$	because $\frac{3}{4}(\frac{4}{3}) = \frac{12}{12} = 1$
17.	$\frac{2}{5}$	$\frac{5}{2}$	because $\frac{2}{5}(\frac{5}{2}) = \frac{10}{10} = 1$ ▲

Although we will not develop multiplication with negative numbers until later in the chapter, you should know that the reciprocal of a negative number is also a negative number. For example, the reciprocal of -4 is $-\frac{1}{4}$. Likewise, the reciprocal of $-\frac{2}{3}$ is $-\frac{3}{2}$. Their products are 1, but we will have to wait until we develop multiplication with negative numbers to see why.

Every real number, except zero, has a reciprocal. Any time we multiply by 0, the result is 0. This is a special property 0 has. In symbols it looks like this:

For any number *a*,

$$a \cdot 0 = 0 \cdot a = 0$$

Multiplying by 0 always results in 0.

To summarize, real numbers are numbers associated with points on the real number line. Each real number has an absolute value, an opposite, and, with the exception of 0, a reciprocal.

Problem Set 1.2

Draw a number line that extends from -10 to $+10$. Label the points with the following coordinates

1.	5	**2.**	6
3.	-4	**4.**	-3
5.	1.5	**6.**	9.25
7.	-3.5	**8.**	-5.5
9.	$\frac{9}{4}$	**10.**	$\frac{8}{3}$
11.	$-\frac{12}{3}$	**12.**	$-\frac{14}{3}$

For each of the following numbers, give the opposite, the reciprocal, and the absolute value.

13.	10	**14.**	8
15.	$\frac{1}{4}$	**16.**	$\frac{1}{3}$
17.	$\frac{3}{4}$	**18.**	$\frac{2}{7}$
19.	$\frac{11}{2}$	**20.**	$\frac{15}{3}$
21.	-3	**22.**	-5
23.	$-\frac{1}{4}$	**24.**	$-\frac{1}{8}$
25.	$-\frac{2}{5}$	**26.**	$-\frac{3}{8}$
27.	x	**28.**	a

Place the correct inequality or equality symbol between each of the following.

29.	$-5 \quad -3$	**30.**	$-8 \quad -1$
31.	$-3 \quad -7$	**32.**	$-6 \quad 5$
33.	$-5 \quad 10$	**34.**	$|5| \quad -|-5|$
35.	$|-4| \quad -|-4|$	**36.**	$3 \quad -|-3|$
37.	$7 \quad -|-7|$	**38.**	$-7 \quad |-7|$
39.	$-12 \quad -12$	**40.**	$-|15| \quad |-15|$
41.	$-|20| \quad |-20|$	**42.**	$-1.5 \quad |-1.5|$
43.	$-\frac{1}{4} \quad -\frac{1}{5}$	**44.**	$-\frac{1}{2} \quad -\frac{1}{3}$
45.	$-\frac{3}{2} \quad -\frac{3}{4}$	**46.**	$-\frac{8}{3} \quad -\frac{17}{3}$

Identify the following statements as true or false.

47.	$(\frac{3}{5})(\frac{5}{3}) = 1$	**48.**	$(\frac{4}{3})(\frac{3}{4}) = 1$
49.	$-10(0) = 1$	**50.**	$0 \cdot 2 = 1$
51.	$-(-3) = -3$	**52.**	$-(-\frac{1}{2}) = \frac{1}{2}$
53.	$-(-15) = 15$	**54.**	$-(-3) = 3$

55. $-|-5| = 5$ **56.** $-|-8| = -8$
57. $-|-9| = -9$ **58.** $-|-3| = 3$
59. $-3 + 3 = 6$ **60.** $-5 + 5 = 0$

Multiply the following:

61. $\dfrac{2}{3} \cdot \dfrac{4}{5}$ **62.** $\dfrac{1}{4} \cdot \dfrac{3}{5}$

63. $\frac{5}{9}(\frac{4}{3})$ **64.** $\frac{7}{8}(\frac{9}{2})$

65. $\frac{2}{3}(\frac{3}{2})$ **66.** $\frac{5}{8}(\frac{8}{5})$

67. $\dfrac{4}{3} \cdot \dfrac{3}{4}$ **68.** $\dfrac{5}{7} \cdot \dfrac{7}{5}$

69. $6(\frac{1}{6})$ **70.** $8(\frac{1}{8})$

71. $3 \cdot \dfrac{1}{3}$ **72.** $4 \cdot \dfrac{1}{4}$

73. The number -7 can be used to represent a temperature of 7 degrees below zero. Give the opposite and the reciprocal of -7.

74. A football team gains 6 yards on one play and then loses 8 yards on the next play. To what number on the number line does a loss of 8 yards correspond? The total yards gained or lost on the two plays corresponds to what negative number?

75. The city with the highest elevation in the world is Wenchuan, which is on the Chinghai-Tibet road. Its elevation is 16,732 feet above sea level. The lowest city in the world is Ein Bokek on the shores of the Dead Sea. Its elevation is 1299 feet below sea level. Write each number using a $+$ or $-$ sign.

76. A woman has a balance of $20 in her checking account. If she writes a check for $30, what negative number can be used to represent the new balance in her checking account?

77. If the distance between two numbers on the number line is 5 and one of the numbers is 3, what are the two possibilities for the other number?

78. The distance between two numbers on the number line is 10. If one of the numbers is 6, what are the two possibilities for the other number?

We begin this section by defining addition of real numbers in terms of the real number line. Once we have the idea of addition from the number line, we can summarize our results by writing a rule for addition of real numbers.

Since real numbers have both a distance from zero (absolute value) and a direction from zero (direction), we can think of addition of two numbers in terms of distance and direction from zero.

**1.3
Addition of
Real Numbers**

Let's look at a problem we know the answer to. Suppose we want to add the numbers 3 and 4. The problem is written $3 + 4$. To put it on the number line, we read the problem as follows:

    1. The 3 tells us to "start at the origin and move 3 units in the positive direction."

    2. The $+$ sign is read "and then move."

    3. The 4 means "4 units in the positive direction."

To summarize, $3 + 4$ means to start at the origin, move 3 units in the *positive* direction and then 4 units in the *positive* direction.

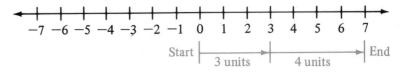

We end up at 7, which is the answer to our problem: $3 + 4 = 7$.

    Let's try other combinations of positive and negative 3 and 4 on the number line.

▼ **Example 1** Add $3 + (-4)$

SOLUTION Starting at the origin, move 3 units in the *positive* direction and then 4 units in the *negative* direction.

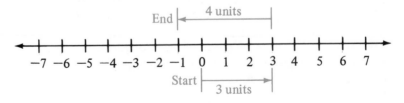

We end up at -1; therefore $3 + (-4) = -1$.

▼ **Example 2** Add $-3 + 4$

SOLUTION Starting at the origin, move 3 units in the *negative* direction and then 4 units in the *positive* direction.

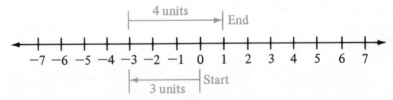

We end up at $+1$; therefore $-3 + 4 = 1$.

▼ **Example 3** Add $-3 + (-4)$

SOLUTION Starting at the origin, move 3 units in the *negative* direction and then 4 units in the *negative* direction.

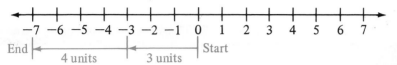

We end up at -7; therefore $-3 + (-4) = -7$. Here is a summary of what we have just completed:

$$3 + 4 = 7$$
$$3 + (-4) = -1$$
$$-3 + 4 = 1$$
$$-3 + (-4) = -7$$ ▲

▼ **Examples** Add

4. $5 + 7 = 12$

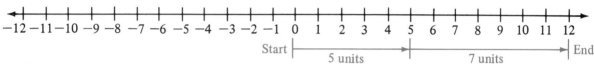

5. $5 + (-7) = -2$

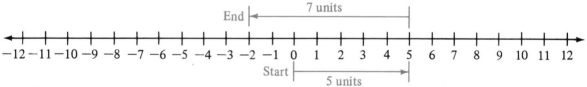

6. $-5 + 7 = 2$

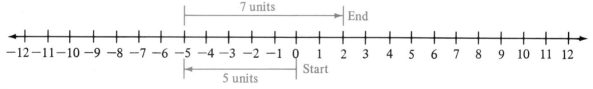

7. $-5 + (-7) = -12$

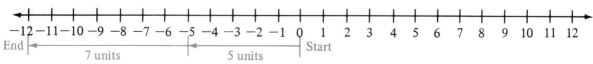

If we look closely at the results of the above addition problems, we can see that they support (or justify) the following rule.

RULE   To add two real numbers with

1. the *same* sign: Simply add their absolute values and use the common sign. (Both numbers positive, the answer is positive. Both numbers negative, the answer is negative.)

2.   *different* signs: Subtract the smaller absolute value from the larger. The answer will have the sign of the number with the larger absolute value.

This rule covers all possible combinations of addition with real numbers. You must memorize it. After you have worked a number of problems, it will seem almost automatic.

Here are some more examples. Check to see that the answers to them are consistent with the rule for addition of real numbers.

▼   **Example 8**   Add all combinations of positive and negative 10 and 13.

SOLUTION

$$10 + 13 = 23$$
$$10 + (-13) = -3$$
$$-10 + 13 = 3$$
$$-10 + (-13) = -23$$
▲

▼   **Example 9**   Add all possible combinations of positive and negative 12 and 17.

SOLUTION

$$12 + 17 = 29$$
$$12 + (-17) = -5$$
$$-12 + 17 = 5$$
$$-12 + (-17) = -29$$
▲

▼   **Examples**   Add

10. $-3 + 2 + (-4) = -1 + (-4)$
$$= -5$$

11. $-8 + [2 + (-5)] + (-1) = -8 + (-3) + (-1)$
$$= -11 + (-1)$$
$$= -12$$

1. Add all positive and negative combinations of 3 and 5. (Look back to Examples 8 and 9).

2. Add all positive and negative combinations of 6 and 4.
3. Add all positive and negative combinations of 15 and 20.
4. Add all positive and negative combinations of 18 and 12.

Work the following problems. You may want to begin by doing a few on the number line.

5. $6 + (-3)$ **6.** $7 + (-8)$
7. $10 + (-5)$ **8.** $12 + (-6)$
9. $18 + (-12)$ **10.** $20 + (-5)$
11. $13 + (-20)$ **12.** $15 + (-25)$
13. $18 + (-32)$ **14.** $6 + (-9)$
15. $-6 + 3$ **16.** $-8 + 7$
17. $-9 + 1$ **18.** $-10 + 4$
19. $-30 + 5$ **20.** $-18 + 6$
21. $-6 + (-6)$ **22.** $-5 + (-5)$
23. $-9 + (-10)$ **24.** $-8 + (-6)$
25. $-10 + (-15)$ **26.** $-18 + (-30)$
27. $-12 + (-13)$ **28.** $-6 + (-5)$
29. $-7 + (-8)$ **30.** $-9 + (-11)$

Work the following problems using the rule for addition of real numbers. You may want to refer back to the rule for order of operation.

31. $5 + (-6) + (-7)$ **32.** $6 + (-8) + (-10)$
33. $-7 + 8 + (-5)$ **34.** $-6 + 9 + (-3)$
35. $5 + [6 + (-2)] + (-3)$ **36.** $10 + [8 + (-5)] + (-20)$
37. $(-18) + (-3) + (-9)$ **38.** $-25 + (-15) + (-30)$
39. $[6 + (-2)] + [3 + (-1)]$ **40.** $[18 + (-5)] + [9 + (-10)]$
41. $20 + (-6) + [3 + (-9)]$ **42.** $18 + (-2) + [9 + (-13)]$
43. $-3 + (-2) + [5 + (-4)]$ **44.** $-6 + (-5) + [-4 + (-1)]$
45. $-17 + 19 + (-30) + 8$ **46.** $-20 + 6 + (-30) + 4$

The following problems will give you experience working with parentheses and negative signs. After you have worked these problems, see if you can state a rule using the letters a and b to summarize your results. Simplify each expression.

47. $-3 + (-4)$ **48.** $-2 + (-5)$
49. $-(3 + 4)$ **50.** $-(2 + 5)$
51. $-8 + (-2)$ **52.** $-6 + (-1)$

53. $-(8 + 2)$ **54.** $-(6 + 1)$
55. $-10 + (-15)$ **56.** $-20 + (-10)$
57. $-(10 + 15)$ **58.** $-(20 + 10)$

Recall that the word "sum" indicates addition. Write the numerical expression that is equivalent to each of the following phrases and then simplify.

59. The sum of 5 and 9
60. The sum of 6 and -3
61. Four added to the sum of -7 and -5
62. Six added to the sum of -9 and 1
63. The sum of -2 and -3 increased by 10
64. The sum of -4 and -12 increased by 2

Answer the following questions.

65. What number do you add to -8 to get -5?
66. What number do you add to 10 to get 4?
67. The sum of what number and -6 is -9?
68. The sum of what number and -12 is 8?

69. The temperature at noon is 12 degrees below zero. By 1:00 it has risen 4 degrees. Write an expression using the numbers -12 and 4 to describe this situation.
70. On Monday a certain stock gains 2 points. On Tuesday it loses 3 points. Write an expression using positive and negative numbers with addition to describe this situation and then simplify it.
71. On three consecutive hands of draw poker a gambler wins $10, loses $6, and then loses another $8. Write an expression using positive and negative numbers and addition to describe this situation and then simplify it.
72. You know from your past experience with numbers that subtracting 5 from 8 results in 3. $(8 - 5 = 3)$. What addition problem that starts with the number 8 gives the same result?

1.4
Subtraction of
Real Numbers

In the last section we spent some time developing the rule for addition of real numbers. Since we want to make as few rules as possible, we can define subtraction in terms of addition. Then we can use the rule for addition to solve our subtraction problems.

RULE   To subtract one real number from another, simply add its opposite.

Algebraically the rule is written like this: If a and b represent two real numbers, then it is always true that

$$\underbrace{a - b}_{\text{To subtract}} = \underbrace{a + (-b)}_{\text{add the opposite.}}$$

This is how subtraction is defined in algebra. This definition of subtraction will not conflict with what you already know about subtraction, but it will allow you to do subtraction with negative numbers.

▼ **Example 1** Subtract all possible combinations of positive and negative 7 and 2.

SOLUTION

$$7 - 2 = \quad 7 + (-2) = 5 \qquad \text{Subtracting 2 is the same}$$
$$-7 - 2 = -7 + (-2) = -9 \qquad \text{as adding } -2$$

$$7 - (-2) = \quad 7 + 2 = 9 \qquad \text{Subtracting } -2 \text{ is the same}$$
$$-7 - (-2) = -7 + 2 = -5 \qquad \text{as adding } +2$$

Notice that each subtraction problem is first changed to an addition problem. The rule for addition is then used to get the answer. ▲

We have defined subtraction in terms of addition and still get answers consistent with the answers we are used to getting with subtraction. Moreover, we can now do subtraction problems involving both positive and negative numbers.

As you proceed through the following examples and the problem set, you will begin to notice shortcuts you can use in working the problems. You will not always have to change subtraction to addition of the opposite to be able to get the answer quickly. Use all the shortcuts you wish as long as you can consistently get the correct answers.

▼ **Example 2** Subtract all combinations of positive and negative 8 and 13.

SOLUTION

$$8 - 13 = \quad 8 + (-13) = -5 \qquad \text{Subtracting } +13 \text{ is the}$$
$$-8 - 13 = -8 + (-13) = -21 \qquad \text{same as adding } -13$$

$$8 - (-13) = \quad 8 + 13 = 21 \qquad \text{Subtracting } -13 \text{ is the}$$
$$-8 - (-13) = -8 + 13 = 5 \qquad \text{same as adding } +13 \qquad ▲$$

The next examples contain both addition and subtraction. Notice in each case that we first change all subtraction to addition of the opposite.

▼ **Examples**   Simplify each expression as much as possible.

3. $7 + (-3) - 5 = 7 + (-3) + (-5)$ Begin by changing all
                                             subtractions to additions
$$= 4 + (-5)$$ Then add left to right
$$= -1$$

4. $8 - (-2) - 6 = 8 + 2 + (-6)$ Begin by changing all
                                         subtractions to additions
$$= 10 + (-6)$$ Then add left to right
$$= 4$$

5. $-2 - (-3 + 1) - 5 = -2 - (-2) - 5$ Do what is in the
                                               parentheses first
$$= -2 + 2 + (-5)$$ Then add and subtract
$$= -5$$ left to right   ▲

Our last examples show how subtraction problems written in words are changed to subtraction problems written with symbols.

▼ **Example 6**   Subtract 7 from −3.

SOLUTION   First, we write the problem in terms of subtraction. We then change to addition of the opposite.

$$-3 - 7 = -3 + (-7)$$
$$= -10 \qquad ▲$$

▼ **Example 7**   Subtract −5 from 2.

SOLUTION   Subtracting −5 is the same as adding +5.

$$2 - (-5) = 2 + 5$$
$$= 7 \qquad ▲$$

Problem Set 1.4

The following problems are intended to give you practice at subtraction. Besides working the problems correctly, make sure you understand that subtraction is the same as addition of the opposite.

1. Subtract all combinations of positive and negative 8 and 2 (see Examples 1 and 2).
2. Subtract all combinations of positive and negative 7 and 3.
3. Subtract all combinations of positive and negative 12 and 20.
4. Subtract all combinations of positive and negative 13 and 17.

Work the following problems.

5.	$5 - 8$	**6.**	$6 - 7$
7.	$18 - (-3)$	**8.**	$19 - (-10)$
9.	$37 - (-4)$	**10.**	$21 - (-11)$
11.	$-8 - (-1)$	**12.**	$-6 - (-2)$
13.	$-15 - (-10)$	**14.**	$-20 - (-5)$
15.	$-32 - (-12)$	**16.**	$-31 - (-1)$
17.	$3 - 2 - 5$	**18.**	$4 - 8 - 6$
19.	$9 - 2 - 3$	**20.**	$8 - 7 - 12$
21.	$-3 + 9 - 7$	**22.**	$-18 + 20 - 4$
23.	$-22 + 4 - 10$	**24.**	$-13 + 6 - 5$
25.	$10 - (-20) - 5$	**26.**	$15 + (-3) + (+8)$
27.	$-13 - 4 - (-6)$	**28.**	$-21 + 5 - (-10)$
29.	$-121 + 169 - (-144)$	**30.**	$36 + (-49) - (-25)$
31.	$8 - (2 - 3) - 5$	**32.**	$10 - (4 - 6) - 8$
33.	$5 + (-8 - 6) + 2$	**34.**	$4 - (-3 - 2) - 1$
35.	$16 - [(4 - 5) - 1]$	**36.**	$15 - [(4 - 2) - 3]$
37.	$5 - [(2 - 3) - 4]$	**38.**	$6 - [(4 - 1) - 5] - 10$
39.	$21 - [-(3 - 4) - 2] - 5$	**40.**	$30 - [-(10 - 5) - 15] - 25$

Work the following problems.

41. Subtract 4 from -7.
42. Subtract 5 from -19.
43. Subtract 16 from -32.
44. Subtract 23 from -61.
45. Subtract -8 from 12.
46. Subtract -2 from 10.
47. Subtract -7 from -5.
48. Subtract -9 from -3.
49. Subtract 17 from the sum of 4 and -5.
50. Subtract -6 from the sum of 6 and -3.

Recall that the word "difference" indicates subtraction. The difference of a and b is $a - b$, in that order. Write a numerical expression that is equivalent to each of the following phrases and then simplify.

51. The difference of 8 and 5.
52. The difference of 5 and 8.
53. The difference of -8 and 5.

54. The difference of −5 and 8.
55. The difference of 8 and −5.
56. The difference of 5 and −8.
57. Seven more than the difference of 3 and −5.
58. Nine more than the difference of −4 and −3.

Answer the following questions.

59. What number do you subtract from 8 to get −2?
60. What number do you subtract from 1 to get −5?
61. What number do you subtract from 8 to get 10?
62. What number do you subtract from 1 to get 5?

63. A man with $1500 in a savings account makes a withdrawal of $730. Write an expression using subtraction that describes this situation.

64. The temperature inside a space shuttle is 73 degrees before reentry. During reentry the temperature inside the craft increases 10 degrees. Upon landing it drops 8 degrees. Write an expression using the numbers 73, 10, and 8 to describe this situation. What is the temperature inside the shuttle upon landing?

65. A man who has lost $35 playing roulette in Las Vegas wins $15 playing blackjack. He then loses $20 playing the wheel of fortune. Write an expression using the numbers −35, 15, and 20 to describe this situation and then simplify it.

66. An airplane flying at 10,000 feet lowers its altitude by 1500 feet to avoid other air traffic. Then it increases its altitude by 3000 feet to clear a mountain range. Write an expression that describes this situation and then simplify it.

1.5 Properties of Real Numbers

In this section we will list all the facts (properties) you know from past experience are true about numbers in general. We will give each property a name so we can refer to it later in the book. Mathematics is very much like a game. The game involves numbers. The rules of the game are the properties and rules we are developing in this chapter. The goal of the game is to extend the basic rules to as many new situations as possible.

You know from past experience with numbers that it makes no difference in which order you add two numbers. That is, 3 + 5 is the same as 5 + 3. This fact about numbers is called the *commutative property of addition*. We say addition is a commutative operation. Changing the order of the numbers does not change the answer.

There is one other basic operation that is commutative. Since 3(5) is the same as 5(3), we say multiplication is a commutative operation. Changing the order of the two numbers you are multiplying does not change the answer.

For all the properties listed in this section, *a, b,* and *c* represent real numbers.

Commutative Property of Addition

In symbols: $a + b = b + a$

In words: Changing the *order* of the numbers in a sum will not change the result.

Commutative Property of Multiplication

In symbols: $a \cdot b = b \cdot a$

In words: Changing the *order* of the numbers in a product will not change the result.

▼   **Examples**

1. The statement $5 + 8 = 8 + 5$ is an example of the commutative property of addition.

2. The statement $2 \cdot y = y \cdot 2$ is an example of the commutative property of multiplication.

3. The expression $5 + x + 3$ can be simplified using the commutative property of addition.

    $$5 + x + 3 = x + 5 + 3 \qquad \text{Commutative property of addition}$$
    $$= x + 8 \qquad \text{Addition}$$

    ▲

The other two basic operations are not commutative. The order in which we subtract or divide two numbers makes a difference in the answer.

Another property of numbers that you have used many times has to do with grouping. You know that when we add three numbers it makes no difference which two we add first. When adding $3 + 5 + 7$ we can add the 3 and 5 first and then the 7 or we can add the 5 and 7 first and then the 3. Mathematically it looks like this: $(3 + 5) + 7 = 3 + (5 + 7)$. This property is true of multiplication as well. Operations that behave in this manner are called *associative* operations. The answer will not change when we change the association (or grouping) of the numbers.

Associative Property of Addition

In symbols: $a + (b + c) = (a + b) + c$
In words: Changing the *grouping* of the numbers in a sum will not change the result.

Associative Property of Multiplication

In symbols: $a(bc) = (ab)c$
In words: Changing the *grouping* of the numbers in a product will not change the result.

The following examples illustrate how the associative properties can be used to simplify expressions that involve both numbers and variables.

▼ **Examples**   Simplify.

4. $4 + (5 + x) = (4 + 5) + x$ Associative property of addition
     $= 9 + x$ Addition

5. $5(2x) = (5 \cdot 2)x$ Associative property
     $= 10x$ Multiplication

6. $\frac{1}{5}(5x) = (\frac{1}{5} \cdot 5)x$ Associative property
     $= 1x$ Multiplication
     $= x$ ▲

The associative and commutative properties apply to problems that are either all multiplication or all addition. There is a third basic property that involves both addition and multiplication. It is called the distributive property and looks like this:

Distributive Property

In symbols: $a(b + c) = ab + ac$
In words: Multiplication *distributes* over addition.

You will see as we progress through the book that the distributive property is used very frequently in algebra. To see that the distributive property is true, compare the following expressions:

$$
\begin{array}{ll}
5(3 + 4) & 5(3) + 5(4) \\
5(7) & 15 + 20 \\
35 & 35
\end{array}
$$

In both cases the result is 35. Since the results are the same, the original two expressions must be equal. Or $5(3 + 4) = 5(3) + 5(4)$. We can either add the 3 and the 4 first and then multiply that sum by 5, or we can multiply the 3 and the 4 separately by 5 and then add the results. In either case we get the same answer.

▼ **Examples** Apply the distributive property to each expression and then simplify the result.

7. $2(x + 3) = 2(x) + 2(3)$ Distributive property
       $= 2x + 6$ Multiplication

8. $5(2x - 8) = 5(2x) - 5(8)$ Distributive property
       $= 10x - 40$ Multiplication

Notice in this example that multiplication distributes over subtraction as well as addition.

9. $4(x + y) = 4x + 4y$ Distributive property   ▲

In addition to the three properties mentioned so far, we want to include in our list two special numbers that have unique properties. They are the numbers zero and one.

Special Numbers

Additive Identity Property
There exists a unique number 0 such that:
In symbols: $a + 0 = a$ and $0 + a = a$
In words: Zero preserves identities under addition. (The identity of the number is unchanged after addition with 0.)

Multiplicative Identity Property
There exists a unique number 1 such that:
In symbols: $a(1) = a$ and $1(a) = a$
In words: The number 1 preserves identities under multiplication. (The identity of the number is unchanged after multiplication by 1.)

Additive Inverse Property
For each real number a, there exists a unique number $-a$ such that:
In symbols: $a + (-a) = 0$
In words: Opposites add to 0.

Multiplicative Inverse Property

For every real number *a*, except 0, there exists a unique real number $\frac{1}{a}$ such that:

> *In symbols:* $a\left(\dfrac{1}{a}\right) = 1$
>
> *In words:* Reciprocals multiply to 1.

Of all the basic properties listed, the commutative, associative, and distributive properties are the ones we will use most often. They are important because they are used as justifications or reasons for many of the things we will do in the future.

The following example illustrates how we use the properties listed above. Each line contains an algebraic expression that has been changed in some way. The property that justifies the change is written to the right.

▼ **Examples**

10. $x + 5 = 5 + x$ Commutative property of addition

11. $2 \cdot 3 \cdot x = 2 \cdot x \cdot 3$ Commutative property of multiplication

12. $(2 + x) + y = 2 + (x + y)$ Associative property of addition

13. $6(x + 3) = 6x + 18$ Distributive property

14. $2 + (-2) = 0$ Additive inverse property

15. $3(\frac{1}{3}) = 1$ Multiplicative inverse property

16. $(2 + 0) + 3 = 2 + 3$ 0 is the identity element for addition

17. $(2 + 3) + 4 = 3 + (2 + 4)$ Commutative and associative properties

18. $(x + 2) + y = (x + y) + 2$ Commutative and associative properties   ▲

As a final note on the properties of real numbers we should mention that although some of the properties are stated for only two or three real numbers, they hold for as many numbers as needed. For example, the distributive property holds for expressions like $3(x + y + z + 5 + 2)$. That is,

$$3(x + y + z + 5 + 2) = 3x + 3y + 3z + 15 + 6$$

It is not important how many numbers are contained in the sum, only that it is a sum. Multiplication, you see, distributes over addition, whether there are two numbers in the sum or two hundred.

State the property or properties that justify the following.

1. $3 + 2 = 2 + 3$

2. $5 + 0 = 5$

3. $4(\frac{1}{4}) = 1$

4. $10(0.1) = 1$

5. $4 + x = x + 4$

6. $3(x - 10) = 3x - 30$

7. $2(y + 8) = 2y + 16$

8. $3 + (4 + 5) = (3 + 4) + 5$

9. $(3 + 1) + 2 = 1 + (3 + 2)$

10. $(5 + 2) + 9 = (2 + 5) + 9$

11. $(8 + 9) + 10 = (8 + 10) + 9$

12. $(7 + 6) + 5 = (5 + 6) + 7$

13. $3(x + 2) = 3(2 + x)$

14. $2(7y) = (7 \cdot 2)y$

15. $x(3y) = 3(xy)$

16. $a(5b) = 5(ab)$

17. $4(xy) = 4(yx)$

18. $3[2 + (-2)] = 3(0)$

19. $8[7 + (-7)] = 8(0)$

20. $7(1) = 7$

Each of the following problems has a mistake in it. Correct the right-hand side.

21. $3(x - 2) = 3x - 2$

22. $5(4 - x) = 4 - 5x$

23. $9(a + b) = 9a + b$

24. $2(y + 1) = 2y - 2$

25. $3(0) = 3$

26. $5(\frac{1}{5}) = 5$

27. $3 + (-3) = 1$

28. $8(0) = 8$

29. $10(1) = 0$

30. $3 - (4 + 5) = (3 - 4) + 5$

Use the associative property to rewrite each of the following expressions, and then simplify the result. (See Examples 4, 5, and 6.)

31. $4 + (2 + x)$

32. $5 + (6 + x)$

33. $(x + 2) + 7$

34. $(x + 8) + 2$

35. $3(5x)$

36. $5(3x)$

37. $9(6y)$

38. $6(9y)$

39. $2(3a)$

40. $4(2a)$

41. $\frac{1}{3}(3x)$

42. $\frac{1}{4}(4x)$

43. $\frac{1}{2}(2y)$

44. $\frac{1}{7}(7y)$

45. $\frac{3}{4}(\frac{4}{3}x)$

46. $\frac{3}{2}(\frac{2}{3}x)$

47. $\frac{6}{5}(\frac{5}{6}a)$

48. $\frac{2}{5}(\frac{5}{2}a)$

Apply the distributive property to each of the following expressions. Simplify when possible.

49. $8(x + 2)$

50. $5(x + 3)$

51. $8(x - 2)$

52. $5(x - 3)$

53. $4(y + 1)$

54. $4(y - 1)$

55. $3(6x + 5)$	56. $3(5x + 6)$
57. $2(3a + 7)$	58. $5(3a + 2)$
59. $9(6y - 8)$	60. $2(7y - 4)$
61. $3(4 + 2x)$	62. $5(2 + 3x)$
63. $7(3 - a)$	64. $2(5 - a)$
65. $3(x + y)$	66. $2(x - y)$
67. $8(a - b)$	68. $7(a + b)$

69. While getting dressed for work, a man puts on his socks and puts on his shoes. Are the two statements "putting on your socks" and "putting on your shoes" commutative?

70. Are the statements "put on your left shoe" and "put on your right shoe" commutative?

71. A skydiver flying over the jump area is about to do two things: jump out of the plane and pull the rip cord. Are the two events "jump out of the plane" and "pull the rip cord" commutative? That is, will changing the order of the events always produce the same result?

72. Give an example of two events in your daily life that are commutative.

73. Give an example that shows that division is not a commutative operation. That is, find two numbers for which changing the order of division gives two different answers.

74. Simplify the expression $10 - (5 - 2)$ and the expression $(10 - 5) - 2$ to show that subtraction is not an associative operation.

1.6 Multiplication of Real Numbers

From our past experience counting numbers, we know that multiplication is simply repeated addition. That is, $3(5) = 5 + 5 + 5$. We will use this fact, along with our knowledge of negative numbers, to develop the rule for multiplication of any two real numbers. The following examples illustrate multiplication with three of the possible combinations of positive and negative numbers.

▼ **Examples** Multiply

1. Two positives:
$$3(5) = 5 + 5 + 5$$
$$= 15 \qquad \text{Positive answer}$$

2. One positive:
$$3(-5) = -5 + (-5) + (-5)$$
$$= -15 \qquad \text{Negative answer}$$

3. One negative:
$$-3(5) = 5(-3) \qquad \text{Commutative property}$$
$$= -3 + (-3) + (-3) + (-3) + (-3)$$
$$= -15 \qquad \text{Negative answer}$$

4. Two negatives:
$$-3(-5) = ? \qquad\qquad\qquad ▲$$

With two negatives, $-3(-5)$, it is not possible to work the problem in terms of repeated addition. (It doesn't "make sense" to write -5 down a -3 number of times.) The answer is probably $+15$ (that's just a guess), but we need some justification for saying so. We will solve a different problem and in so doing get the answer to the problem $(-3)(-5)$.

Here is a problem we know the answer to. We will work it two different ways.

$$-3[5 + (-5)] = -3(0) = 0$$

The answer is zero. We can also work the problem using the distributive property.

$$-3[5 + (-5)] = -3(5) + (-3)(-5) \qquad \text{Distributive property}$$
$$= -15 + \quad ?$$

Since the answer to the problem is 0, our ? must be $+15$. (What else could we add to -15 to get 0? Only $+15$.)

We will do another example like this one later. Let us first summarize our results so far and look for similarities between the signs of the numbers in the problem and the answers we get.

Original numbers have		The answer is
the same signs	$3(5) = 15$	positive
different signs	$3(-5) = -15$	negative
different signs	$-3(5) = -15$	negative
the same signs	$-3(-5) = 15$	positive

Here are more examples involving repeated addition and the commutative property.

▼ **Examples** Multiply

5. $2(3) = 3 + 3$
     $= 6$

6. $2(-3) = -3 + (-3)$
     $= -6$

7. $-2(3) = 3(-2)$ \qquad Commutative property
     $= -2 + (-2) + (-2)$
     $= -6$

8. $-2(-3) = ?$ \qquad ▲

Here it is again. To find the answer to $-2(-3)$ we will work the problem $-2[3 + (-3)]$ two different ways:

$$-2[3 + (-3)] = -2(0) = 0$$

Or, using the distributive property, we have

$$-2[3 + (-3)] = -2(3) + (-2)(-3)$$
$$= -6 \quad + \quad ?$$

The missing number must be 6, because we already know the answer to the problem is 0, and 6 is the only number we can add to -6 to get 0.

Here is the summary from the last examples.

Original numbers have		The answer is
the same signs	$2(3) = 6$	positive
different signs	$2(-3) = -6$	negative
different signs	$-2(3) = -6$	negative
the same signs	$-2(-3) = 6$	positive

The above examples and their summaries illustrate the following rule, which you should memorize.

RULE   To multiply any two real numbers, simply multiply their absolute values. The sign of the answer is

1. positive if both numbers had the same sign (both $+$ or both $-$).

2. negative if the numbers had opposite signs (one $+$, the other $-$).

▼   **Examples**   Simplify as much as possible.

9. $5(-7) = -35$

10. $-7(6) = -42$

11. $-8(-3) = 24$

12. $-9(-1) = 9$

13. $2(-3)(-4) = -6(-4)$
$$= 24$$

14. $-5(-1)(-8) = 5(-8)$
$$= -40$$

15. $3(-2) + 5(-4) = -6 + (-20)$
$$= -26$$

16. $4(-3) + 6(-5) - 10 = -12 + (-30) + (-10)$
$$= -42 + (-10)$$
$$= -52 \qquad \blacktriangle$$

Previously we mentioned that to multiply two fractions we multiply numerators and multiply denominators. We can apply the rule for multiplication of positive and negative numbers to fractions in the same way we apply it to other numbers. We multiply absolute values: the product is positive if both fractions have the same sign and negative if they have different signs. Here are some examples.

▼ **Examples** Multiply

17. $-\dfrac{3}{4}\left(\dfrac{5}{7}\right) = -\dfrac{3 \cdot 5}{4 \cdot 7}$ Different signs give
                       a negative answer
$$= -\dfrac{15}{28}$$

18. $-6\left(\dfrac{1}{2}\right) = -\dfrac{6}{1}\left(\dfrac{1}{2}\right)$ Different signs give
                       a negative answer
$$= -\dfrac{6}{2}$$
$$= -3$$

19. $-\dfrac{2}{3}\left(-\dfrac{3}{2}\right) = \dfrac{2 \cdot 3}{3 \cdot 2}$ Same signs give a
                       positive answer
$$= \dfrac{6}{6}$$
$$= 1$$

20. $-2\left(-\dfrac{1}{2}\right) = \dfrac{2}{1}\left(\dfrac{1}{2}\right)$ Same signs give a
                       positive answer
$$= \dfrac{2}{2}$$
$$= 1 \qquad \blacktriangle$$

Notice in Examples 19 and 20 that the product is 1. This means that the two original numbers we multiplied in each problem were reciprocals.

We can use the rule for multiplication of real numbers, along with the associative property, to multiply expressions that contain numbers and variables.

▼ **Examples**   Apply the associative property and then multiply.

21. $-3(2x) = (-3 \cdot 2)x$ Associative property
            $= -6x$ Multiplication

22. $6(-5y) = [6(-5)]y$ Associative property
            $= -30y$ Multiplication

23. $-2(-\frac{1}{2}x) = [(-2)(-\frac{1}{2})]x$ Associative property
            $= 1x$ Multiplication
            $= x$ Multiplication   ▲

The following examples show how we can use the distributive property and multiplication with real numbers together.

▼ **Examples**   Apply the distributive property to each expression.

24. $-2(a + 3)\ = -2a + (-2)(3)$ Distributive property
            $= -2a + (-6)$ Multiplication
            $= -2a - 6$

25. $-3(2x + 1) = -3(2x) + (-3)(1)$ Distributive property
            $= -6x + (-3)$ Multiplication
            $= -6x - 3$

26. $-5(6y - 2) = -5(6y) - (-5)(2)$ Distributive property
            $= -30y - (-10)$ Multiplication
            $= -30y + 10$ ▲

Problem Set 1.6

Find the following products (multiply).

1.	$7(-6)$	**2.**	$8(-4)$
3.	$-7(3)$	**4.**	$-5(4)$
5.	$-8(2)$	**6.**	$-16(3)$
7.	$-3(-1)$	**8.**	$-7(-1)$
9.	$-11(-11)$	**10.**	$-12(-12)$
11.	$-3(2)(-1)$	**12.**	$-2(3)(-4)$
13.	$-3(-4)(-5)$	**14.**	$-5(-6)(-7)$
15.	$-6(-1)(2)$	**16.**	$-12(5)(-1)$
17.	$3(8)(-2)$	**18.**	$4(9)(-3)$
19.	$-2(-4)(-3)(-1)$	**20.**	$-1(-3)(-2)(-1)$

Work the following problems. You may refer to the rule for order of operation.

21. $3(5 - 4)$ 22. $4(7 - 2)$
23. $-2(2 - 5)$ 24. $-3(3 - 7)$
25. $-5(8 - 10)$ 26. $-4(6 - 12)$
27. $-4(6) - 7$ 28. $-5(7) - 2$
29. $-3(-2) - 5$ 30. $-4(-6) - 10$
31. $-3(-6) + 4(-1)$ 32. $-4(-5) + 8(-2)$
33. $-3(-2) + (-8)(-6)$ 34. $-8(-9) - (-6)(-7)$
35. $2(3) - 3(-4) + 4(-5)$ 36. $5(4) - (-2)(-1) - 8$
37. $4(2 + 3) - 5(4 + 7)$ 38. $3(5 - 7) - 4(2 - 3)$
39. $2(9 - 8) - 7(5 - 6)$ 40. $5(3 - 6) - 4(2 - 5)$

Multiply the following fractions. (See Examples 17–20.)

41. $-\dfrac{2}{3} \cdot \dfrac{5}{7}$ 42. $-\dfrac{6}{5} \cdot \dfrac{2}{7}$
43. $-8(\frac{1}{2})$ 44. $-12(\frac{1}{3})$
45. $-\frac{3}{4}(-\frac{4}{3})$ 46. $-\frac{5}{8}(-\frac{8}{5})$
47. $-3(-\frac{1}{3})$ 48. $-5(-\frac{1}{5})$

Find the following products. (See Examples 21, 22, and 23.)

49. $-2(4x)$ 50. $-8(7x)$
51. $5(-9y)$ 52. $3(-6y)$
53. $-7(-6x)$ 54. $-8(-9x)$
55. $-\frac{1}{3}(-3x)$ 56. $-\frac{1}{5}(-5x)$
57. $-4(-\frac{1}{4}x)$ 58. $-2(-\frac{1}{2}x)$

Apply the distributive property to each expression and then simplify the result. (See Examples 24, 25, and 26.)

59. $-4(a + 2)$ 60. $-7(a + 6)$
61. $-2(4x + 1)$ 62. $-8(7x + 3)$
63. $-5(3x + 4)$ 64. $-4(2x + 3)$
65. $-7(6x - 3)$ 66. $-8(2x - 3)$
67. $-1(3x - 4)$ 68. $-1(5x - 6)$
69. Five added to the product of 3 and -10 is what number?
70. If the product of -8 and -2 is decreased by 4, what number results?
71. Write an expression for twice the product of -4 and x and then simplify it.
72. Write an expression for twice the product of -2 and $3x$ and then simplify it.
73. What number results if 8 is subtracted from the product of -9 and 2?
74. What number results if -8 is subtracted from the product of -9 and 2?

1.7
Division of
Real Numbers

The last of the four basic operations is division. We will use the same approach to define division as we did for subtraction. That is, we will define division in terms of rules we already know. Recall that we developed the rule for subtraction by defining subtraction in terms of addition. We changed our subtraction problems to addition problems. Since we already have a rule for multiplication of real numbers and division is the inverse operation of multiplication, we will simply define division in terms of multiplication.

We know that division by the number 2 is the same as multiplication by $\frac{1}{2}$. That is, 6 divided by 2 is 3, which is the same as $\frac{1}{2}$ of 6. Dividing a number by 5 gives the same result as multiplication by $\frac{1}{5}$. From these examples we can generalize the following rule.

RULE If a and b represent any two real numbers (b cannot be 0), then it is always true that

$$a \div b = \frac{a}{b} = a\left(\frac{1}{b}\right)$$

We cannot use 0 for b since 0 has no reciprocal (Section 1.2).

Division by a number is the same as multiplication by its reciprocal. Since every division problem can be written as a multiplication problem and we have the rule for multiplication of two real numbers, we do not have to write a new rule for division of real numbers. We will simply replace our division problem by multiplication and use the rule we already have.

▼ **Examples**

1. $\dfrac{6}{2} = 6\left(\dfrac{1}{2}\right) = 3$ The product of two positives is positive

2. $\dfrac{6}{-2} = 6\left(-\dfrac{1}{2}\right) = -3$

3. $\dfrac{-6}{2} = -6\left(\dfrac{1}{2}\right) = -3$ The product of a positive and a negative is a negative

4. $\dfrac{-6}{-2} = -6\left(-\dfrac{1}{2}\right) = 3$ The product of two negatives is positive ▲

The second step in the examples above is only used to show that we *can* write division in terms of multiplication. [In actual practice we wouldn't write $\frac{6}{2}$ as $6(\frac{1}{2})$.] The results, therefore, must follow from the rule for multiplication. Like signs produce a positive answer. Unlike signs produce a negative answer.

Here are some more examples. This time we will not show division as multiplication by the reciprocal. We will simply divide. If the original numbers have the same signs the answer will be positive. If the original numbers have different signs the answer will be negative.

▼ **Examples** Divide

5. $\dfrac{12}{6} = 2$ Like signs give a positive answer

6. $\dfrac{12}{-6} = -2$ Unlike signs give a negative answer

7. $\dfrac{-12}{6} = -2$ Unlike signs give a negative answer

8. $\dfrac{-12}{-6} = 2$ Like signs give a positive answer

9. $\dfrac{15}{-3} = -5$ Unlike signs give a negative answer

10. $\dfrac{-40}{-5} = 8$ Like signs give a positive answer

11. $\dfrac{-14}{2} = -7$ Unlike signs give a negative answer ▲

The answers in the next examples are fractions. The last step in each of these examples involves reducing a fraction to lowest terms. To reduce a fraction to lowest terms we divide the numerator and denominator by the largest number that divides each of them exactly. For example, to reduce $\frac{15}{20}$ to lowest terms we divide both 15 and 20 by 5 to get $\frac{3}{4}$.

▼ **Examples** Simplify as much as possible.

12. $\dfrac{-4(5)}{6} = \dfrac{-20}{6}$ Simplify numerator

      $= -\dfrac{10}{3}$ Reduce to lowest terms by dividing numerator and denominator by 2

13. $\dfrac{30}{-4-5} = \dfrac{30}{-9}$ Simplify denominator

      $= -\dfrac{10}{3}$ Reduce to lowest terms by dividing numerator and denominator by 3 ▲

The next two examples are somewhat more involved. In each case we simplify the numerator and denominator separately.

▼ **Examples** Simplify

14. $\dfrac{2(-3) + 4}{12} = \dfrac{-6 + 4}{12}$ Multiplication before addition

$= \dfrac{-2}{12}$ Addition

$= -\dfrac{1}{6}$ Reduce to lowest terms

15. $\dfrac{5(-4) + 6(-1)}{2(3) - 4(1)} = \dfrac{-20 + (-6)}{6 - 4}$ Multiplication before addition

$= \dfrac{-26}{2}$ Simplify numerator and denominator

$= -13$ Reduce to lowest terms ▲

Problem Set 1.7

Find the following quotients (divide).

1. $\dfrac{8}{-4}$ 2. $\dfrac{10}{-5}$ 3. $\dfrac{15}{-3}$

4. $\dfrac{20}{-10}$ 5. $\dfrac{-11}{33}$ 6. $\dfrac{-12}{48}$

7. $\dfrac{-48}{16}$ 8. $\dfrac{-32}{4}$ 9. $\dfrac{-27}{3}$

10. $\dfrac{-18}{9}$ 11. $\dfrac{-7}{21}$ 12. $\dfrac{-25}{100}$

13. $\dfrac{-39}{-13}$ 14. $\dfrac{-18}{-6}$ 15. $\dfrac{-24}{-8}$

16. $\dfrac{-4}{-28}$ 17. $\dfrac{-7}{-35}$ 18. $\dfrac{-21}{-28}$

19. $\dfrac{0}{-32}$ 20. $\dfrac{0}{17}$

The following problems involve more than one operation. Simplify as much as possible.

21. $\dfrac{3(-2)}{-10}$

22. $\dfrac{4(-3)}{24}$

23. $\dfrac{-5(-5)}{-15}$

24. $\dfrac{-7(-3)}{-35}$

25. $\dfrac{-8(-7)}{-28}$

26. $\dfrac{-3(-9)}{-6}$

27. $\dfrac{27}{4-13}$

28. $\dfrac{27}{13-2}$

29. $\dfrac{-50}{5-10}$

30. $\dfrac{-60}{6-10}$

31. $\dfrac{2(-3)-6}{6-4}$

32. $\dfrac{3(-2)+4}{6-4}$

33. $\dfrac{15(-5)-25}{2(-10)}$

34. $\dfrac{10(-3)-20}{5(-2)}$

35. $\dfrac{27-2(-4)}{-3(3)}$

36. $\dfrac{20-5(-3)}{10(-3)}$

37. $\dfrac{12-6(-2)}{12(-2)}$

38. $\dfrac{3(-4)+5(-6)}{10-6}$

39. $\dfrac{6(-4)-2(5-8)}{-6-3-5}$

40. $\dfrac{3(-4)-5(9-11)}{-9-2-3}$

41. $\dfrac{3(-5-3)+4(7-9)}{5(-2)+3(-4)}$

42. $\dfrac{-2(6-10)-3(8-5)}{6(-3)-6(-2)}$

We can apply the definition of division to fractions. Since dividing by a number is the same as multiplying by its reciprocal, to divide by the fraction $\frac{3}{4}$ we can multiply by its reciprocal $\frac{4}{3}$. For example,

$$\frac{2}{5} \div \frac{3}{4} = \frac{2}{5} \cdot \frac{4}{3} = \frac{8}{15}$$

You may have learned this rule in previous math classes. In some math classes, multiplication by the reciprocal is referred to as "inverting the divisor and multiplying." Divide the following fractions. Reduce all answers to lowest terms.

43. $\dfrac{4}{5} \div \dfrac{3}{4}$

44. $\dfrac{6}{8} \div \dfrac{3}{4}$

45. $-\dfrac{5}{6} \div \left(-\dfrac{5}{8}\right)$ **46.** $-\dfrac{7}{9} \div \left(-\dfrac{1}{6}\right)$

47. $\dfrac{10}{13} \div \left(-\dfrac{5}{4}\right)$ **48.** $\dfrac{5}{12} \div \left(-\dfrac{10}{3}\right)$

49. $-\dfrac{5}{6} \div \dfrac{5}{6}$ **50.** $-\dfrac{8}{9} \div \dfrac{8}{9}$

Answer the following questions.

51. What is the quotient of -12 and -4?
52. The quotient of -4 and -12 is what number?
53. What number do we divide by -5 to get 2?
54. Fifteen divided by what number is -3?
55. If the quotient of -20 and 4 is decreased by 3, what number results?
56. If -4 is added to the quotient of 24 and -8, what number results?

**1.8
Subsets of
Real Numbers**

Section 1.2 introduced real numbers. In this section we will classify the different subsets of the real numbers and look at some relationships among them.

DEFINITION Set A is called a *subset* of set B if set A is contained in set B. That is, if each and every element in set A is also a member of set B.

The set of *counting numbers* (the numbers we use to count) is the set $\{1, 2, 3, . . .\}$ (remember, the dots mean "and so on"). Since each of these numbers is also a real number, we say the counting numbers are a subset of the real numbers.

$$\text{Counting numbers} = \{1, 2, 3, . . .\}$$

▼ **Example 1** Each of the following is a counting number:

$$7, \quad 25, \quad 78, \quad 1000, \quad 100,000 \qquad \blacktriangle$$

If we extend the set of counting numbers to include the number zero, we get the set of *whole numbers:*

$$\text{Whole numbers} = \{0, 1, 2, 3, . . .\}$$

The set formed by taking the opposites of all counting numbers together with the whole numbers is the set of *integers:*

$$\text{Integers} = \{. . ., -3, -2, -1, 0, 1, 2, 3, . . .\}$$

▼ **Example 2** Each of the following is an integer:

$$-59{,}000, \quad -8, \quad 0, \quad 10, \quad 37, \quad 10{,}000 \qquad ▲$$

The set of *rational numbers* is the set of numbers commonly called "fractions" together with the integers. The set of rational numbers is difficult to list in the same way we have listed the other sets, so we will use a different kind of notation:

$$\text{Rational numbers} = \left\{ \frac{a}{b} \;\middle|\; a \text{ and } b \text{ are integers } (b \neq 0) \right\}$$

This notation is read "The set of elements $\frac{a}{b}$ such that a and b are integers." If a number can be put in the form $\frac{a}{b}$, where a and b are both from the set of integers, then it is called a rational number. A rational number is the ratio of two integers. We cannot have zero in the denominator, since division by zero is undefined.

▼ **Examples**

3. $\frac{5}{8}$ is a rational number because it is the ratio of the two integers 5 and 8.

4. $-\frac{2}{3}$ is a rational number because it can be thought of as the ratio of -2 to 3. (It can also be thought of as the ratio of 2 to -3.)

5. -8 is a rational number because it can be thought of as the ratio of -8 to 1:

$$-8 = \frac{-8}{1}$$

6. 10 is a rational number because it can be thought of as the ratio of 10 to 1:

$$10 = \frac{10}{1} \qquad ▲$$

As you can see from Examples 5 and 6, any counting number, whole number, or integer is also a rational number. It is true that a number may belong to many different sets. For example, the number 8 is a counting number, a whole number, an integer, and a rational number. As we will see later in this section, 8 is also a real number, and in Chapter 8 we will find that 8 is also a complex number.

Many numbers written in decimal form are rational numbers. If you are familiar with decimals, you may recall that 0.5 means $\frac{1}{2}$. Since $\frac{1}{2}$ is a rational number, so is any number equal to it. A number doesn't have to be written in the form $\frac{a}{b}$ to be considered a rational number. If it *can* be written in that form, then it is a rational number.

▼ **Example 7** 0.75 is a rational number because it can be written as $\frac{75}{100}$. ▲

▼ **Example 8** The number 0.666 . . . is a rational number because it is equal to $\frac{2}{3}$. To see that this is true, we can divide 2 by 3.

$$
\begin{array}{r}
0.66666 \\
3\overline{)2.00000} \\
\underline{1\ 8} \\
20 \\
\underline{18} \\
20 \\
\underline{18} \\
20 \\
\underline{18} \\
20 \\
\underline{18} \\
\end{array}
$$

As you can see, the 6s repeat forever. ▲

In general, any decimal that terminates after a certain number of places (such as 0.75) or any decimal that repeats a sequence of digits indefinitely (such as 0.666 . . .) can be written as the ratio of two integers and is therefore a rational number.

Still other numbers, each of which is associated with a point on the real number line, cannot be written as the ratio of two integers. In decimal form they never terminate and never repeat a sequence of digits indefinitely. They are called *irrational numbers* (because they are not rational).

Irrational numbers = {nonrational numbers; nonrepeating, nonterminating decimals}

We cannot write any irrational number in a form that is familiar to us, because they are all nonterminating, nonrepeating decimals. Since they are not rational, they cannot be written as the ratio of two integers. They have to be represented in other ways. One irrational number we have seen before is π. It is not 3.14. Rather, 3.14 is an approximation to π. It cannot be written as a decimal number. Other representations for irrational numbers are $\sqrt{2}$, $\sqrt{3}$, $\sqrt{5}$, $\sqrt{6}$, and, in general, the square root of any number that is not itself a perfect square. (If you are not familiar with square roots, you will be after Chapter 7.) Right now it is enough to know some numbers on the number line cannot be written as the ratio of two integers or in decimal form. We call them irrational numbers.

The set of real numbers is the set of numbers that are either rational or irrational. That is, a real number is either rational or irrational.

Real numbers = {all rational numbers and all irrational numbers}

▼ **Example 9** For the set $\{-9, -4.2, -\frac{1}{3}, 0, \frac{5}{6}, \sqrt{3}, \pi, 19\}$ which numbers are (a) counting numbers, (b) whole numbers, (c) integers, (d) rational numbers, (e) irrational numbers, and (f) real numbers.

SOLUTION

(a)   The counting number is 19.

(b)   The whole numbers are 0 and 19.

(c)   The integers are −9, 0, and 19.

(d)   The rational numbers are $-9, -4.2, -\frac{1}{3}, 0, \frac{5}{6}$, and 19.

(e)   The irrational numbers are $\sqrt{3}$ and π.

(f)   All of them are real numbers.   ▲

In a final note, we should mention the empty set. The empty set is the set without members. It is empty. We use the symbol Ø to represent the empty set. It is considered a subset of every set. We will use it mostly when solving equations.

Given the numbers in the set $\{-3, -2.5, 0, 1, \frac{3}{2}, \sqrt{15}\}$:   Problem Set 1.8

1.   List all the whole numbers.   **2.**   List all the integers.
3.   List all the rational numbers.   **4.**   List all the irrational numbers.
5.   List all the real numbers.

Given the numbers in the set $\{-10, -8, -0.333 \ldots, -2, 9, \frac{27}{3}, \pi\}$:

6.   List all the whole numbers.   **7.**   List all the integers.
8.   List all the rational numbers.   **9.**   List all the irrational numbers.
10. List all the real numbers.

Identify the following statements as either true or false.

11. Every whole number is also an integer.

12. The set of whole numbers is a subset of the set of integers.
13. A number can be both rational and irrational.
14. The set of rational and the set of irrational numbers have some elements in common.
15. Some whole numbers are also negative integers.
16. Every rational number is also a real number.
17. All integers are also rational numbers.
18. The set of integers is a subset of the set of rational numbers.

List all the sets the following numbers belong to. (Some numbers belong to more than one set.)

19. 5	20. 18	21. -6	22. -12
23. $-\frac{3}{4}$	24. $-\frac{7}{12}$	25. $\sqrt{3}$	26. $\sqrt{5}$
27. $\sqrt{25}$	28. $\sqrt{36}$	29. 1.32	30. 11.78
31. $-\frac{17}{25}$	32. $-\frac{34}{8}$	33. $\frac{27}{3}$	34. $\frac{30}{10}$
35. π	36. $-1\frac{2}{3}$	37. $-3\frac{3}{4}$	38. $5\frac{2}{3}$

Chapter 1 Summary and Review

Examples We will use the margins in the chapter summaries to give examples that correspond to the topic being reviewed whenever it is appropriate.

The number(s) in brackets next to each heading indicates the section(s) in which that topic is discussed.

SYMBOLS [1.1]

$a = b$ a is equal to b
$a \neq b$ a is not equal to b
$a < b$ a is less than b
$a \not< b$ a is not less than b
$a > b$ a is greater than b
$a \not> b$ a is not greater than b

The inequality symbols always point to the smaller quantity.

1. $3 + 5(3 + 4)$
    $= 3 + 5(7)$
    $= 3 + 35$
    $= 38$

ORDER OF OPERATION [1.1]

1. Do what is in the parentheses first, if you can. (In some cases it is not possible to do what is in the parentheses, as is the case when one of the quantities is a number and the other is a variable.)
2. Then perform all multiplications and divisions left to right (in the same direction in which you read.)
3. Perform all additions and subtractions left to right.

ADDITION OF REAL NUMBERS [1.3]

To add two real numbers with

1. the same sign: Simply add their absolute values and use the common sign.
2. different signs: Subtract the smaller absolute value from the larger absolute value. The answer has the same sign as the number with larger absolute value.

2. Add all combinations of positive and negative 10 and 13:

$$10 + 13 = 23$$
$$10 + (-13) = -3$$
$$-10 + 13 = 3$$
$$-10 + (-13) = -23$$

SUBTRACTION OF REAL NUMBERS [1.4]

To subtract one number from another, simply add the opposite of the number you are subtracting. That is, if a and b represent real numbers, then

$$a - b = a + (-b)$$

3. Subtracting 2 is the same as adding -2:

$$7 - 2 = 7 + (-2) = 5$$

Subtracting -2 is the same as adding $+2$:

$$7 - (-2) = 7 + 2 = 9$$

MULTIPLICATION OF REAL NUMBERS [1.6]

To multiply two real numbers, simply multiply their absolute values. Like signs give a positive answer. Unlike signs give a negative answer.

4.
$$3(5) = 15$$
$$3(-5) = -15$$
$$-3(5) = -15$$
$$-3(-5) = 15$$

DIVISION OF REAL NUMBERS [1.7]

Division by a number is the same as multiplication by its reciprocal. Like signs give a positive answer. Unlike signs give a negative answer.

5.
$$\frac{-6}{2} = -6\left(\frac{1}{2}\right) = -3$$

$$\frac{-6}{-2} = -6\left(-\frac{1}{2}\right) = 3$$

ABSOLUTE VALUE [1.2]

The absolute value of a real number is its distance from zero on the real number line. Absolute value is never negative.

6.
$$|5| = 5$$
$$|-5| = 5$$

OPPOSITES [1.2, 1.5]

Any two real numbers the same distance from zero on the number line but in opposite directions from zero are called opposites. Opposites always add to zero.

7. The numbers 3 and -3 are opposites; their sum is 0:

$$3 + (-3) = 0$$

8. The numbers 2 and $\frac{1}{2}$ are reciprocals; their product is 1:

$$2\left(\frac{1}{2}\right) = 1$$

RECIPROCALS [1.2, 1.5]

Any two real numbers whose product is one are called reciprocals. Every real number has a reciprocal except zero.

PROPERTIES OF REAL NUMBERS [1.5]

	For addition	*For multiplication*
Commutative:	$a + b = b + a$	$a \cdot b = b \cdot a$
Associative:	$a + (b + c) = (a + b) + c$	$a \cdot (b \cdot c) = (a \cdot b) \cdot c$
Identity:	$a + 0 = a$	$a \cdot 1 = a$
Inverse:	$a + (-a) = 0$	$a\left(\dfrac{1}{a}\right) = 1$
Distributive:	$a(b + c) = ab + ac$	

SUBSETS OF THE REAL NUMBERS [1.8]

Counting numbers:	$\{1, 2, 3, \ldots\}$
Whole numbers:	$\{0, 1, 2, 3, \ldots\}$
Integers:	$\{\ldots -3, -2, -1, 0, 1, 2, 3, \ldots\}$
Rational numbers:	{all numbers that can be expressed as the ratio of two integers}
Irrational numbers:	{all numbers on the number line that cannot be expressed as the ratio of two integers}
Real numbers:	{all numbers that are either rational or irrational}

COMMON MISTAKES

1. Interpreting absolute value as changing the sign of the number inside the absolute value symbols. $|-5| = +5$, $|+5| = -5$. (The first expression is correct, the second one is not.) To avoid this mistake, remember: Absolute value is a distance and distance is always measured in positive units.

2. Confusing $-(-5)$ with $-|-5|$. The first is the opposite of -5. The second expression is the opposite of the absolute value of -5. The two expressions are not the same. The first is $+5$. The second is -5.

3. Using the phrase "two negatives make a positive." This only works with multiplication. With addition, two

negative numbers produce a negative answer. It is best not to use the phrase "two negatives make a positive" at all.

Perform the indicated operations and simplify your answers as much as possible.

1. $3 + (-7)$

2. $-8 + 1$

3. $-4 + (-3)$

4. $-9 + (-6)$

5. $6(-3)$

6. $-6(5)$

7. $-5(-3)$

8. $-3(4) + 5$

9. $-4(-3) - 6$

10. $-5(-1) - 4$

11. $\dfrac{-3(-1) + 2}{8 - 9}$

12. $\dfrac{-2(-5) - 3(-4)}{6 - 12}$

13. $\dfrac{-9(-8) + (-3)(2)}{2(4) - 5}$

14. $8 - 3(6 - 4)$

15. $|-2|(5) - 4$

Answer the following true or false.

16. Subtraction is a commutative operation.
17. Multiplication distributes over addition.
18. Addition distributes over multiplication.
19. $-|-3| = 3$
20. The reciprocal of $-\frac{3}{4}$ is $\frac{4}{3}$.
21. Some numbers are both rational and irrational.

Given the set of numbers $\{1, 1.5, \sqrt{2}, \frac{3}{4}, -8\}$:

22. List all the integers.
23. List all the rational numbers.
24. List all irrational numbers.
25. List all real numbers.

For each expression on the left list the letter(s) that justify the property used.

26. $(x + y) + z = x + (z + y)$ A. Commutative property
27. $3(x - 5) = 3x - 15$ of addition
28. $a(3 \cdot b) = 3(a \cdot b)$ B. Commutative property
                                                            of multiplication

29. $(x + 5) + 7 = 7 + (x + 5)$ C. Associative property of
30. $3 \cdot 5 - 4 \cdot 5 = (3 - 4)5$ addition
31. $(8 + x)5 = 5(8 + x)$ D. Associative property
                                          of multiplication
                                 E. Distributive property

Write an expression in symbols that is equivalent to each of the following and then simplify.

32. The sum of 8 and -3
33. The difference of -24 and 2
34. The product of -5 and -4
35. The quotient of -24 and -2
36. Five added to the product of 6 and -1
37. Twice the sum of -9 and 2
38. The sum of twice 3 and -4

Apply the associative property to the following expressions and then simplify.

39. $3 + (5 + x)$ **40.** $-4 + (2 + x)$
41. $-\frac{1}{3}(-3x)$ **42.** $5(-4x)$

Apply the distributive property to each expression and then simply the result.

43. $2(3x + 5)$ **44.** $-3(x + 4)$
45. $-5(2x - 1)$ **46.** $-1(5x - 9)$

Linear Equations and Inequalities

Much of what we do in algebra is concerned with solving equations. In this chapter we will develop most of the properties necessary to solve many different types of equations. It may seem surprising, but most equations can be solved by applying just two main properties— the *addition property of equality* and the *multiplication property of equality*. To be successful in this chapter you must know the following concepts from Chapter 1:

1. Addition, subtraction, multiplication, and division of positive and negative real numbers
2. The commutative, associative, and distributive properties
3. That reciprocals multiply to 1 and opposites add to zero

As you will see in the next few sections, the first step in solving an equation is to simplify both sides as much as possible. In the first part of this section, we will practice simplifying expressions by combining what are called similar (or like) terms.

For our immediate purposes, a term is a number or a number and one or more variables multiplied together. For example, the number 5 is a term, as are the expressions $3x$, $-7y$, and $15xy$.

2.1 Simplifying Expressions

DEFINITION   Two or more terms with the same variable part are called *similar* (or *like*) terms.

The terms $3x$ and $4x$ are similar since their variable parts are identical. Likewise, the terms $18y$, $-10y$, and $6y$ are similar terms.

To simplify an algebraic expression, we simply reduce the number of terms in the expression. We accomplish this by applying the distributive property along with our knowledge of addition and subtraction of positive and negative real numbers. The following examples illustrate the procedure.

▼   **Examples**   Simplify by combining similar terms.

1. $3x + 4x = (3 + 4)x$ Distributive property
               $= 7x$ Addition of 3 and 4

2. $7a - 10a = (7 - 10)a$ Distributive property
               $= -3a$ Addition of 7 and -10

3. $18y - 10y + 6y = (18 - 10 + 6)y$ Distributive property
                       $= 14y$ Addition of 18, -10,
                                            and 6   ▲

When the expression we intend to simplify is more complicated, we use the commutative and associative properties first.

▼   **Examples**   Simplify each expression.

4. $3x + 5 + 2x - 3 = 3x + 2x + 5 - 3$ Commutative property
                     $= (3x + 2x) + (5 - 3)$ Associative property
                     $= (3 + 2)x + (5 - 3)$ Distributive property
                     $= 5x + 2$ Addition

5. $4a - 7 - 2a + 3 = (4a - 2a) + (-7 + 3)$ Commutative and
                                                  associative
                                                  properties
                     $= (4 - 2)a + (-7 + 3)$ Distributive property
                     $= 2a - 4$ Addition

6. $5x + 8 - x - 6 = (5x - x) + (8 - 6)$ Commutative and
                                                  associative
                                                  properties
                     $= (5 - 1)x + (8 - 6)$ Distributive property
                     $= 4x + 2$ Addition   ▲

Notice that in each case the result has fewer terms than the original expression. Since there are fewer terms, the resulting expression is said to be simpler than the original expression.

If an expression contains parentheses, it is often necessary to apply the distributive property to remove the parentheses before combining similar terms.

▼ **Example 7** Simplify the expression $5(2x - 8) - 3$.

SOLUTION We begin by distributing the 5 across $2x - 8$. We then combine similar terms:

$$
\begin{aligned}
5(2x - 8) - 3 &= 10x - 40 - 3 \qquad \text{Distributive property} \\
&= 10x - 43
\end{aligned}
$$
▲

▼ **Example 8** Simplify $7 - 3(2y + 1)$.

SOLUTION By the rule for order of operation we must multiply before we add or subtract. For that reason, it would be incorrect to subtract 3 from 7 first. Instead we multiply -3 and $2y + 1$ to remove the parentheses and then combine similar terms:

$$
\begin{aligned}
7 - 3(2y + 1) &= 7 - 6y - 3 \qquad \text{Distributive property} \\
&= -6y + 4
\end{aligned}
$$
▲

▼ **Example 9** Simplify $5(x - 2) - (3x + 4)$.

SOLUTION We begin by applying the distributive property to remove the parentheses. The expression $-(3x + 4)$ can be thought of as $-1(3x + 4)$. Thinking of it in this way allows us to apply the distributive property:

$$
\begin{aligned}
-1(3x + 4) &= -1(3x) + (-1)(4) \\
&= -3x - 4
\end{aligned}
$$

The complete solution looks like this:

$$
\begin{aligned}
5(x - 2) - (3x + 4) &= 5x - 10 - 3x - 4 \qquad \text{Distributive} \\
&\qquad\qquad\qquad\qquad\qquad\quad \text{property} \\
&= 2x - 14 \qquad\qquad\qquad \text{Combine similar} \\
&\qquad\qquad\qquad\qquad\qquad\quad \text{terms}
\end{aligned}
$$
▲

The last topic we will cover in this section will be useful later for checking solutions to equations. We will see how to find the value of an expression when the variable in the expression has been replaced by a number.

The Value of
an Expression

An expression like $3x + 2$ will have a certain value depending on what number we assign to x. For instance, when x is 4, $3x + 2$ becomes $3(4) + 2$, or 14. When x is -8, $3x + 2$ becomes $3(-8) + 2$, or -22. The value of an expression is found by replacing the variable with a given number.

▼ **Examples** Find the value of the following expressions by replacing the variable with the given number.

Expression	Value of the Variable	Value of the Expression
10. $3x - 1$	$x = 2$	$3(2) - 1 = 6 - 1$ $= 5$
11. $7a + 4$	$a = -3$	$7(-3) + 4 = -21 + 4$ $= -17$
12. $2x - 3 + 4x$	$x = -1$	$2(-1) - 3 + 4(-1)$ $= -2 - 3 + (-4)$ $= -9$
13. $2x - 5 - 8x$	$x = 5$	$2(5) - 5 - 8(5)$ $= 10 - 5 - 40$ $= -35$

▲

Simplifying an expression should not change its value. That is, if an expression has a certain value when x is 3, then it will always have that value no matter how much it has been simplified as long as x is 3. If we were to simplify the expression in Example 13 first, it would look like $2x - 5 - 8x = -6x - 5$. When x is 5, the simplified expression $-6x - 5$ is $-6(5) - 5 = -30 - 5 = -35$. It has the same value as the original expression when x is 5.

Problem Set 2.1

Simplify the following expressions.

1. $3x - 6x$
2. $7x - 5x$
3. $-2a + a$
4. $3a - a$
5. $7x + 3x + 2x$
6. $8x - 2x - x$
7. $3a - 2a + 5a$
8. $7a - a + 2a$
9. $4x - 3 + 2x$
10. $5x + 6 - 3x$
11. $3a + 4a + 5$
12. $6a + 7a + 8$
13. $2x - 3 + 3x - 2$
14. $6x + 5 - 2x + 3$
15. $3a - 1 + a + 3$
16. $-a + 2 + 8a - 7$

17. $-4x + 8 - 5x - 10$	**18.** $-9x - 1 + x - 4$
19. $7a + 3 + 2a + 3a$	**20.** $8a - 2 + a + 5a$
21. $5(2x - 1) + 4$	**22.** $2(4x - 3) + 2$
23. $7(3y + 2) - 8$	**24.** $6(4y + 2) - 7$
25. $-3(2x - 1) + 5$	**26.** $-4(3x - 2) - 6$
27. $5 - 2(a + 1)$	**28.** $7 - 8(2a + 3)$
29. $6 - 4(x - 5)$	**30.** $12 - 3(4x - 2)$
31. $-9 - 4(2 - y) + 1$	**32.** $-10 - 3(2 - y) + 3$
33. $-6 + 2(2 - 3x) + 1$	**34.** $-7 - 4(3 - x) + 1$
35. $3(x - 2) + 2(x - 3)$	**36.** $2(2x + 1) - 3(x + 4)$
37. $4(2y - 8) - (y + 7)$	**38.** $5(y - 3) - (y - 4)$
39. $-9(2x + 1) - 3(x + 5)$	**40.** $-3(3x - 2) - 4(2x + 3)$

Evaluate the following expressions when x is 2. (Find the value of the expressions if x is 2.)

41. $3x - 1$	**42.** $4x + 3$
43. $7x - 8$	**44.** $8x - 7$
45. $3x + 2x + 1$	**46.** $4x + 3x + 2$
47. $-2x - 5$	**48.** $-3x + 6$
49. $-2x + 6 + 3x$	**50.** $-6x + 7 - 2x$

Find the value of the following expressions when x is -5. Then simplify the expression and check to see that it has the same value for $x = -5$.

51. $2x - 5x + 1$	**52.** $4x - 7x + 5$
53. $5x + 6 + 2x + 1$	**54.** $2x - 3 - 5x - 1$
55. $7x - 4 - x - 3$	**56.** $3x + 4 + 7x - 6$
57. $3(x - 2) + 4$	**58.** $2(x - 5) + 3$
59. $5(2x + 1) + 4$	**60.** $2(3x - 10) + 5$

When we use words to describe a mathematical expression that includes parentheses, we sometimes use the phrase "the quantity" to show where the parentheses start and a comma to show where they end. For instance, the expression $2(3x + 4) - 5$ can be read "2 times the quantity $3x$ plus 4, minus 5." Write each of the following expressions in words and include the phrase "the quantity."

61. $3(5x + 1) - 6$	**62.** $4(2x - 6) + 7$
63. $4(2x + 3y) - 1$	**64.** $8(3x - 5y) + 4$
65. $4(2x + 3y - 1)$	**66.** $8(3x - 5y + 4)$
67. $2(4x + 1) + 3(2x + 1)$	**68.** $5(7x - 8) + 2(3x + 1)$

Review Problems From here on, each problem set will end with a series of review problems. In mathematics it is very important to review. The more you review, the better you will understand the topics we cover and the longer you will remember them. Also, there are times when material that seemed confusing earlier will be less confusing the second time around.

The following problems review material we covered in Sections 1.6 and 1.7.

Simplify each expression.

69. $-2(4 - 9)$ **70.** $-6(3 - 10)$

71. $-4(-5) - 3$ **72.** $-2(6) - 8$

73. $3 - 5(6 - 2)$ **74.** $7 - 4(1 - 5)$

75. $\dfrac{8}{2 - 4}$ **76.** $\dfrac{25}{5 - 10}$

77. $\dfrac{2(-6) - 3(-2)}{4 - 7}$ **78.** $\dfrac{-3(5) - 7(-5)}{2 - 6}$

2.2 Addition Property of Equality

In this section we will solve some simple equations. To solve an equation, we must find all replacements for the variable that make the equation a true statement.

DEFINITION The *solution set* for an equation is the set of all numbers that when used in place of the variable make the equation a true statement.

For example, the equation $x + 2 = 5$ has solution set $\{3\}$ because when x is 3 the equation becomes the true statement $3 + 2 = 5$, or $5 = 5$.

The important thing about an equation is its solution set. We therefore make the following definition in order to classify together all equations with the same solution set.

DEFINITION Two or more equations with the same solution set are said to be *equivalent equations*.

Equivalent equations may look different but must have the same solution set.

▼ **Examples**

1. $x + 2 = 5$ and $x = 3$ are equivalent equations, since both have solution set $\{3\}$.

2. $a - 4 = 3$, $a - 2 = 5$, and $a = 7$ are equivalent equations, since they all have solution set $\{7\}$.

3. $y + 3 = 4$, $y - 8 = -7$, and $y = 1$ are equivalent equations, since they all have solution set $\{1\}$.　　　　　　　　　▲

If two numbers are equal and we increase (or decrease) both of them by the same amount, the resulting quantities are also equal. We can apply this concept to equations. Adding the same amount to both sides of an equation always produces an equivalent equation—one with the same solution set. This fact about equations is called the *addition property of equality* and can be stated more formally as follows.

Addition Property of Equality

For any three algebraic expressions A, B, and C,

$$\text{If} \qquad A = B$$
$$\text{then } A + C = B + C$$

In words: Adding the same quantity to both sides of an equation will not change the solution set.

This property is just as simple as it seems. We can add any amount to both sides of an equation and always be sure we have not changed the solution set.

Consider the equation $x + 6 = 5$. We want to solve this equation for the value of x that makes it a true statement. We want to end up with x on one side of the equal sign and a number on the other side. Since we want x by itself, we will add -6 to both sides.

$$x + 6 + (-6) = 5 + (-6) \qquad \text{Addition property of equality}$$
$$x + 0 = -1 \qquad\qquad\quad \text{Addition}$$
$$x = -1$$

All three equations say the same thing about x. They all say that x is -1. All three equations are equivalent. The last one is just easier to read.

Here are some further examples of how the addition property of equality can be used to solve equations.

▼   **Examples**   Solve the following equations.

4. $x - 5 = 12$

SOLUTION   Since we want x alone on the left side, we choose to add $+5$ to both sides.

$$x - 5 + 5 = 12 + 5 \qquad \text{Addition property of equality}$$
$$x + 0 = 17$$
$$x = 17$$

5. $a + 7 = -3$

   SOLUTION   Since we want a to be by itself, we will add the opposite of 7 to both sides.

$$a + 7 + (-7) = -3 + (-7)$$ Addition property of
   equality

$$a + 0 = -10$$
$$a = -10$$

6. $y - 8 = -3$

   SOLUTION   Again, we want to isolate y so we add the opposite of -8 to both sides.

$$y - 8 + 8 = -3 + 8$$ Addition property of
   equality

$$y + 0 = 5$$
$$y = 5$$ ▲

Sometimes it is necessary to simplify each side of an equation before using the addition property of equality. The reason we simplify both sides first is that we want as few terms as possible on each side of the equation before we use the addition property of equality. The following examples illustrate this procedure.

▼   Examples   Solve for x.

7. $-x + 2 + 2x = 7 + 5$

    SOLUTION

$$x + 2 = 12$$ Simplify both sides first
$$x + 2 + (-2) = 12 + (-2)$$ Addition property of
   equality

$$x + 0 = 10$$
$$x = 10$$

8. $3x - 4 - 2x = 11 - 4$

    SOLUTION

$$x - 4 = 7$$ Simplify both sides first
$$x - 4 + 4 = 7 + 4$$ Addition property of
   equality

$$x + 0 = 11$$
$$x = 11$$ ▲

So far in this section we have used the addition property of equality to add only numbers to both sides of an equation. It is often necessary to add a term involving a variable to both sides of an equation, as the following examples indicate.

▼   **Examples**   Solve for x.

9. $5x = 4x + 2$

SOLUTION   To isolate x on the left side of the equation, we add $-4x$ to both sides:

$$5x + (-4x) = 4x + (-4x) + 2 \qquad \text{Add } -4x \text{ to both sides}$$
$$x = 2 \qquad \text{Simplify}$$

10. $3x - 5 = 2x + 7$

SOLUTION   We can solve this equation in two steps. First, we add $-2x$ to both sides of the equation. When this has been done, x will appear on the left side only. Second, we add 5 to both sides:

$$3x + (-2x) - 5 = 2x + (-2x) + 7 \qquad \text{Add } -2x \text{ to both sides}$$
$$x - 5 = 7 \qquad \text{Simplify each side}$$
$$x - 5 + 5 = 7 + 5 \qquad \text{Add 5 to both sides}$$
$$x = 12 \qquad \text{Simplify each side} \qquad ▲$$

To check your work, substitute your solution for x into the original equation.

▼   **Example 11**   Check the solution for the equation in Example 10.

SOLUTION   We replace x with 12 in the equation $3x - 5 = 2x + 7$:

$$3(12) - 5 = 2(12) + 7$$
$$36 - 5 = 24 + 7$$
$$31 = 31$$

Since $x = 12$ makes the original equation a true statement, it must be the solution to the equation.   ▲

Once you have become proficient at solving equations, it will not be necessary to check every solution to every equation you solve. What is important is that you know how to check your solutions when you think it is necessary—as on a test.

Many of the equations in the problem set that follows will seem very simple. You will be able to recognize solutions to some of them without doing any work. It won't always be this way. The idea here is to develop a method of solving equations. To do so, you must show all your work until you can consistently solve equations correctly using the method shown in the examples in this section.

Problem Set 2.2

Find the solution set for the following equations. Be sure to show when you have used the addition property of equality.

1. $x - 3 = 8$
2. $x - 2 = 7$
3. $x + 2 = 6$
4. $x + 5 = 4$
5. $a + 7 = -2$
6. $a + 8 = -1$
7. $x + 5 = -3$
8. $y + 10 = -4$
9. $y + 11 = -6$
10. $x - 3 = -1$
11. $x - 5 = -2$
12. $x - 3 = -5$
13. $m - 6 = -10$
14. $m - 10 = -6$
15. $5 + x = 4$
16. $4 + x = 7$
17. $5 = a + 4$
18. $12 = a - 3$
19. $-3 = x - 8$
20. $-5 = x - 4$

Simplify both sides of the following equations as much as possible and then solve.

21. $4x + 2 - 3x = 4 + 1$
22. $5x + 2 - 4x = 7 - 3$
23. $8a - 5 - 7a = 3 + 5$
24. $9a - 18 - 8a = 20 - 7$
25. $-3 - 4x + 5x = 18$
26. $10 - 3x + 4x = 20$
27. $-11x + 2 + 10x + 2x = 9$
28. $-10x + 5 - 4x + 15x = 0$
29. $-3 + 4 = 8x - 5 - 7x$
30. $-2 + 5 = 7x - 4 - 6x$
31. $2y - 10 + y - 4y = 18 - 6$
32. $3y - 20 + 4y - 8y = 21$
33. $15 - 21 = 8x + x - 10x$
34. $23 - 17 = -7x - x + 9x$
35. $24 - 3 + 8a - 5a - 2a = 21$
36. $30 - 4 + 7a - 2 - 6a = 30$

The following equations contain parentheses. Apply the distributive property to remove the parentheses, then simplify each side before using the addition property of equality.

37. $2(x + 3) - x = 4$
38. $5(x + 1) - 4x = 2$
39. $-3(x - 4) + 4x = 3 - 7$
40. $-2(x - 5) + 3x = 4 - 9$
41. $5(2a + 1) - 9a = 8 - 6$
42. $4(2a - 1) - 7a = 9 - 5$
43. $-(x + 3) + 2x - 1 = 6$
44. $-(x - 7) + 2x - 8 = 4$
45. $4y - 3(y - 6) + 2 = 8$
46. $7y - 6(y - 1) + 3 = 9$
47. $2(3x + 1) - 5(x + 2) = 1 - 10$
48. $4(2x + 1) - 7(x - 1) = 2 - 6$
49. $-3(2m - 9) + 7(m - 4) = 12 - 9$
50. $-5(m - 3) + 2(3m + 1) = 15 - 8$

Solve the following equations by the method used in Examples 9 and 10 in this section.

51.	$4x = 3x + 2$	**52.**	$6x = 5x - 4$
53.	$8a = 7a - 5$	**54.**	$9a = 8a - 3$
55.	$2x = 3x + 1$	**56.**	$4x = 3x + 5$
57.	$3y + 4 = 2y + 1$	**58.**	$5y + 6 = 4y + 2$
59.	$2m - 3 = m + 5$	**60.**	$8m - 1 = 7m - 3$
61.	$4x - 7 = 5x + 1$	**62.**	$3x - 7 = 4x - 6$
63.	$-2x = -3x + 5$	**64.**	$-5x = -6x + 3$
65.	$4 - 2a = -7 - 3a$	**66.**	$2 - 3a = -5 - 4a$

Review Problems The problems below review material we covered in Section 1.5. Reviewing this material will help you in the next section.

Apply the associative property to each expression and then simplify the result.

67.	$3(6x)$	**68.**	$5(4x)$
69.	$\frac{1}{5}(5x)$	**70.**	$\frac{1}{3}(3x)$
71.	$8(\frac{1}{8}y)$	**72.**	$6(\frac{1}{6}y)$
73.	$-2(-\frac{1}{2}x)$	**74.**	$-4(-\frac{1}{4}x)$

In the previous section we found that adding the same number to both sides of an equation never changed the solution set. The same idea holds for multiplication by numbers other than zero. We can multiply both sides of an equation by the same nonzero number and always be sure we have not changed the solution set. (The reason we cannot multiply both sides by zero will become apparent later.) This fact about equations is called the *multiplication property of equality*, and can be stated formally as follows.

**2.3
Multiplication
Property of Equality**

Multiplication Property of Equality

For any three algebraic expressions A, B, and C, where $C \neq 0$,

$$\text{If} \quad A = B$$
$$\text{then } AC = BC$$

In words: Multiplying both sides of an equation by the same nonzero number will not change the solution set.

Suppose we want to solve the equation $5x = 30$. We have $5x$ on the left side but would like to have just x. We choose to multiply both sides by $\frac{1}{5}$ since $(\frac{1}{5})(5) = 1$. Here is the solution:

$$5x = 30$$
$$(\tfrac{1}{5})(5x) = (\tfrac{1}{5})(30) \qquad \text{Multiplication property of equality}$$
$$(\tfrac{1}{5} \cdot 5)x = (\tfrac{1}{5})(30) \qquad \text{Associative property of multiplication}$$
$$1x = 6$$
$$x = 6$$

We chose to multiply by $\frac{1}{5}$ because it is the reciprocal of 5. We can see that multiplication by any number but zero will not change the solution set. If, however, we were to multiply both sides by zero, the result would always be $0 = 0$, since multiplication by zero always results in zero. The statement $0 = 0$ is nice, but we have lost our variable and cannot solve the equation. This is the only restriction on the multiplication property of equality. We are free to multiply both sides of an equation by any number except zero.

Here are some more examples that use the multiplication property of equality.

▼   **Examples**   Solve each equation.

1. $3x = 18$

   SOLUTION   Since we want x alone on the left side, we choose to multiply by $\frac{1}{3}$:

   $$(\tfrac{1}{3})(3x) = (\tfrac{1}{3})(18) \qquad \text{Multiplication property of equality}$$
   $$(\tfrac{1}{3} \cdot 3)x = (\tfrac{1}{3})(18) \qquad \text{Multiplication is associative}$$
   $$1x = 6$$
   $$x = 6$$

2. $-4a = 24$

   SOLUTION   Since we want a alone on the left side, we choose to multiply both sides by $-\frac{1}{4}$:

   $$(-\tfrac{1}{4})(-4a) = (-\tfrac{1}{4})(24) \qquad \text{Multiplication property of equality}$$
   $$[-\tfrac{1}{4}(-4)]a = (-\tfrac{1}{4})(24) \qquad \text{Associative property}$$
   $$a = -6$$

3. $\dfrac{x}{6} = 7$

   SOLUTION   Since $\dfrac{x}{6} = \left(\dfrac{1}{6}\right)x$, we will multiply both sides by 6 and come up with $1x$ on the left side:

$$6\left(\frac{x}{6}\right) = 6(7) \qquad \text{Multiplication property of equality}$$

$$\left(6 \cdot \frac{1}{6}\right)x = 6(7) \qquad \text{Associative property}$$

$$x = 42 \qquad \text{Simplify}$$

4. $\dfrac{y}{8} = 2$

SOLUTION Since $\dfrac{y}{8} = \left(\dfrac{1}{8}\right)y$, we can multiply both sides by 8 and have $1y$ on the left side:

$$8\left(\frac{y}{8}\right) = 8(2) \qquad \text{Multiplication property of equality}$$

$$\left(8 \cdot \frac{1}{8}\right)y = 8(2) \qquad \text{Associative property}$$

$$y = 16 \qquad \text{Simplify} \qquad \blacktriangle$$

Notice in the above examples that if the variable is being multiplied by a number like -4 or $\frac{1}{6}$, we always multiply by the number's reciprocal, $-\frac{1}{4}$ or 6, to end up with just x on one side of the equation.

▼ **Examples** Solve for x.

5. $-3x + 5x + 2x = 8 + 2$

    SOLUTION

$$
\begin{aligned}
4x &= 10 && \text{Simplify both sides first} \\
(\tfrac{1}{4})(4x) &= (\tfrac{1}{4})(10) && \text{Multiplication property of equality} \\
x &= \tfrac{10}{4} && \text{Multiplication} \\
x &= \tfrac{5}{2} && \text{Reduce to lowest terms}
\end{aligned}
$$

6. $7x - 4x - x = -11 - 3$

    SOLUTION

$$
\begin{aligned}
2x &= -14 && \text{Simplify both sides first} \\
(\tfrac{1}{2})(2x) &= (\tfrac{1}{2})(-14) && \text{Multiplication property of equality} \\
x &= -7 && \text{Multiply}
\end{aligned}
$$

7. $5 + 8 = 10x + 20x - 4x$

SOLUTION

$13 = 26x$	Simplify both sides first
$(\frac{1}{26})(13) = (\frac{1}{26})(26x)$	Multiplication property of equality
$\frac{13}{26} = x$	Multiplication
$\frac{1}{2} = x$	Reduce to lowest terms ▲

   In the last three examples in this section we will use both the addition property of equality and the multiplication property of equality. As a general rule, we use the addition property of equality before the multiplication property of equality.

▼ **Example 8**   Solve for x: $6x + 5 = -13$.

SOLUTION   We begin by adding -5 to both sides of the equation:

$6x + 5 + (-5) = -13 + (-5)$	Add -5 to both sides
$6x = -18$	Simplify
$\frac{1}{6}(6x) = \frac{1}{6}(-18)$	Multiply both sides by $\frac{1}{6}$
$x = -3$	▲

▼ **Example 9**   Solve for x: $5x = 2x + 12$.

SOLUTION   We begin by adding $-2x$ to both sides of the equation:

$5x + (-2x) = 2x + (-2x) + 12$	Add $-2x$ to both sides
$3x = 12$	Simplify
$\frac{1}{3}(3x) = \frac{1}{3}(12)$	Multiply both sides by $\frac{1}{3}$
$x = 4$	Simplify ▲

   Notice that in Example 9 we used the addition property of equality first in order to combine all the terms containing x on the left side of the equation. Once this had been done, we used the multiplication property to get just x on the left side.

▼ **Example 10**   Solve for x: $3x - 4 = -2x + 6$.

SOLUTION   We begin by adding $2x$ to both sides:

$3x + 2x - 4 = -2x + 2x + 6$	Add $2x$ to both sides
$5x - 4 = 6$	Simplify

Now we add 4 to both sides:

$$5x - 4 + 4 = 6 + 4 \qquad \text{Add 4 to both sides}$$
$$5x = 10 \qquad \text{Simplify}$$
$$\tfrac{1}{5}(5x) = \tfrac{1}{5}(10) \qquad \text{Multiply by } \tfrac{1}{5}$$
$$x = 2 \qquad \text{Simplify} \qquad \blacktriangle$$

As you can see in the three examples above, both properties can be used in the process of solving an equation. The addition property is used first to combine all the terms containing the variable together on one side of the equation and terms without the variable on the other side. Once that has been accomplished, the multiplication property of equality is used to obtain just one of whatever variable is being solved for.

Solve the following equations. Be sure to show your work.

1. $5x = 10$ **2.** $6x = 12$

3. $7a = 28$ **4.** $4a = 36$

5. $-2x = 8$ **6.** $-3x = 15$

7. $-8x = 4$ **8.** $-6x = 2$

9. $8m = -16$ **10.** $5m = -25$

11. $-3x = -9$ **12.** $-9x = -36$

13. $-7y = -28$ **14.** $-15y = -30$

15. $-20x = -20$ **16.** $-2x = -10$

17. $2x = 0$ **18.** $7x = 0$

19. $-5x = 0$ **20.** $-3x = 0$

21. $\dfrac{x}{3} = 2$ **22.** $\dfrac{x}{4} = 3$

23. $\dfrac{m}{5} = 10$ **24.** $\dfrac{m}{7} = 1$

25. $\dfrac{x}{2} = -3$ **26.** $\dfrac{x}{3} = -2$

27. $\dfrac{a}{-5} = -3$ **28.** $\dfrac{a}{-6} = -10$

29. $-\dfrac{x}{5} = 11$ **30.** $-\dfrac{x}{8} = 4$

Simplify both sides as much as possible and then solve.

31. $3x - 4x - 2x = 6$ **32.** $-2x - 3x + x = 8$

33. $-8x + x - 2x = 18$ **34.** $-x - 2x - 3x = 12$
35. $-4x - 2x + 3x = 24$ **36.** $7x - 5x + 8x = 20$
37. $4x + 8x - 2x = 15 - 10$ **38.** $5x + 4x + 3x = 4 + 8$
39. $-3 - 5 = 3x + 5x - 10x$ **40.** $10 - 16 = 12x - 6x - 3x$
41. $18 - 3 = 8a - 5a + 2a$ **42.** $20 - 4 = 7a + a - 4a$

Solve each of the following equations by multiplying both sides by -1.

43. $-x = 4$ **44.** $-x = -3$
45. $-x = -4$ **46.** $-x = 3$
47. $15 = -a$ **48.** $-15 = -a$
49. $-y = \frac{1}{2}$ **50.** $-y = -\frac{3}{4}$

Solve each of the following equations using the method shown in Examples 8, 9, and 10 in this section.

51. $3x - 2 = 7$ **52.** $2x - 3 = 9$
53. $3a - 7 = 14$ **54.** $9a - 7 = 29$
55. $2a + 1 = 3$ **56.** $5a - 3 = 7$
57. $1 + 4x = 2$ **58.** $7 + 3x = -8$
59. $6x = 2x - 12$ **60.** $8x = 3x - 10$
61. $2y = -4y + 18$ **62.** $3y = -2y - 15$
63. $-7x = -3x - 8$ **64.** $-5x = -2x - 12$
65. $3x + 4 = 2x - 5$ **66.** $5x + 6 = 3x - 6$
67. $6m - 3 = 5m + 2$ **68.** $2m - 5 = m + 5$
69. $7y + 2 = 6y - 4$ **70.** $3y + 14 = 2y - 2$

Review Problems The problems below review material we covered in Section 2.1. Reviewing this material will help you in the next section.

Simplify each expression.

71. $5(2x - 8) - 3$ **72.** $4(3x - 1) + 7$
73. $-2(3x + 5) + 3(x - 1)$ **74.** $6(x + 3) - 2(2x + 4)$
75. $7 - 3(2y + 1)$ **76.** $8 - 5(3y - 4)$
77. $4x - (9x - 3) + 4$ **78.** $x - (5x + 2) - 3$

2.4
Solving Linear
Equations

We will now use the material we have developed in the first three sections of this chapter to build a method for solving any linear equation.

DEFINITION A *linear equation* in one variable is any equation that can be put in the form $ax + b = 0$, where a and b are real numbers and a is not zero.

Each of the equations we will solve in this section is a linear equation in one variable. The steps we use to solve a linear equation in one variable are listed below.

Step 1:   Use the distributive property to separate terms, if necessary, and simplify both sides of the equation as much as possible.

Step 2:   Use the addition property of equality to get all variable terms on one side and all constant terms on the other. A variable term is a term that contains the variable (for example, 5x). A constant term is a term that does not contain the variable (the number 3, for example).

Step 3:   Use the multiplication property of equality to get x by itself on one side.

Step 4:   Check your solution in the original equation if you think it is necessary.

▼   **Example 1**   Solve: $2(x + 3) = 10$.

SOLUTION

$$2x + 6 = 10 \qquad \text{Distributive property}$$
$$2x + 6 + (-6) = 10 + (-6) \qquad \text{Addition property of equality}$$
$$2x = 4$$
$$(\tfrac{1}{2})(2x) = (\tfrac{1}{2})(4) \qquad \text{Multiplication property of}$$
$$x = 2 \qquad\qquad\qquad \text{equality}$$
▲

The general method of solving linear equations is actually very simple. It is based on the properties we developed in Chapter 1 and two very simple new properties. We can add any number to both sides of the equation and multiply both sides by any nonzero number. The equation may change in form, but the solution set will not. If we look back to Example 1, each equation looks a little different from each preceding equation. What is interesting and useful is that each equation says the same thing about x. They all say x is 2. The last equation, of course, is the easiest to read, and that is why our goal is to end up with x by itself.

▼   **Examples**   Solve

2. $3(x - 5) + 4 = 13$

SOLUTION

$$3x - 15 + 4 = 13 \qquad \text{Distributive property}$$
$$3x - 11 = 13 \qquad \text{Simplify the left side}$$

$$3x - 11 + 11 = 13 + 11 \qquad \text{Add } 11 \text{ to both sides}$$
$$3x = 24$$
$$(\tfrac{1}{3})(3x) = (\tfrac{1}{3})(24) \qquad \text{Multiply both sides by } \tfrac{1}{3}$$
$$x = 8$$

3. $5(2a - 5) = 3a - 4$

SOLUTION

$$10a - 25 = 3a - 4 \qquad \text{Distributive property}$$
$$10a - 25 + 25 = 3a - 4 + 25 \qquad \text{Add } 25 \text{ to both sides}$$
$$10a = 3a + 21$$
$$10a + (-3a) = 3a + (-3a) + 21 \qquad \text{Add } -3a \text{ to both sides}$$
$$7a = 21$$
$$(\tfrac{1}{7})(7a) = (\tfrac{1}{7})(21) \qquad \text{Multiply both sides by } \tfrac{1}{7}$$
$$a = 3$$

4. $5(x - 3) + 2 = 5(2x - 8) - 3$

SOLUTION

$$5x - 15 + 2 = 10x - 40 - 3 \qquad \text{Distributive property}$$
$$5x - 13 = 10x - 43 \qquad \text{Simplify both sides}$$
$$5x + (-5x) - 13 = 10x + (-5x) - 43 \qquad \text{Add } -5x \text{ to both sides}$$
$$-13 = 5x - 43$$
$$-13 + 43 = 5x - 43 + 43 \qquad \text{Add } 43 \text{ to both sides}$$
$$30 = 5x$$
$$(\tfrac{1}{5})(30) = (\tfrac{1}{5})(5x) \qquad \text{Multiply both sides by } \tfrac{1}{5}$$
$$6 = x \qquad \blacktriangle$$

It makes no difference which side of the equal sign x ends up on. Most people prefer to have x on the left side because we read from left to right and it seems to sound better to say x is 6 rather than 6 is x. Both expressions, however, have exactly the same meaning.

▼ **Example 5**   Solve: $7 - 3(2y + 1) = 16$.

SOLUTION

$$7 - 6y - 3 = 16 \qquad \text{Distributive property}$$
$$-6y + 4 = 16 \qquad \text{Simplify the left side}$$
$$-6y + 4 + (-4) = 16 + (-4) \qquad \text{Add } -4 \text{ to both sides}$$
$$-6y = 12$$
$$(-\tfrac{1}{6})(-6y) = (-\tfrac{1}{6})(12) \qquad \text{Multiply both sides by } -\tfrac{1}{6}$$
$$y = -2 \qquad \blacktriangle$$

▼ **Example 6** Check the solution from Example 5 in the original equation.

SOLUTION When $y = -2$
the equation $7 - 3(2y + 1) = 16$
becomes $7 - 3[2(-2) + 1] = 16$
$$7 - 3(-4 + 1) = 16$$
$$7 - 3(-3) = 16$$
$$7 + 9 = 16$$
$$16 = 16$$

Since replacing y with -2 in the original equation yields a true statement, we know $y = -2$ is the solution.   ▲

Solve each of the following equations using the four steps shown in this section.

1. $2(x + 3) = 12$
2. $3(x - 2) = 6$
3. $6(x - 1) = -18$
4. $4(x + 5) = 16$
5. $2(4a + 1) = -6$
6. $3(2a - 4) = 12$
7. $14 = 2(5x - 3)$
8. $-25 = 5(3x + 4)$
9. $-2(3y + 5) = 14$
10. $-3(2y - 4) = -6$
11. $-5(2a + 4) = 0$
12. $-3(3a - 6) = 0$
13. $6 = 3(4x + 2)$
14. $12 = 4(2x + 3)$
15. $3(t - 4) + 5 = -4$
16. $5(t - 1) + 6 = -9$
17. $4(2y + 1) - 7 = 1$
18. $6(3y + 2) - 8 = -2$
19. $4(x - 3) = 2(x + 1)$
20. $2(x - 4) = 3(x - 6)$
21. $-7(2x - 7) = 3(11 - 4x)$
22. $-3(2x - 5) = 7(3 - x)$
23. $5(3x - 1) = 4(2x + 1) - 30$
24. $4(2x + 3) = -3(x - 1) + 20$
25. $-2(3y + 1) = 3(1 - 6y) - 9$
26. $-5(4y - 3) = 2(1 - 8y) + 11$
27. $8 - 4(x + 5) = -12$
28. $7 - 3(x + 2) = 1$
29. $6 - 5(2a - 3) = 1$
30. $-8 - 2(3 - a) = 0$
31. $2x - 5 = 5 - 2(2x - 13)$
32. $4x - 1 = 7 - 3(6 - 2x)$
33. $2(t - 3) + 3(t - 2) = 28$
34. $-3(t - 5) - 2(2t + 1) = -8$
35. $5(x - 2) - (3x + 4) = 3(6x - 8) + 10$
36. $3(x - 1) - (4x - 5) = 2(5x - 1) - 7$

None of the following equations has a single number for its solution. The solution set for each one is either all real numbers or the empty set (in which case we say there are no solutions). In each case, if you try to collect all the variable terms on one side, you will eliminate the variable completely. Once the variable has been eliminated, you will be left with either a true statement or a false statement. A true statement indicates the solution set is all real numbers, in which case every real number satisfies the equation. A false statement indicates there are no solutions to the equations; any number you substitute for x will give a false statement.

Solve each equation if possible.

37. $2(x + 3) = 2x + 6$ **38.** $3x - 9 = 3(x - 3)$

39. $x + 5 = x + 7$ **40.** $2x + 4 = 2x - 1$

41. $8 - 4x = 2(4 - 2x)$ **42.** $12 - 10x = 2(6 - 5x)$

43. $4x + 6 = 2(2x - 3)$ **44.** $2x - 14 = 2(x + 3)$

Review Problems The problems below review material we covered in Chapter 1. Reviewing these problems will help you in the next section.

Write an equivalent expression in English. Include the words "sum" and "difference" when possible.

45. $4 + 1 = 5$ **46.** $7 + 3 = 10$

47. $6 - 2 = 4$ **48.** $8 - 1 = 7$

For each of the following expressions, write an equivalent expression with numbers.

49. Twice the sum of 6 and 3.

50. Four added to the product of 5 and −1.

51. The sum of twice 5 and 3 is 13.

52. Twice the difference of 8 and 2 is 12.

2.5
Word Problems

It seems that the major difference between those people who are good at working word problems and those who are not is confidence. The people with confidence know that no matter how long it takes them, they will eventually be able to solve the problem. Those without confidence begin by saying to themselves, "I'll never be able to work this problem." Are you like that? If you are, what you need to do is put your old ideas about you and word problems aside for awhile and make a decision to be successful. Sometimes that's all it takes. Instead of telling yourself that you can't do word problems, that you don't like them, or that they're not good for anything anyway, decide to do whatever it takes to master them. I think you'll find your attitude toward word problems will change once you've experienced success solving them. Let's begin our study of word problems by listing the steps we will use in the examples that follow.

Step 1: Let x (or any letter you choose) represent the unknown quantity. It is usually the quantity asked for in the problem.

Step 2: Write an equation, in x, that describes the situation.

Step 3: Solve the equation you wrote in Step 2.

Step 4: Check your solution in the original problem.

Step 2 is the most difficult part. It may be necessary to read each problem a number of times before you can do this step correctly. You may make some mistakes along the way. There will be times when your first attempt at Step 2 results in the wrong equation. Mistakes are part of the process of learning to do things correctly. Many times the correct equation will become obvious after you have written an equation that is partially wrong. In any case, it is better to write an equation that is partially wrong and be actively involved with the problem than to write nothing at all. Word problems, like other problems in algebra, are not always solved correctly the first time.

Here are some common English words and phrases and their mathematical translation.

English	Algebra
the difference of a and b	$a - b$
the product of a and b	$a \cdot b$
the quotient of a and b	$\dfrac{a}{b}$
of	$\cdot$ (multiply)
is	$=$ (equals)
a number	x
4 more than x	$x + 4$
4 times x	$4x$
4 less than x	$x - 4$

Look over the following examples and maybe try to solve some on your own. The examples are similar to the problems in the problem set.

▼ **Example 1** The sum of twice a number and three is seven. Find the number.

   SOLUTION

   Step 1: Let $x =$ the number asked for.
   Step 2: "The sum of twice a number and three" translates to

$2x + 3$, "is" translates to $=$, and "seven" is 7. An equation that describes the situation, then, is

$$2x + 3 = 7$$

Step 3:   Solve the equation:

$$2x + 3 = 7$$
$$2x + 3 + (-3) = 7 + (-3)$$
$$2x = 4$$
$$(\tfrac{1}{2})(2x) = (\tfrac{1}{2})(4)$$
$$x = 2$$

Step 4:   Check the solution in the original problem.

The sum of twice two and three is seven.   ▲

▼   **Example 2**   If twice the difference of a number and three were decreased by five, the result would be three. Find the number.

SOLUTION

Step 1:   Let $x =$ the number asked for.

Step 2:   "Twice the difference of a number and three" translates to $2(x - 3)$; "is decreased by five" translates to -5 (subtract 5); "the result is" translates to $=$; and "three" is 3. An equation that describes the situation is

$$2(x - 3) - 5 = 3$$

Step 3:   Solve the equation:

$$2(x - 3) - 5 = 3$$
$$2x - 6 - 5 = 3$$
$$2x - 11 = 3$$
$$2x - 11 + 11 = 3 + 11$$
$$2x = 14$$
$$x = 7$$

Step 4:   Try it and see.   ▲

▼   **Example 3**   Bill is six years older than Tom. Three years ago Bill's age was four times Tom's age. Find the age of each boy now.

SOLUTION   It is sometimes very helpful to make a table for age problems. The people are arranged in rows and the different times are arranged in columns. For example,

	Three years ago	Now
Bill		
Tom		

Proceeding as usual we have:

Step 1:   Let x = Tom's age now. That makes Bill $x + 6$ years old now.

Step 2:   If Tom is x years old now, three years ago he was $x - 3$ years old. If Bill is $x + 6$ years old now, three years ago he was $x + 6 - 3 = x + 3$ years old.

Using the information from Steps 1 and 2, we fill in the table as follows:

	Three years ago	Now
Bill	$x + 3$	$x + 6$
Tom	$x - 3$	x

Reading the problem again, we see that three years ago Bill's age was four times Tom's age. Writing this as an equation, we have

$$x + 3 = 4(x - 3)$$
$$\text{Bill's age} = 4(\text{Tom's age})$$

Step 3:   Solve the equation:

$$x + 3 = 4(x - 3)$$
$$x + 3 = 4x - 12$$
$$x + (-x) + 3 = 4x + (-x) - 12$$
$$3 = 3x - 12$$
$$3 + 12 = 3x - 12 + 12$$
$$15 = 3x$$
$$x = 5$$

Tom is 5 years old. Bill is 11 years old.   ▲

▼   **Example 4**   The length of a rectangle is five more than twice the width. The perimeter is 34 inches. Find the length and width.

SOLUTION

Step 1:   Let x = the width of the rectangle.

Step 2: The length is 5 more than twice the width, so it must be $2x + 5$. It is a good idea when working problems that involve geometric figures (circles, squares, rectangles, triangles, etc.) to visualize the problem by drawing a picture. A picture that describes the information given in this problem looks like this:

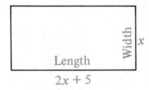

Since the perimeter of a rectangle is the total length of all the sides, we can find it by taking twice the length and twice the width. The equation that describes the situation in this problem is

$$2(2x + 5) \ + \quad 2x \quad = \quad 34$$

twice length + twice width = perimeter

Step 3: Solve the equation:

$$2(2x + 5) + 2x = 34$$
$$4x + 10 + 2x = 34$$
$$6x + 10 = 34$$
$$6x + 10 + (-10) = 34 + (-10)$$
$$6x = 24$$
$$x = 4$$

The width is 4 inches. The length is $2(4) + 5 = 13$ inches.

Step 4: If the length is 13 and the width is 4, then the perimeter must be $2(13 + 4) = 2(17) = 34$, which checks with the original problem. ▲

When working the problems in the problem set, you may have to read each problem many times. Read the problems as many times as it takes for you to understand them. Remember also that word problems are not always solved correctly the first time.

Problem Set 2.5

Solve the following word problems. Follow the four steps given in the text and be sure to show the equation you use.

1. The sum of a number and five is thirteen. Find the number.
2. The difference of ten and a number is negative eight. Find the number.

3. The sum of twice a number and four is fourteen. Find the number.
4. The difference of four times a number and eight is sixteen. Find the number.
5. Five times the sum of a number and seven is thirty. Find the number.
6. Five times the difference of twice a number and six is negative twenty. Find the number.
7. One number is two more than another. Their sum is eight. Find both numbers.
8. One number is three less than another. Their sum is fifteen. Find the numbers.
9. One number is four less than three times another. If their sum is increased by five, the result is twenty-five. Find the numbers.
10. One number is five more than twice another. If their sum is decreased by ten, the result is twenty-two. Find the numbers.
11. The length of a rectangle is five more than the width. The perimeter is 34 inches. Find the length and width.
12. The width of a rectangle is three less than the length. The perimeter is 10 feet. Find the length and width.
13. The perimeter of a square is 48 meters. Find the length of one side.
14. One side of a triangle is twice the smallest side. The third side is seven more than the shortest side. The perimeter is 15 feet. Find all three sides.
15. The length of a rectangle is three less than twice the width. The perimeter is 54 inches. Find the length and width.
16. The length of a rectangle is four less than three times the width. The sum of the length and width is 14 more than the width. Find the width.
17. Fred is four years older than Barney. Five years ago the sum of their ages was forty-eight. How old are they now?
18. John is four times as old as Martha. Five years ago he was seven times as old as she was. How old are they now?
19. Jack is twice as old as Lacy. In three years the sum of their ages will be 54. How old are they now?
20. Tim is five years older than JoAnn, and JoAnn is four years older than Stacey. The sum of their ages is forty-three. Find Stacey's age.

Review Problems   The problems below review material we covered in Sections 1.1 and 1.2. Reviewing these problems will help you understand the material in the next section.

Write an equivalent statement in English.

21. $4 < 10$
22. $4 \leq 10$
23. $9 \geq -5$
24. $x - 2 > 4$

Copy each problem and place the symbol $<$ or the symbol $>$ between the quantities in each expression.

25. 12   20
26. -12 20

27. $-4(-3)$ $-4(5)$ **28.** $-4(-5)$ $-4(-3)$
29. -4 $2(6-3)$ **30.** $-2(3)$ $5(4-7)$

2.6
Linear Inequalities

Linear inequalities are solved by a method similar to the one used in solving linear equations. The only real differences between the methods are in the multiplication property for inequalities and in graphing the solution set.

An inequality differs from an equation only with respect to the comparison symbol between the two quantities being compared. In place of the equal sign, we use $<$ (less than), $\leq$ (less than or equal to), $>$ (greater than), or $\geq$ (greater than or equal to). The addition property for inequalities is almost identical to the addition property for equality.

Addition Property for Inequalities

For any three algebraic expressions A, B, and C,

$$\text{If} \qquad A < B$$
$$\text{then } A + C < B + C$$

In words: Adding the same quantity to both sides of an inequality will not change the solution set.

It makes no difference which inequality symbol we use to state the property. Adding the same amount to both sides always produces an inequality equivalent to the original inequality.

▼ **Example 1**   Solve the inequality $x + 5 < 7$.

SOLUTION

$$x + 5 < 7$$
$$x + 5 + (-5) < 7 + (-5) \qquad \text{Addition property for inequalities}$$
$$x < 2$$
▲

We can go one step further here and graph the solution set. The solution set is all real numbers less than 2. To graph this set, we simply draw a straight line and label the center 0 (zero) for reference. Then we label the 2 on the right side of zero and extend an arrow beginning at 2 and pointing to the left. We use an open circle at 2, since it is not included in the solution set. Here is the graph.

▼ **Example 2** Solve: $x - 6 \leq -3$.

SOLUTION

$$x - 6 \leq -3$$
$$x - 6 + 6 \leq -3 + 6 \qquad \text{Add 6 to both sides}$$
$$x \leq 3$$

The graph of the solution set is

▲

Notice that the 3 is darkened because 3 is included in the solution set. We will always use open circles on the graph of solution sets with $<$ or $>$ and closed (darkened) circles on the graphs of solution sets with $\leq$ or $\geq$.

To see the idea behind the multiplication property of inequalities, we will consider four true inequality statements and explore what happens when we multiply both sides by a positive number and then what happens when we multiply by a negative number.

Consider the following three true statements:

$$3 < 5 \qquad\qquad -3 < 5 \qquad\qquad -5 < -3$$

Now multiply both sides by the positive number 4:

$$4(3) < 4(5) \qquad 4(-3) < 4(5) \qquad 4(-5) < 4(-3)$$
$$12 < 20 \qquad\quad -12 < 20 \qquad\quad -20 < -12$$

In each case, the inequality symbol in the result points in the same direction it did in the original inequality. We say the "sense" of the inequality doesn't change when we multiply both sides by a positive quantity.

Notice what happens when we go through the same process but multiply both sides by -4 instead of 4:

$$3 < 5 \qquad\qquad -3 < 5 \qquad\qquad -5 < -3$$
$$\downarrow \qquad\qquad\quad \downarrow \qquad\qquad\quad \downarrow$$
$$-4(3) > -4(5) \qquad -4(-3) > -4(5) \qquad -4(-5) > -4(-3)$$
$$-12 > -20 \qquad\quad 12 > -20 \qquad\quad 20 > 12$$

In each case, we have to change the direction in which the inequality symbol points to keep each statement true. Multiplying both sides of an inequality by a negative quantity *always* reverses the sense of the inequality. Our results are summarized in the multiplication property for inequalities.

Multiplication Property for Inequalities

For any three algebraic expressions A, B, and C,

If $\quad A < B$
then $AC < BC \quad$ when C is positive
and $\quad AC > BC \quad$ when C is negative

In words: Multiplying both sides of an inequality by a positive number does not change the solution set. When multiplying both sides of an inequality by a negative number, it is necessary to reverse the inequality symbol in order to produce an equivalent inequality.

We can multiply both sides of an inequality by any nonzero number we choose. If that number happens to be negative, we must also reverse the sense of the inequality.

▼ **Examples**   Solve each inequality and graph the solution.

3. $3a < 15$

SOLUTION

$$\tfrac{1}{3}(3a) < \tfrac{1}{3}(15) \qquad \text{Multiply by } \tfrac{1}{3}$$
$$a < 5$$

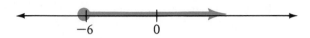

4. $-3x \leq 18$

SOLUTION

$$-\tfrac{1}{3}(-3x) \geq -\tfrac{1}{3}(18) \qquad \text{Reverse the sense of the inequality}$$
$$x \geq -6 \qquad\qquad\qquad \text{because we are multiplying both}$$
sides by a negative number.

5. $-\dfrac{x}{4} > 2$

SOLUTION

$$-4\left(-\dfrac{x}{4}\right) < -4(2) \qquad \text{Multiply by } -4 \text{ and reverse the sense}$$
of the inequality.
$$x < -8$$

▲

To solve more complicated inequalities we use the following four steps.

Step 1: Use the distributive property, if necessary, to separate terms. Simplify both sides as much as possible.

Step 2: Use the addition property of inequalities to get all variable terms on one side and all constant terms on the other.

Step 3: Use the multiplication property of inequalities to get x on just one side.

Step 4: Graph the solution set.

▼ **Examples** Use the four steps given above to solve the following inequalities.

6. $3(x - 4) \geq -2$

   SOLUTION

   $$3x - 12 \geq -2 \qquad \text{Distributive property}$$
   $$3x - 12 + 12 \geq -2 + 12 \qquad \text{Add 12 to both sides}$$
   $$3x \geq 10$$
   $$\tfrac{1}{3}(3x) \geq \tfrac{1}{3}(10) \qquad \text{Multiply both sides by } \tfrac{1}{3}$$
   $$x \geq \tfrac{10}{3}$$

7. $2(1 - 3x) + 4 < 4x - 14$

   SOLUTION

   $$2 - 6x + 4 < 4x - 14 \qquad \text{Distributive property}$$
   $$-6x + 6 < 4x - 14 \qquad \text{Simplify}$$
   $$-6x + 6 - 6 < 4x - 14 - 6 \qquad \text{Add } -6 \text{ to both sides}$$
   $$-6x < 4x - 20$$
   $$-6x - 4x < 4x - 4x - 20 \qquad \text{Add } -4x \text{ to both sides}$$
   $$-10x < -20$$
   $$\downarrow$$
   $$(-\tfrac{1}{10})(-10x) > (-\tfrac{1}{10})(-20) \qquad \text{Multiply by } -\tfrac{1}{10}, \text{ reverse}$$
   $$x > 2 \qquad \qquad \qquad \text{the sense of the inequality}$$

Solve the following inequalities using the addition property of inequalities. Graph each solution set.   Problem Set 2.6

1. $x - 5 < 7$ 2. $x + 3 < -5$

3. $a - 4 \leq 8$ 4. $a + 3 \leq 10$
5. $x - 5 > 8$ 6. $x - 2 > 11$
7. $y + 6 \geq 10$ 8. $y + 3 \geq 12$
9. $2 < x - 7$ 10. $3 < x + 8$

Solve the following inequalities using the multiplication property of inequalities. If you multiply both sides by a negative number, be sure to change the sense of the inequality symbol. Graph the solution set.

11. $3x < 6$ 12. $2x < 14$
13. $5a \leq 25$ 14. $4a \leq 16$

15. $\dfrac{x}{3} > 5$ 16. $\dfrac{x}{7} > 1$

17. $\dfrac{y}{5} \geq 2$ 18. $\dfrac{y}{4} \geq 3$

19. $-2x > 6$ 20. $-3x \geq 9$
21. $-4a < 28$ 22. $-10a \leq 30$
23. $-3x > -18$ 24. $-8x \geq -24$

25. $-\dfrac{x}{5} \leq 10$ 26. $-\dfrac{x}{9} \geq -1$

Solve the following inequalities. Graph the solution set in each case.

27. $2x - 3 < 9$ 28. $3x - 4 < 17$
29. $-3y - 5 \leq 10$ 30. $-2y - 6 \leq 8$
31. $-4x + 1 > -11$ 32. $-6x - 1 > 17$
33. $3(a + 1) \leq 12$ 34. $4(a - 2) \leq 4$
35. $2(2x - 5) > 20$ 36. $7(2x - 8) > -28$
37. $3x - 5 > 8x$ 38. $8x - 4 > 6x$
39. $2y - 3 \leq 5y + 3$ 40. $7y + 8 \leq 11y - 7$
41. $-2x + 3 < -10x - 5$ 42. $-6x - 2 < -2x + 10$
43. $3(m - 2) - 4 \geq 7m + 14$ 44. $2(3m - 1) + 5 \geq 8m - 7$
45. $3 - 4(x - 2) \leq -5x + 6$ 46. $8 - 6(x - 3) \leq -4x + 12$

47. The sum of twice a number and six is less than ten. Find all solutions.
48. Twice the difference of a number and three is greater than or equal to the number increased by five. Find all solutions.
49. The product of a number and four is greater than the number minus eight. Find the solution set.
50. The quotient of a number and five is less than the sum of seven and two. Find the solution set.
51. Twice the sum of a number and five is less than or equal to twelve. Find all solutions.
52. Three times the difference of a number and four is greater than twice the number. Find all solutions.
53. The difference of three times a number and five is less than the sum of the number and seven. Find all solutions.

54. If twice a number is added to three, the result is less than the sum of three times the number and two. Find all solutions.

Review Problems   The problems below review material we covered in Section 1.5.

Match each expression on the left with one or more of the properties on the right.

55. $x + 4 = 4 + x$ A.   Distributive property

56. $2(3x) = (2 \cdot 3)x$ B.   Commutative property of addition

57. $5(x - 3) = 5x - 15$ C.   Associative property of addition

58. $x + (y + 4) = (x + y) + 4$ D.   Commutative property of multiplication

59. $x + (y + 4) = (x + 4) + y$ E.   Associative property of multiplication

60. $7 \cdot 5 = 5 \cdot 7$

The *union* of two sets A and B is the set of all elements that are in A or in B. The word "or" is the key word in the definition. The intersection of two sets A and B is the set of elements contained in both A and B. The key word in this definition is the word "and." We can put the words "and" and "or" together with our methods of graphing inequalities to find the solution sets for compound inequalities.

**2.7
Compound
Inequalities**

DEFINITION   A compound inequality is two or more inequalities connected by the words "and" or "or."

▼   **Example 1**   Graph the solution set for the compound inequality

$$x < -1 \quad \text{or} \quad x \geq 3$$

SOLUTION   Graphing each inequality separately, we have:

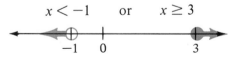

Since the two inequalities are connected by "or," we want to graph their union.

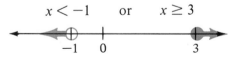

▼   Example 2   Graph the solution set for the compound inequality

$$x > -2 \quad \text{and} \quad x < 3$$

SOLUTION   Graphing each inequality separately, we have

$x > -2$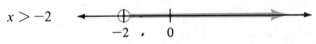

$x < 3$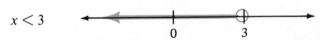

Since the two inequalities are connected by the word "and," we will graph their intersection, which consists of all points that are common to both graphs:

▲

▼   Example 3   Solve and graph the solution set for

$$2x - 1 \geq 3 \quad \text{and} \quad -3x > -12$$

SOLUTION   Solving the two inequalities separately, we have

$$2x - 1 \geq 3 \quad \text{and} \quad -3x > -12$$
$$2x \geq 4 \qquad -\tfrac{1}{3}(-3x) < -\tfrac{1}{3}(-12)$$
$$x \geq 2 \quad \text{and} \qquad x < 4$$

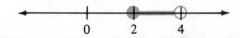

Since the word "and" connects the two graphs, we will graph their intersection—the points they have in common:

▲

Notation: Sometimes compound inequalities that use the word "and" can be written in a shorter form. For example, the compound inequality $-2 < x$ and $x < 3$ can be written as $-2 < x < 3$. The word "and" does not appear when an inequality is written in this form. It is implied. The solution set for $-2 < x$ and $x < 3$ is

It is all the numbers between -2 and 3 on the number line. It seems reasonable, then, that this graph should be the graph of

$$-2 < x < 3$$

In both the graph and the inequality, x is said to be between -2 and 3.

▼ **Example 4** Solve and graph $-3 \leq 2x - 1 \leq 9$.

SOLUTION To solve for x we must add 1 to the center expression and then divide the result by 2. Whatever we do to the center expression, we must also do to the two expressions on the ends. In this way, we can be sure we are producing equivalent inequalities. The solution set will not be affected.

$$
\begin{aligned}
-3 &\leq 2x - 1 \leq 9 \\
-2 &\leq \quad 2x \quad \leq 10 \qquad \text{Add 1 to each expression} \\
-1 &\leq \quad x \quad \leq 5 \qquad \text{Multiply each by } \tfrac{1}{2}
\end{aligned}
$$

▲

▼ **Example 5** Solve and graph $-5 < 3x + 1 < 10$.

SOLUTION

$$
\begin{aligned}
-5 &< 3x + 1 < 10 \\
-6 &< \quad 3x \quad < 9 \qquad \text{Add } -1 \text{ to each member} \\
-2 &< \quad x \quad < 3 \qquad \text{Multiply through by } \tfrac{1}{3}
\end{aligned}
$$

The solution set is all real numbers between -2 and 3. We must keep in mind that the notation $-2 < x < 3$ is equivalent to the expression $-2 < x$ and $x < 3$. This notation always implies the word "and"; it never implies the word "or." ▲

Graph the following compound inequalities. Problem Set 2.7

1. $x < -1$ or $x > 5$
2. $x \leq -2$ or $x \geq -1$
3. $x < -3$ or $x \geq 0$
4. $x < 5$ and $x > 1$
5. $x \leq 6$ and $x > -1$
6. $x \leq 7$ and $x > 0$
7. $x > 2$ and $x < 4$
8. $x < 2$ or $x > 4$
9. $x \geq 2$ and $x \leq 4$
10. $x \leq 2$ or $x \geq 4$
11. $x < 5$ and $x > -1$
12. $x > 5$ or $x < -1$
13. $-1 < x < 3$
14. $-1 \leq x \leq 3$

15. $-3 < x \le -2$ **16.** $-5 \le x \le 0$

Solve the following. Graph the solution set in each case.

17. $3x - 1 < 5$ or $5x - 5 > 10$
18. $x + 1 < -3$ or $x - 2 > 6$
19. $x - 2 > -5$ and $x + 7 < 13$
20. $3x + 2 \le 11$ and $2x + 2 \ge 0$
21. $11x < 22$ or $12x > 36$
22. $-5x < 25$ and $-2x \ge -12$
23. $3x - 5 < 10$ and $2x + 1 > -5$
24. $5x + 8 < -7$ or $3x - 8 > 10$
25. $2x - 3 < 8$ and $3x + 1 > -10$
26. $11x - 8 > 3$ or $12x + 7 < -5$
27. $2x - 1 < 3$ and $3x - 2 > 1$
28. $3x + 9 < 7$ or $2x - 7 > 11$

Solve and graph each of the following.

29. $-1 \le x - 5 \le 2$ **30.** $0 \le x + 2 \le 3$
31. $-4 \le 2x \le 6$ **32.** $-5 < 5x < 10$
33. $-3 < 2x + 1 < 5$ **34.** $-7 \le 2x - 3 \le 7$
35. $0 \le 3x + 2 \le 7$ **36.** $2 \le 5x - 3 \le 12$
37. $-7 < 2x + 3 < 11$ **38.** $-5 < 6x - 2 < 8$
39. $-1 \le 4x + 5 \le 9$ **40.** $-8 \le 7x - 1 \le 13$

Review Problems   The problems below review material we covered in Section 1.8

For the set $\{-10, -\sqrt{5}, -\frac{3}{4}, 0, \frac{5}{4}, 3, \sqrt{15}\}$, list all the numbers that are:

41. Counting numbers   **42.** Whole numbers
43. Integers   **44.** Rational numbers
45. Irrational numbers   **46.** Real numbers

Chapter 2 Summary and Review

Examples

1. The terms $2x$, $5x$, and $-7x$ are all similar since their variable parts are the same.

SIMILAR TERMS [2.1]

A *term* is a number or a number and one or more variables multiplied together. *Similar terms* are terms with the same variable part.

2. Simplify $3x + 4x$

$$3x + 4x = (3 + 4)x$$
$$= 7x$$

SIMPLIFYING EXPRESSIONS [2.1]

In this chapter we simplified expressions that contained variables by using the distributive property to combine similar terms.

SOLUTION SET [2.2]

The *solution set* for an equation (or inequality) is all the numbers that, when used in place of the variable, make the equation a true statement.

3. The solution set for the equation $x + 2 = 5$ is $\{3\}$ because when x is 3 the equation is $3 + 2 = 5$ or $5 = 5$.

EQUIVALENT EQUATIONS [2.2]

Two equations are called *equivalent* if they have the same solution set.

4. The equations $a - 4 = 3$ and $a - 2 = 5$ are equivalent since both have solution set $\{7\}$.

ADDITION PROPERTY OF EQUALITY [2.2]

When the same quantity is added to both sides of an equation, the solution set for the equation is unchanged. Adding the same amount to both sides of an equation produces an equivalent equation.

5. Solve $x - 5 = 12$

$$x - 5 + 5 = 12 + 5$$
$$x + 0 = 17$$
$$x = 17$$

MULTIPLICATION PROPERTY OF EQUALITY [2.3]

If both sides of an equation are multiplied by the same nonzero number, the solution set is unchanged. Multiplying both sides of an equation by a nonzero quantity produces an equivalent equation.

6. Solve $3x = 18$

$$\tfrac{1}{3}(3x) = \tfrac{1}{3}(18)$$
$$x = 6$$

TO SOLVE A LINEAR EQUATION [2.4]

Step 1: Use the distributive property, if necessary, to separate terms. Simplify both sides as much as possible.
Step 2: Use the addition property of equality to get all variable terms on one side and all constant terms on the other.
Step 3: Use the multiplication property of equality to get x by itself.
Step 4: Check your solution, if necessary.

7. Solve $2(x + 3) = 10$

$$2x + 6 = 10$$
$$2x + 6 + (-6) = 10 + (-6)$$
$$2x = 4$$
$$\tfrac{1}{2}(2x) = \tfrac{1}{2}(4)$$
$$x = 2$$

TO SOLVE A WORD PROBLEM [2.5]

Step 1: Let x represent the unknown quantity.
Step 2: Using x, write an equation that describes the situation.
Step 3: Solve the equation.
Step 4: Check your solution with the original problem.

8. The sum of twice a number and three is seven. Find the number. Let $x =$ the number:

$$2x + 3 = 7$$
$$x = 2$$

The sum of twice 2 and 3 is 7.

9. Solve $x + 5 < 7$

$$x + 5 + (-5) < 7 + (-5)$$
$$x < 2$$

ADDITION PROPERTY OF INEQUALITY [2.6]

Adding the same quantity to both sides of an inequality produces an equivalent inequality, one with the same solution set.

10. Solve $-3a \leq 18$

$$-\tfrac{1}{3}(-3a) \geq -\tfrac{1}{3}(18)$$
$$a \geq -6$$

MULTIPLICATION PROPERTY OF INEQUALITY [2.6]

Multiplying both sides of an inequality by a positive number never changes the solution set. If both sides are multiplied by a negative number, the sense of the inequality must be reversed to produce an equivalent inequality.

11. Solve $3(x - 4) \geq -2$

$$3x - 12 \geq -2$$
$$3x - 12 + 12 \geq -2 + 12$$
$$3x \geq 10$$
$$\tfrac{1}{3}(3x) \geq \tfrac{1}{3}(10)$$
$$x \geq \tfrac{10}{3}$$

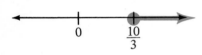

TO SOLVE A LINEAR INEQUALITY [2.6]

Step 1: Use the distributive property, if necessary, to separate terms. Simplify both sides as much as possible.

Step 2: Use the addition property of inequality to get all variable terms on one side and all constant terms on the other.

Step 3: Use the multiplication property of inequality to get just one x. (Remember to reverse the direction of the inequality symbol if you multiply both sides by a negative number.)

Step 4: Graph the solution set.

12. $x < -3$ or $x > 1$

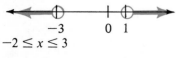

$-2 \leq x \leq 3$

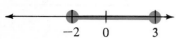

COMPOUND INEQUALITIES [2.7]

Two inequalities connected by the word "and" or "or" form a compound inequality. If the connecting word is "or," we graph all points that are on either graph. If the connecting word is "and," we graph only those points that are common to both graphs. The inequality $-2 \leq x \leq 3$ is equivalent to the compound inequality $-2 \leq x$ and $x \leq 3$.

COMMON MISTAKES

1. Trying to subtract away coefficients (the number in front of variables) when solving equations. For example:

$$4x = 12$$
$$4x - 4 = 12 - 4$$
$$x = 8 \quad \longleftarrow \text{ Mistake}$$

It is not incorrect to add (-4) to both sides, it's just that $4x - 4$ is not equal to x. Both sides should be multiplied by $\frac{1}{4}$ to solve for x.

2. Forgetting to reverse the direction of the inequality symbol when multiplying both sides of an inequality by a negative number. For instance:

$$-3x < 12 \qquad 12$$
$$-\tfrac{1}{3}(-3x) < -\tfrac{1}{3}(12) \quad \longleftarrow \quad \text{Mistake}$$
$$x < -4 \qquad -4$$

It is not incorrect to multiply both sides by $-\frac{1}{3}$. But if we do, we must also reverse the sense of the inequality.

Simplify each of the following algebraic expressions.   Chapter 2 Test

  1. $3x + 2 - 7x + 3$ **2.** $2(3x + 1) + 4x$
  3. $4a - 5 - a + 1$ **4.** $3(2 - x) + 8x$
  5. $7 - 3(y + 5) - 4$ **6.** $8(2x + 1) - 5(x - 4)$
  7. $2(4 - a) - 2(3 - a)$ **8.** $11y - 2(3y + 1) + 5$

Solve the following equations.

  9. $2x - 5 = 7$ **10.** $3a + 2 = 17$
11. $2x + 4 = 3x$ **12.** $5x - 1 = 2x + 5$
13. $2(x + 8) = 20$ **14.** $-5(2x + 1) - 6 = 19$
15. $4 - 3(3a - 5) = 1$ **16.** $2(x - 4) + 3(x + 5) = 2x - 2$
17. $2x - 4(x + 1) = 3x + 1$ **18.** $2a - 3(4a + 2) = 4a + 22$

Solve the following inequalities. Graph the solution set.

19. $2x + 3 < 5$ **20.** $4 - 2x \geq 10$
21. $-5a > 20$ **22.** $2(4x - 5) > 6$
23. $3(x - 8) \leq 0$ **24.** $4 - 5(m + 1) < 9$
25. $x < -2$ or $x > 1$ **26.** $-3 \leq 2x + 1 \leq 9$

Solve the following word problems.

27. The sum of twice a number and four is ten. Find the number.
28. One number is five times another number. Their sum is twelve more than twice the smaller. Find the two numbers.
29. Dave is twice as old as Rick. Ten years ago Dave was three times as old as Rick. Find their ages, now.
30. A rectangle is twice as long as it is wide. The perimeter is 60 inches. What are the length and width?

Graphing and Linear Systems

In the last chapter we spent most of our time developing and using the method of solving linear equations in one variable. The equations we worked with had the form $ax + b = 0$. (If they didn't have this form, they could be put into this form.) In this chapter we will expand our work with equations to include linear equations in *two* variables. We are going to include another variable so that most of the equations in this chapter will have the form $ax + by = c$, where a, b, and c are constants and x and y are variables. We will also extend the technique of graphing to include points associated with two number lines instead of just one. The background material needed for this chapter is in Chapter 2. You need to know how to solve a linear equation in one variable.

If we solve the equation $3x - 2 = 10$, the solution is $x = 4$. If we were to graph this solution, we would simply draw the real number line and place a dot at the point whose coordinate is 4. The relationship between linear equations in one variable, their solutions, and the graphs of those solutions looks like this:

3.1 Solutions to Linear Equations in Two Variables

Equation	Solution	Graph of Solution Set
$3x - 2 = 10$	$x = 4$	

$$x + 5 = 7 \qquad x = 2$$

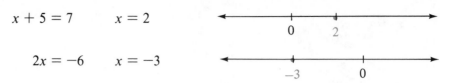

$$2x = -6 \qquad x = -3$$

When the equation has one variable, the solution is a single number whose graph is a point on a line.

Now consider the equation $2x + y = 3$. The first thing we notice is that there are two variables instead of one. Therefore a solution to the equation $2x + y = 3$ will not be a single number but a pair of numbers, one for x and one for y, that make the equation a true statement. One pair of numbers that works is $x = 2$, $y = -1$, because when we substitute them for x and y in the equation, we get a true statement. That is:

$$2(2) - 1 = 3$$

$$4 - 1 = 3$$

$$3 = 3 \qquad \text{A true statement}$$

The pair of numbers $x = 2$, $y = -1$ is written as $(2, -1)$ in a kind of shorthand form. $(2, -1)$ is called an *ordered pair* because it is a pair of numbers written in a specific order. The first number is always associated with the variable x, and the second number is always associated with the variable y. We call the first number in the ordered pair the *x-coordinate* (or x component) and the second number the *y-coordinate* (or y component) of the ordered pair.

Let's look back to the equation $2x + y = 3$. The ordered pair $(2, -1)$ is not the only solution. Another solution is $(0, 3)$, because when we substitute 0 for x and 3 for y we get

$$2(0) + 3 = 3$$

$$0 + 3 = 3$$

$$3 = 3 \qquad \text{A true statement}$$

Still another solution is the ordered pair $(5, -7)$, because

$$2(5) - 7 = 3$$

$$10 - 7 = 3$$

$$3 = 3 \qquad \text{A true statement}$$

As a matter of fact, for any number we want to use for x there is another number we can use for y that will make the equation a true statement. There is an infinite number of ordered pairs that satisfy (are solutions to) the equation $2x + y = 3$; we have listed just a few of them.

▼ **Example 1** Given the equation $2x + 3y = 6$, complete the following ordered pairs so they will be solutions to the equation: $(0, \ )$, $(\ , 1)$, $(3, \ )$.

SOLUTION To complete the ordered pair $(0, \ )$, we substitute 0 for x in the equation and then solve for y:

$$2(0) + 3y = 6$$
$$3y = 6$$
$$y = 2$$

The ordered pair is $(0, 2)$.

To complete the ordered pair $(\ , 1)$, we substitute 1 for y in the equation and solve for x:

$$2x + 3(1) = 6$$
$$2x + 3 = 6$$
$$2x = 3$$
$$x = \tfrac{3}{2}$$

The ordered pair is $(\tfrac{3}{2}, 1)$.

To complete the ordered pair $(3, \ )$, we substitute 3 for x in the equation and solve for y:

$$2(3) + 3y = 6$$
$$6 + 3y = 6$$
$$3y = 0$$
$$y = 0$$

The ordered pair is $(3, 0)$.   ▲

Notice in each case that once we have used a number in place of one of the variables, the equation becomes a linear equation in one variable. We then use the method explained in Chapter 2 to solve for that variable.

▼ **Example 2** Complete the following table for the equation $2x - 5y = 20$.

x	y
0	
	2
	0
−5	

SOLUTION   Filling in the table is equivalent to completing the following ordered pairs: $(0, \quad), (\quad, 2), (\quad, 0), (-5, \quad)$. So we proceed as in Example 1.

When $x = 0$, we have

$$2(0) - 5y = 20$$
$$0 - 5y = 20$$
$$-5y = 20$$
$$y = -4$$

When $y = 2$, we have

$$2x - 5(2) = 20$$
$$2x - 10 = 20$$
$$2x = 30$$
$$x = 15$$

When $y = 0$, we have

$$2x - 5(0) = 20$$
$$2x - 0 = 20$$
$$2x = 20$$
$$x = 10$$

When $x = -5$, we have

$$2(-5) - 5y = 20$$
$$-10 - 5y = 20$$
$$-5y = 30$$
$$y = -6$$

The completed table looks like this:

x	y
0	−4
15	2
10	0
−5	−6

which is equivalent to the ordered pairs $(0, -4)$, $(15, 2)$, $(10, 0)$, and $(-5, -6)$.   ▲

▼ **Example 3** Complete the following table for the equation $y = 2x - 1$.

x	y
0	
5	
	7
	3

SOLUTION   When $x = 0$, we have

$$y = 2(0) - 1$$
$$y = 0 - 1$$
$$y = -1$$

When $x = 5$, we have

$$y = 2(5) - 1$$
$$y = 10 - 1$$
$$y = 9$$

When $y = 7$, we have

$$7 = 2x - 1$$
$$8 = 2x$$
$$4 = x$$

When $y = 3$, we have

$$3 = 2x - 1$$
$$4 = 2x$$
$$2 = x$$

The completed table is:

x	y
0	-1
5	9
4	7
2	3

which means the ordered pairs $(0, -1)$, $(5, 9)$, $(4, 7)$, and $(2, 3)$ are among the solutions to the equation $y = 2x - 1$.   ▲

▼ **Example 4**   Which of the ordered pairs $(2, 3)$, $(1, 5)$, and $(-2, -4)$ are solutions to the equation $y = 3x + 2$?

SOLUTION If an ordered pair is a solution to the equation, then it must satisfy the equation. That is, when the coordinates are used in place of the variables in the equation, the equation becomes a true statement.

Try $(2, 3)$ in $y = 3x + 2$:

$$3 = 3(2) + 2$$
$$3 = 6 + 2$$
$$3 = 8 \qquad \text{A false statement}$$

Try $(1, 5)$ in $y = 3x + 2$:

$$5 = 3(1) + 2$$
$$5 = 3 + 2$$
$$5 = 5 \qquad \text{A true statement}$$

Try $(-2, -4)$ in the equation $y = 3x + 2$:

$$-4 = 3(-2) + 2$$
$$-4 = -6 + 2$$
$$-4 = -4 \qquad \text{A true statement}$$

The ordered pairs, $(1, 5)$ and $(-2, -4)$ are solutions to the equation $y = 3x + 2$, and $(2, 3)$ is not. ▲

Problem Set 3.1

For each equation, complete the given ordered pairs.

1. $2x + y = 6$ $(0, \ )$, $(3, \ )$, $(\ , -6)$
2. $3x - y = 5$ $(0, \ )$, $(1, \ )$, $(\ , 5)$
3. $3x + 4y = 12$ $(0, \ )$, $(\ , 0)$, $(-4, \ )$
4. $5x - 5y = 20$ $(0, \ )$, $(\ , -4)$, $(1, \ )$
5. $y = 4x - 3$ $(1, \ )$, $(\ , 0)$, $(5, \ )$
6. $y = 3x - 5$ $(\ , 13)$, $(0, \ )$, $(-2, \ )$
7. $y = 7x - 1$ $(2, \ )$, $(\ , 6)$, $(0, \ )$
8. $y = 8x + 2$ $(3, \ )$, $(\ , 0)$, $(\ , -6)$
9. $x = -5$ $(\ , 4)$, $(\ , -3)$, $(\ , 0)$
10. $y = 2$ $(5, \ )$, $(-8, \ )$, $(\frac{1}{2}, \ )$

For each of the following equations complete the given table.

11. $y = 3x$

x	y
1	
-3	
	12
	18

12. $y = -2x$

x	y
-4	
0	
	10
	12

13. $y = 4x$

x	y
0	
	-2
-3	
	12

14. $y = -5x$

x	y
3	
	0
-2	
	-20

15. $x + y = 5$

x	y
2	
3	
	0
	-4

16. $x - y = 8$

x	y
0	
4	
	-3
	-2

17. $2x - y = 4$

x	y
	0
	2
3	
-3	

18. $3x - y = 9$

x	y
	0
	-9
5	
-4	

19. $y = 6x - 1$

x	y
0	
-1	
-3	
	8

20. $y = 5x + 7$

x	y
0	
-2	
-4	
	-8

For the following equations, tell which of the given ordered pairs are solutions.

21. $2x - 5y = 10$ $(2, 3)$, $(0, -2)$, $(\frac{5}{2}, 1)$

22. $3x + 7y = 21$ $(0, 3)$, $(7, 0)$, $(1, 2)$

23. $y = 7x - 2$ $(1, 5)$, $(0, -2)$, $(-2, -16)$

24. $y = 8x - 3$ $(0, 3)$, $(5, 16)$, $(1, 5)$

25. $y = 6x$ $(1, 6)$, $(-2, -12)$, $(0, 0)$

26. $y = -4x$ $(0, 0)$, $(2, 4)$, $(-3, 12)$

27. $x + y = 0$ $(1, 1)$, $(2, -2)$, $(3, 3)$

28. $x - y = 1$ $(0, 1)$, $(0, -1)$, $(1, 2)$

29. $x = 3$ $(3, 0)$, $(3, -3)$, $(5, 3)$

30. $y = -4$ $(3, -4)$, $(-4, 4)$, $(0, -4)$

Find four solutions for the following equations.

31. $2x + 2y = 4$ **32.** $3x - 2y = 6$
33. $4x - 6y = 24$ **34.** $7x - 2y = 14$
35. $y = -6x$ **36.** $y = \frac{1}{2}x$
37. $y = 3x - 7$ **38.** $y = -5x - 2$
39. $y = -7$ **40.** $x = -1$

41. If the perimeter of a rectangle is 30 inches, then the relationship between the length l and the width w is given by the equation

$$2l + 2w = 30$$

What is the length when the width is 3 inches?

42. The relationship between the perimeter P of a square and the length of its side s is given by the formula

$$P = 4s$$

If each side of a square is 5 inches, what is the perimeter? If the perimeter of a square is 28 inches, how long is a side?

43. If every ordered pair that satisfies an equation has a y-coordinate that is twice as large as its x-coordinate, then the equation is $y = 2x$ and every ordered pair has the form $(x, 2x)$. Write the ordered pair that has 5 as its x-coordinate.

44. If every ordered pair that satisfies an equation has a y-coordinate that is three times as large as its x-coordinate, then the equation is $y = 3x$. What is the form of each ordered pair? (*Hint:* Fill in the second coordinate in $(x,\ )$; see Problem 43.) If x is 4, what is y?

45. For the equation $y = |x|$, y is equal to the absolute value of x. Complete the following ordered pairs so that they are solutions to $y = |x|$.

$$(3,\ ) \quad (-3,\ ) \quad (5,\ ) \quad (-5,\ )$$

46. If $y = |x|$ and y is 4, there are two possible x values: either 4 or -4, because both the ordered pairs $(4, 4)$ and $(-4, 4)$ satisfy the equation. What two ordered pairs that are solutions to $y = |x|$ have a y-coordinate of 6?

Review Problems   The problems below review material we covered in Section 2.3.

Solve each equation.

47. $\dfrac{x}{3} = -4$ **48.** $-\dfrac{x}{4} = 2$
49. $3x + 2x - 6x = 4$ **50.** $9x - 6x - 4x = -3$
51. $4x + 2 = 2x - 8$ **52.** $3x + 1 = 7x - 7$
53. $7x - 5 = 2x + 10$ **54.** $6x + 8 = x - 7$

In this section we will graph ordered pairs. Since linear equations in two variables have solution sets made up of ordered pairs, we will also examine the graphs of those solution sets.

To graph an ordered pair such as (3, 4), we need two real number lines—one associated with the first coordinate and one associated with the second coordinate. A rectangular (or Cartesian) coordinate system looks like this:

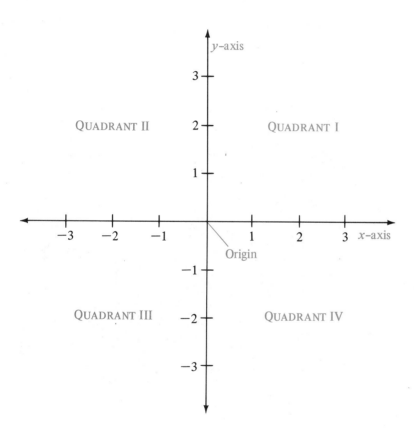

The rectangular coordinate system is built from two number lines oriented perpendicular to each other. The horizontal number line is exactly the same as our real number line and is called the *x*-axis. The vertical number line is also the same as our real number line with the positive direction up and the negative direction down. It is called the *y*-axis. The point where the two axes intersect is called the origin. As you can see from the diagram, the axes divide the plane into four quadrants, which are numbered I through IV in a counterclockwise direction.

Graphing Ordered
Pairs

To graph the ordered pair (*a, b*), we start at the origin and move *a* units forward or back (forward if *a* is positive and back if *a* is negative). Then we move *b* units up or down (up if *b* is positive, down if *b* is negative). The point where we end up is the graph of the ordered pair (*a, b*).

▼ **Example 1** Graph the ordered pairs (3, 4), (3, −4), (−3, 4) and (−3, −4).

SOLUTION

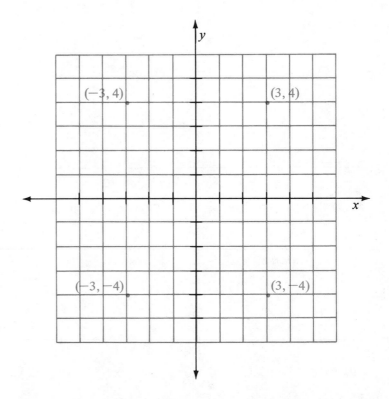

We can see that when graphing ordered pairs the *x*-coordinate corresponds to movement parallel to the *x*-axis (horizontal) and the *y*-coordinate corresponds to movement parallel to the *y*-axis (vertical).   ▲

▼ **Example 2** Graph the following ordered pairs: (−1, 3), (2, 5), (0, 0), (0, −3), (4, 0).

SOLUTION

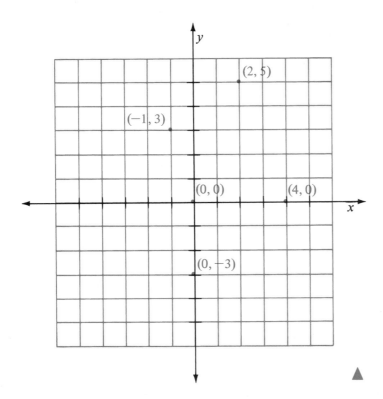

We will now see why linear equations are called linear. It is because their graphs are straight lines.

Graphing Straight Lines

To graph the solution set for an equation in two variables, we simply graph all the points whose coordinates satisfy the equation.

▼   **Example 3**   Graph the solution set for $x + y = 5$.

SOLUTION   We know from the last section that there is an infinite number of ordered pairs that are solutions to the equation $x + y = 5$. We can't possibly list them all. What we can do is list a few of them and see if there is any pattern to their graphs.

Some ordered pairs that are solutions to $x + y = 5$ are $(0, 5)$, $(2, 3)$, $(3, 2)$, $(5, 0)$. If we graph them, we get the following:

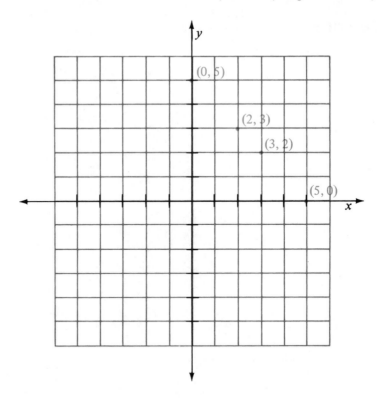

Now, by passing a straight line through these points we can graph the solution set for the equation $x + y = 5$. Linear equations in two variables always have graphs that are straight lines. The graph of the solution set for $x + y = 5$ is on page 95.

Every ordered pair that satisfies $x + y = 5$ has its graph on the line and any point on the line has coordinates that satisfy the equation. That is, there is a one-to-one correspondence between points on the line and solutions to the equation. ▲

Here are some steps to follow when graphing straight lines.

Step 1: Find any three ordered pairs that satisfy the equation. This can be done by using a convenient number for one variable and solving for the other variable.

Step 2: Graph the three ordered pairs found in Step 1. Actually, we only need two points to graph a straight line. The third point serves as a check. If all three points do not line up, there is a mistake in our work.

Step 3: Draw a straight line through the three points graphed in Step 2.

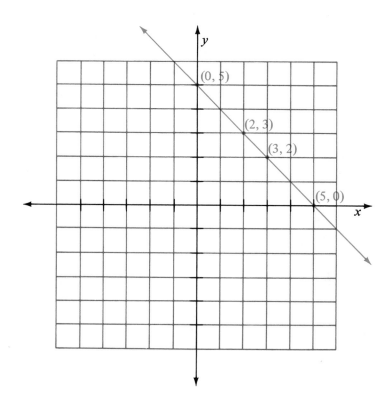

▼ **Example 4**   Graph the equation $y = 3x - 1$.

SOLUTION   We first find three solutions to the equation by choosing convenient numbers for one variable and solving for the other variable.

Let $x = 0$:

$$y = 3(0) - 1$$
$$y = 0 - 1$$
$$y = -1 \qquad (0, -1) \text{ is one solution}$$

Let $x = 2$:

$$y = 3(2) - 1$$
$$y = 6 - 1$$
$$y = 5 \qquad (2, 5) \text{ is another solution}$$

Let $x = -2$:

$$y = 3(-2) - 1$$
$$y = -6 - 1$$
$$y = -7 \qquad \text{(-2, -7) is a third solution}$$

Next we graph the ordered pairs $(0, -1)$, $(2, 5)$, $(-2, -7)$ and draw a straight line through them.

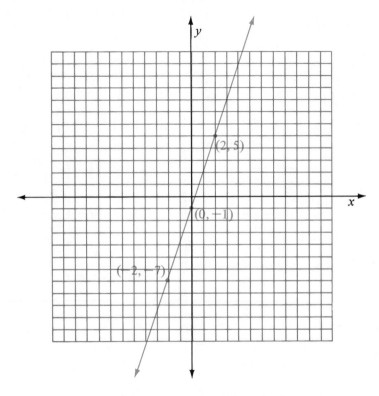

The line we have drawn is the graph of the solution set for $y = 3x - 1$.   ▲

▼ **Example 5** Graph the solution set for $3x - 2y = 6$.

SOLUTION Let $x = 0$:

$$3(0) - 2y = 6$$
$$0 - 2y = 6$$
$$-2y = 6$$
$$y = -3 \qquad \text{(0, -3) is a solution}$$

Let $y = 0$:

$$3x - 2(0) = 6$$
$$3x - 0 = 6$$
$$3x = 6$$
$$x = 2 \qquad (2, 0) \text{ is another solution}$$

Let $y = 3$:

$$3x - 2(3) = 6$$
$$3x - 6 = 6$$
$$3x = 12$$
$$x = 4 \qquad (4, 3) \text{ is a third solution}$$

The graph is:

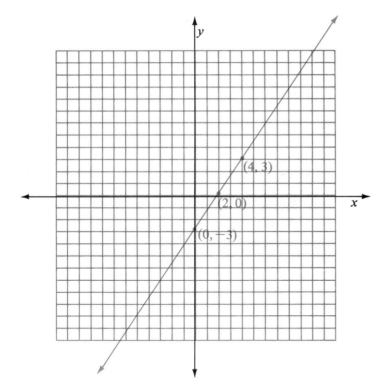

Graph the following ordered pairs.

1. $(3, 2)$ **2.** $(3, -2)$ **3.** $(-3, 2)$

4. $(-3, -2)$ **5.** $(5, 1)$ **6.** $(5, -1)$

    7. $(1, 5)$ **8.** $(1, -5)$ **9.** $(-1, 5)$
10. $(-1, -5)$ **11.** $(2, \frac{1}{2})$ **12.** $(3, \frac{3}{2})$
13. $(-4, -\frac{5}{2})$ **14.** $(-5, -\frac{3}{2})$ **15.** $(-2, -8)$
16. $(-3, -5)$ **17.** $(3, 0)$ **18.** $(-2, 0)$
19. $(0, 5)$ **20.** $(0, 0)$

For the following equations, complete the given ordered pairs and use the results to graph the solution set for the equation.

21. $x + y = 3$ $(0, \ )$, $(2, \ )$, $(\ , -1)$
22. $x - y = 4$ $(1, \ )$, $(-1, \ )$, $(\ , 0)$
23. $y = 2x$ $(0, \ )$, $(-2, \ )$, $(2, \ )$
24. $y = \frac{1}{2}x$ $(0, \ )$, $(-2, \ )$, $(2, \ )$
25. $y = -3x$ $(-1, \ )$, $(\ , 0)$, $(1, \ )$
26. $y = -\frac{1}{3}x$ $(-3, \ )$, $(\ , 0)$, $(3, \ )$
27. $3x + 2y = 6$ $(0, \ )$, $(\ , 0)$, $(-3, \ )$
28. $2x + 3y = 6$ $(0, \ )$, $(\ , 0)$, $(\ , -3)$
29. $2x - 7y = 14$ $(0, \ )$, $(\ , 0)$, $(\ , 2)$
30. $7x - 2y = 14$ $(0, \ )$, $(\ , 0)$, $(-2, \ )$
31. $y = 2x + 1$ $(0, \ )$, $(-1, \ )$, $(1, \ )$
32. $y = -2x + 1$ $(0, \ )$, $(-1, \ )$, $(1, \ )$
33. $y = 4x - 3$ $(0, \ )$, $(\ , 1)$, $(\ , 5)$
34. $y = 4x + 3$ $(\ , 3)$, $(-1, \ )$, $(-2, \ )$
35. $x = -2$ $(\ , 3)$, $(\ , 0)$, $(-2, \ )$
36. $x = 3$ $(3, \ )$, $(\ , 1)$, $(\ , -2)$
37. $y = 1$ $(2, \ )$, $(3, \ )$, $(\ , 1)$
38. $y = -5$ $(0, \ )$, $(4, \ )$, $(\ , -5)$

Find three solutions to each of the following equations and then graph the solution set.

    39. $x + y = 4$ **40.** $x - y = 3$
    41. $2x + y = 6$ **42.** $x - 2y = 5$
    43. $y = -\frac{1}{2}x$ **44.** $y = -2x$
    45. $y = 4x$ **46.** $y = -\frac{1}{4}x$
    47. $2x - 3y = 6$ **48.** $3x - 2y = 6$
    49. $5x - 2y = 10$ **50.** $2x + 5y = 10$
    51. $y = 3x - 1$ **52.** $y = -3x - 1$
    53. $y = 2x + 4$ **54.** $y = 2x - 4$
    55. $x = 5$ **56.** $x = -3$
    57. $y = \frac{1}{2}$ **58.** $y = -6$

59. If the perimeter of a rectangle is 10 inches, then the equation that describes the relationship between the length l and width w is

$$2l + 2w = 10$$

To graph this equation, we use a coordinate system in which the horizontal axis is labeled l and the vertical axis is labeled w.

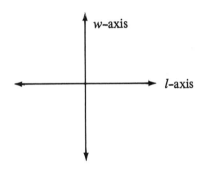

The ordered pairs that satisfy the equation look like (l, w). Complete the ordered pairs $(1, \ )$, $(2, \ )$, and $(3, \ )$ for the equation $2l + 2w = 10$ and use the results to graph the equation. Compare your answer with Example 3 in this section.

60. The perimeter of a rectangle is 6 inches. Graph the equation that describes the relationship between the length l and width w.

61. The solutions to the equation $y = x$ are ordered pairs of the form (x, x). The x- and y-coordinates are equal. Graph the line $y = x$.

62. The solutions to a certain equation are ordered pairs of the form $(x, 2x)$. Find the equation and its graph.

63. The y-coordinate of each ordered pair that satisfies the equation $y = |x|$ will never be negative. Each value of y is the absolute value of the corresponding value of x. Fill in the following ordered pairs so that they are solutions to the equation $y = |x|$. Then graph them and connect their graphs in a way that makes the most sense to you.

$$(-3, \ ) \quad (-2, \ ) \quad (-1, \ ) \quad (0, \ ) \quad (1, \ ) \quad (2, \ ) \quad (3, \ )$$

64. Fill in the following ordered pairs so that they are solutions to the equation $y = |x - 2|$. Graph each point and then connect the graphs in a way that makes the most sense to you.

$$(-1, \ ) \quad (0, \ ) \quad (1, \ ) \quad (2, \ ) \quad (3, \ ) \quad (4, \ ) \quad (5, \ )$$

65. Graph the lines $y = x + 1$ and $y = x - 3$ on the same coordinate system. Can you tell from looking at these first two graphs where the graph of $y = x + 3$ would be?

66. Graph the lines $y = 2x + 2$ and $y = 2x - 1$ on the same coordinate system. Use the similarities between these two graphs to graph the line $y = 2x - 4$.

Review Problems   The problems below review material we covered in Section 2.4.

Solve each equation.

67. $3(x - 2) = 9$ **68.** $-4(x - 3) = -16$

69. $2(3x - 1) + 4 = -10$ **70.** $-5(2x + 3) - 10 = 15$
71. $6 - 2(4x - 7) = -4$ **72.** $5 - 3(2 - 3x) = 8$

**3.3
Solving Linear
Systems by
Graphing**

Two linear equations considered at the same time make up what is called a *system* of linear equations. Both equations contain two variables and, of course, have graphs that are straight lines. The following are systems of linear equations:

$$x + y = 3 \qquad y = 2x + 1 \qquad 2x - y = 1$$
$$3x + 4y = 2 \qquad y = 3x + 2 \qquad 3x - 2y = 6$$

The solution set for a *system* of linear equations is all ordered pairs that are solutions to both equations. Since each linear equation has a graph that is a straight line, we can expect the intersection of the graphs to be a point whose coordinates are solutions to the system. That is, if we graph both equations on the same coordinate system, we can read the coordinates of the point of intersection and have the solution to our system. Here is an example.

▼ Example 1   Solve the following system by graphing:

$$x + y = 4$$
$$x - y = -2$$

SOLUTION   On the same set of coordinate axes we graph each equation separately, using the methods from Section 3.2. Without showing the work necessary to get both graphs, the graphs look like the figure on the following page. We can see from the graphs that they intersect at the point $(1, 3)$. The point $(1, 3)$ must therefore be the solution to our system, since it is the only ordered pair whose graph lies on both lines. Its coordinates satisfy both equations.

We can check our results by substituting the coordinates $x = 1$, $y = 3$ into both equations to see if they work.

When	$x = 1$
and	$y = 3$
the equation	$x + y = 4$
becomes	$1 + 3 = 4$
or	$4 = 4$

When $\qquad x = 1$
and $\qquad y = 3$
the equation $\qquad x - y = -2$
becomes $\qquad 1 - 3 = -2$
or $\qquad -2 = -2$

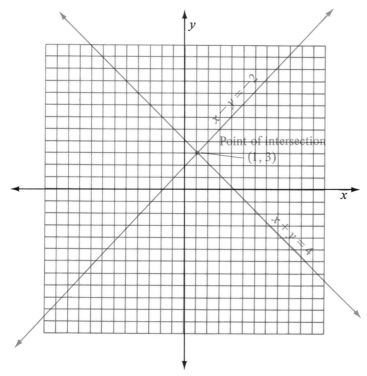

The point $(1, 3)$ satisfies both equations and, therefore, must be the solution to our system. ▲

QUESTION How do we know $(1, 3)$ is the only solution to the system in Example 1?

ANSWER Once we have graphed both equations, we have a picture of each solution set. Obviously the only point on both graphs is $(1, 3)$. Therefore it must be the only solution to our system.

Here are some steps to follow in solving linear systems by graphing.

Step 1: Graph the first equation by the methods described in Section 3.2.

Step 2: Graph the second equation on the same set of axes used for the first equation.

Step 3: Read the coordinates of the point of intersection of the two graphs. The ordered pair is the solution to the system.

Step 4: Check the solution in both equations, if necessary.

▼ **Example 2** Solve the following system by graphing:

$$x + 2y = 8$$
$$2x - 3y = 2$$

SOLUTION We graph each equation on the same set of coordinate axes and get

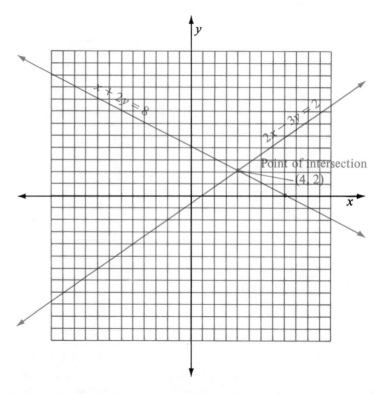

From the graph, we can see the solution for our system is (4, 2). We check this solution as follows.

When	$x = 4$
and	$y = 2$
the equation	$x + 2y = 8$
becomes	$4 + 2(2) = 8$
	$4 + 4 = 8$
	$8 = 8$

When $\qquad\qquad x = 4$
and $\qquad\qquad\qquad y = 2$
the equation $\qquad\qquad 2x - 3y = 2$
becomes $\qquad\qquad 2(4) - 3(2) = 2$
$\qquad\qquad\qquad\qquad\quad 8 - 6 = 2$
$\qquad\qquad\qquad\qquad\qquad\quad 2 = 2$

The point (4, 2) satisfies both equations and, therefore, must be the solution to our system. ▲

▼ **Example 3** Solve this system by graphing:

$$y = 2x - 3$$
$$x = 3$$

SOLUTION Graphing both equations on the same set of axes, we have

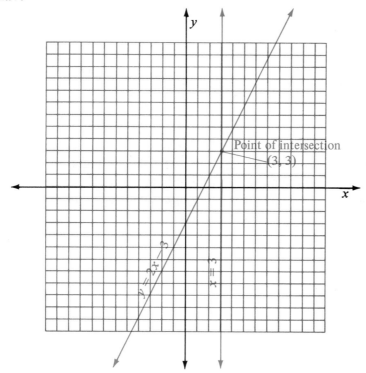

The solution to the system is the point (3, 3). ▲

▼ **Example 4** Solve by graphing:

$$y = x - 2$$
$$y = x + 1$$

SOLUTION   Graphing both equations, we have the following:

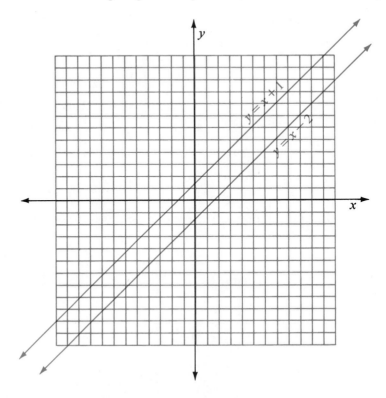

We can see that the lines are parallel and therefore do not intersect. Our system has no ordered pair as a solution, since there is
no ordered pair that satisfies both equations. We say the solution
set is the empty set and write ∅.   ▲

Example 4 is one example of two special cases associated with linear
systems. The other special case happens when the two graphs coincide.
Here is an example.

▼   **Example 5**   Graph the system

$$2x + y = 4$$
$$4x + 2y = 8$$

SOLUTION   When we graph both equations, we come up with the
following:

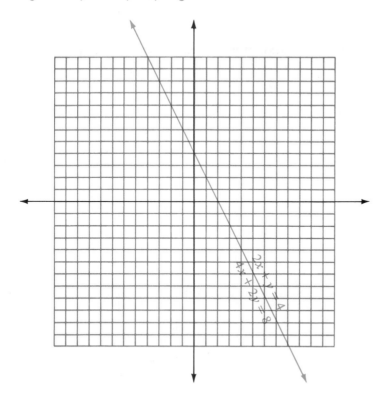

The two graphs coincide. The reason becomes apparent when we multiply both sides of the first equation by 2.

$$2x + y = 4$$
$$2(2x + y) = 2(4) \qquad \text{Multiply both sides by 2}$$
$$4x + 2y = 8$$

The equations have the same solution set. Any ordered pair that is a solution to one is a solution to the system. The system has an infinite number of solutions. (Any point on the line is a solution to the system.) ▲

The two special cases illustrated in the last two examples do not happen often. Usually a system has a single ordered pair as a solution. Solving a system of linear equations by graphing is useful only when the ordered pair in the solution set has integers for coordinates. There are two other solution methods that work well in all cases. We will develop the other two methods in the next two sections.

Problem Set 3.3

Solve the following systems of linear equations by graphing.

1. $x + y = 3$
   $x - y = 1$

2. $x + y = 2$
   $x - y = 4$

3. $x + y = 1$
   $-x + y = 3$

4. $x + y = 1$
   $x - y = -5$

5. $x + y = 10$
   $-x + y = 4$

6. $x + y = 8$
   $-x + y = -4$

7. $3x - 2y = 6$
   $x - y = 1$

8. $5x - 2y = 10$
   $x - y = -1$

9. $6x - 2y = 12$
   $3x + y = -6$

10. $4x - 2y = 8$
    $2x + y = -4$

11. $4x + y = 4$
    $3x - y = 3$

12. $5x - y = 10$
    $2x + y = 4$

13. $x + 2y = 0$
    $2x - y = 0$

14. $3x + y = 0$
    $5x - y = 0$

15. $3x - 5y = 15$
    $-2x + y = 4$

16. $4x - 5y = 20$
    $2x - y = -2$

17. $y = 2x + 1$
    $y = -2x - 3$

18. $y = 3x - 4$
    $y = -2x + 1$

19. $x + 3y = 3$
    $y = x + 5$

20. $2x + y = -2$
    $y = x + 4$

21. $4x - 5y = 3$
    $y = -3x + 7$

22. $4x + 3y = 9$
    $y = -2x + 5$

23. $x + y = 2$
    $x = -3$

24. $x + y = 6$
    $y = 2$

25. $x = -4$
    $y = 6$

26. $x = 5$
    $y = -1$

27. $x + y = 4$
    $2x + 2y = -6$

28. $x - y = 3$
    $2x - 2y = 6$

29. $4x - 2y = 8$
    $2x - y = 4$

30. $3x - 6y = 6$
    $x - 2y = 4$

Review Problems   The problems below review material we covered in Section 2.1.

Find the value of each expression when x is -3.

31. $2x - 9$

32. $-4x + 3$

33. $9 - 6x$

34. $7 - 5x$

35. $7x + 6 - 4x - 3$

36. $4x + 9 - 6x + 1$

37. $4(3x + 2) + 1$

38. $3(6x + 5) + 10$

The addition property states that if equal quantities are added to both sides of an equation the solution set is unchanged. In the past we have used this property to help solve equations in one variable. We will now use it to solve systems of linear equations. Here is another way to state the addition property of equality.

Let A, B, C, and D represent algebraic expressions.

If	$A = B$
and	$C = D$
then	$A + C = B + D$

Since C and D are equal (that is, they represent the same number), what we have done is added the same amount to both sides of the equation $A = B$. Let's see how we can use this form of the addition property of equality to solve a system of linear equations.

▼ **Example 1** Solve the system

$$x + y = 4$$
$$x - y = 2$$

SOLUTION The system is written in the form of the addition property of equality as written above. It looks like

$$A = B$$
$$C = D$$

where A is $x + y$, B is 4, C is $x - y$, and D is 2.

We use the addition property of equality to add the left sides together and the right sides together:

$$\begin{array}{r} x + y = 4 \\ \underline{x - y = 2} \\ 2x + 0 = 6 \end{array}$$

We now solve the resulting equation for x:

$$2x + 0 = 6$$
$$2x = 6$$
$$x = 3$$

The value we get for x is the value of the x-coordinate of the point of intersection of the two lines $x + y = 4$ and $x - y = 2$. To find the y-coordinate we simply substitute $x = 3$ into either of the two original equations and get

$$3 + y = 4$$
$$y = 1$$

The solution to our system is the ordered pair (3, 1). It satisfies both equations.

When	$x = 3$
and	$y = 1$
the equation	$x + y = 4$
becomes	$3 + 1 = 4$
or	$4 = 4$

When	$x = 3$
and	$y = 1$
the equation	$x - y = 2$
becomes	$3 - 1 = 2$
or	$2 = 2$

▲

The most important part of this method of solving linear systems is eliminating one of the variables when we add the left and right sides together. In our first example the equations were written so that the y variable was eliminated when we added the left and right sides together. If the equations are not set up this way to begin with, we have to work on one or both of them separately before we can add them together to eliminate one variable.

▼ Example 2   Solve the following system:

$$x + 2y = 4$$
$$x - y = -5$$

SOLUTION   Notice that if we were to add the equations together as they are, the resulting equation would have terms in both x and y. Let's eliminate the variable x by multiplying both sides of the second equation by -1 before we add the equations together. (As you will see, we can choose to eliminate either the x or the y variable.) Multiplying both sides of the second equation by -1 will not change its solution, so we do not need to be concerned that we have altered the system.

$$x + 2y = 4 \xrightarrow{\text{no change}} x + 2y = 4 \qquad \text{Add left and right}$$
$$x - y = -5 \xrightarrow[\text{multiply by } -1]{} -x + y = 5 \qquad \text{sides to get}$$
$$\overline{0 + 3y = 9}$$
$$3y = 9$$
$$y = 3$$

$\left\{\begin{array}{l} y\text{—coordinate of} \\ \text{the point of} \\ \text{intersection} \end{array}\right.$

Substituting $y = 3$ into either of the two original equations, we get $x = -2$. The solution to the system is $(-2, 3)$. It satisfies both equations. ▲

▼ **Example 3** Solve the system

$$2x - y = 6$$
$$x + 3y = 3$$

SOLUTION Let's eliminate the y variable from the two equations. We can do this by multiplying the first equation by 3 and leaving the second equation unchanged:

$$2x - y = 6 \xrightarrow{\text{3 times both sides}} 6x - 3y = 18$$
$$x + 3y = 3 \xrightarrow[\text{no change}]{} x + 3y = 3$$

The important thing about our system now is that the coefficients (the numbers in front) of the y variables are opposites. When we add the terms on each side of the equal sign, then, the terms in y will add to zero and be eliminated:

$$6x - 3y = 18$$
$$\underline{x + 3y = 3}$$
$$7x = 21 \qquad \text{Add corresponding terms}$$

which gives us $x = 3$. Using this value of x in the second equation of our original system, we have

$$3 + 3y = 3$$
$$3y = 0$$
$$y = 0$$

We could substitute $x = 3$ into any of the equations with both x and y variables and also get $y = 0$. The solution to our system is the ordered pair $(3, 0)$. ▲

▼ **Example 4**   Solve the system

$$2x + 3y = -1$$
$$3x + 5y = -2$$

SOLUTION   Let's eliminate x from the two equations. If we multiply the first equation by 3 and the second by -2, the coefficients of x will be 6 and -6, respectively. The x terms in the two equations will then add to zero:

$$2x + 3y = -1 \xrightarrow{\text{multiply by 3}} 6x + 9y = -3$$
$$3x + 5y = -2 \xrightarrow[\text{multiply by } -2]{} -6x - 10y = 4$$

We now add the left and right sides of our new system together:

$$
\begin{array}{r}
6x + 9y = -3 \\
-6x - 10y = 4 \\
\hline
- y = 1 \\
y = -1
\end{array}
$$

Substituting $y = -1$ into the first equation in our original system, we have

$$2x + 3(-1) = -1$$
$$2x - 3 = -1$$
$$2x = 2$$
$$x = 1$$

The solution to our system is $(1, -1)$. It is the only ordered pair that satisfies both equations.   ▲

▼ **Example 5**   Solve the system

$$3x + 5y = -7$$
$$5x + 4y = 10$$

SOLUTION   Let's eliminate y by multiplying the first equation by -4 and the second equation by 5:

$$3x + 5y = -7 \xrightarrow{\text{multiply by } -4} -12x - 20y = 28$$
$$5x + 4y = 10 \xrightarrow[\text{multiply by 5}]{} 25x + 20y = 50$$
$$
\begin{array}{r}
\hline
13x = 78 \\
x = 6
\end{array}
$$

Substitute $x = 6$ into any equation with both x and y, and the result will be $y = -5$. The solution is therefore $(6, -5)$.   ▲

▼ **Example 6**   Solve the system

$$2x - y = 2$$
$$4x - 2y = 8$$

SOLUTION   Let us choose to eliminate y from the system. We can do this by multiplying the first equation by -2 and leaving the second equation unchanged.

$$2x - y = 2 \xrightarrow{\text{multiply by } -2} -4x + 2y = -4$$
$$4x - 2y = 8 \xrightarrow[\text{no change}] {} 4x - 2y = 8$$

If we add both sides of the resulting system, we have

$$-4x + 2y = -4$$
$$\underline{4x - 2y = 8}$$
$$0 + 0 = 4$$
$$\text{or} \quad 0 = 4 \qquad \text{A false statement}$$

Both variables have been eliminated, and we end up with the false statement $0 = 4$. We have tried to solve a system that consists of two parallel lines. There is no solution, and that is the reason we end up with a false statement.   ▲

If, on the other hand, we had eliminated both variables and ended up with a true statement, it would mean that the lines coincided and every ordered pair whose graph was on the line would be a solution to the system.

Both variables elimi-nated and the resulting statement false	↔	The lines are parallel and there is no solution to the system
Both variables elimi-nated and the resulting statement true	↔	The lines coincide and there is an infinite number of solutions to the system

The main idea in solving a system of linear equations by the elimination method is to use the multiplication property of equality on one or both of the original equations, if necessary, to make the coefficients of either variable opposites. Here are some steps to follow when solving a system of linear equations by the elimination method.

Step 1: Decide which variable to eliminate. (In some cases one variable will be easier to eliminate than the other. With some practice you will notice which one it is.)

Step 2: Use the multiplication property of equality on each equation separately to make the coefficients of the variable that is to be eliminated opposites.

Step 3: Add the respective left and right sides of the system together.

Step 4: Solve for the variable remaining.

Step 5: Substitute the value of the variable from Step 4 into an equation containing both variables and solve for the other variable.

Step 6: Check your solution in both equations, if necessary.

Problem Set 3.4

Solve the following systems of linear equations by elimination.

1. $x + y = 3$
   $x - y = 1$

2. $x + y = -2$
   $x - y = 6$

3. $x + y = 10$
   $-x + y = 4$

4. $x - y = 1$
   $-x - y = -7$

5. $x - y = 7$
   $-x - y = 3$

6. $x - y = 4$
   $2x + y = 8$

7. $x + y = -1$
   $3x - y = -3$

8. $2x - y = -2$
   $-2x - y = 2$

9. $3x + 2y = 1$
   $-3x - 2y = -1$

10. $-2x - 4y = 1$
    $2x + 4y = -1$

Solve each of the following systems by eliminating the y variables.

11. $3x - y = 4$
    $2x + 2y = 24$

12. $2x + y = 3$
    $3x + 2y = 1$

13. $5x - 3y = 1$
    $x - y = -1$

14. $4x - y = 13$
    $2x + 4y = 2$

15. $11x - 4y = 11$
    $5x + y = 5$

16. $3x - y = 7$
    $10x - 5y = 25$

Solve each of the following systems by eliminating the x variable.

17. $3x - 5y = 7$
    $-x + y = -1$

18. $4x + 2y = 32$
    $x + y = -2$

19. $-x - 8y = 3$
    $-2x + y = -11$

20. $-x + 10y = 3$
    $-5x + 11y = -24$

21. $-3x - y = 7$
    $6x + 7y = 11$

22. $-5x + 2y = -6$
    $10x + 7y = 34$

Solve each of the following systems of linear equations by the elimination method.

23. $6x - y = -8$
$2x + y = -16$

24. $5x - 3y = -3$
$3x + 3y = -21$

25. $x + 3y = 9$
$2x - y = 4$

26. $x + 2y = 0$
$2x - y = 0$

27. $x - 3y = -9$
$4x + 3y = 39$

28. $3x + y = 7$
$4x - 5y = 3$

29. $2x + y = 5$
$5x + 3y = 11$

30. $5x + 2y = 11$
$7x + y = 10$

31. $4x + 3y = 14$
$9x - 2y = 14$

32. $7x - 6y = 13$
$6x - 5y = 11$

33. $3x + 2y = -1$
$6x + 4y = 0$

34. $8x - 2y = 2$
$4x - y = 2$

35. $11x + 6y = 17$
$5x - 4y = 1$

36. $3x - 8y = 7$
$10x - 5y = 45$

37. $3x + y = 2$
$-6x - 2y = -1$

38. $2x - 3y = 4$
$-4x + 6y = -8$

39. For some systems of equations it is necessary to apply the addition property of equality to each equation to line up the x variables and y variables before trying to eliminate a variable. Solve the following system by first writing each equation so that the variable terms with x in them come first, the variable terms with y in them come second, and the constant terms are on the right side of each equation.

$$4x - 5y = 17 - 2x$$
$$5y = 3x + 4$$

40. Solve the following system by first writing each equation so that the variable terms with x in them come first, the variable terms with y in them come second, and the constant terms are on the right side of each equation.

$$3x - 6y = -20 + 7x$$
$$4x = 3y - 34$$

41. Multiply both sides of the second equation in the following system by 100 and then solve as usual.

$$x + y = 22$$
$$0.05x + 0.10y = 1.70$$

42. Multiply both sides of the second equation in the following system by 100 and then solve as usual.

$$x + y = 15,000$$
$$0.06x + 0.07y = 980$$

Review Problems   The problems below review material we covered in Section 2.6.

Solve each inequality.

43. $x - 3 < 2$

44. $x + 4 \le 6$

45. $-3x \ge 12$

46. $-2x > 10$

47. $-\dfrac{x}{3} \le -1$

48. $-\dfrac{x}{5} < -2$

49. $-4x + 1 < 17$

50. $-3x + 2 \le -7$

**3.5
The Substitution
Method**

There is a third method of solving systems of equations. It is the substitution method, and like the elimination method, it can be used on any system of linear equations. Some systems, however, lend themselves more to the substitution method than others.

▼   **Example 1**   Solve the following system:

$$x + y = 2$$
$$y = 2x - 1$$

SOLUTION   If we were to solve this system by the methods used in the last section, we would have to rearrange the terms of the second equation so that similar terms would be in the same column. There is no need to do this, however, since the second equation tells us that y is $2x - 1$. We can replace the y variable in the first equation with the expression $2x - 1$ from the second equation. That is, we *substitute* $2x - 1$ from the second equation for y in the first equation. Here is what it looks like:

$$x + y = 2$$
$$y = 2x - 1$$

Substituting $2x - 1$ for y, we have

$$x + (2x - 1) = 2$$

The equation we end up with contains only the variable x. The y variable has been eliminated by substitution.

Solving the resulting equation, we have

$$x + (2x - 1) = 2$$
$$3x - 1 = 2$$
$$3x = 3$$
$$x = 1$$

This is the x-coordinate of the solution to our system. To find the y-coordinate, we substitute $x = 1$ into the second equation of our system. (We could substitute $x = 1$ into the first equation also and have the same result.)

$$y = 2(1) - 1$$
$$y = 2 - 1$$
$$y = 1$$

The solution to our system is the ordered pair $(1, 1)$. It satisfies both of the original equations.   ▲

The key to this method of solving a system of equations is again elimination of one of the variables. In this method we eliminate the variable by substitution rather than addition.

▼ **Example 2** Solve the following system by the substitution method:

$$2x - 3y = 12$$
$$y = 2x - 8$$

SOLUTION Again, the second equation says y *is* $2x - 8$. Since we are looking for the ordered pair that satisfies both equations, the y in the first equation must also be $2x - 8$. Substituting $2x - 8$ from the second equation for y in the first equation, we have

$$2x - 3(2x - 8) = 12$$

This equation can still be read as $2x - 3y = 12$, because $2x - 8$ is the same as y. Solving the equation, we have

$$2x - 3(2x - 8) = 12$$
$$2x - 6x + 24 = 12$$
$$-4x + 24 = 12$$
$$-4x = -12$$
$$x = 3$$

To find the y-coordinate of our solution, we substitute $x = 3$ into the second equation in the original system.

When	$x = 3$
the equation	$y = 2x - 8$
becomes	$y = 2(3) - 8$
	$y = 6 - 8$
	$y = -2$

The solution to our system is $(3, -2)$.   ▲

▼ **Example 3** Solve the following system by solving the first equation for x and then using the substitution method:

$$x - 3y = -1$$
$$2x - 3y = 4$$

SOLUTION We solve the first equation for x by adding $3y$ to both sides to get

$$x = 3y - 1$$

Using this value of x in the second equation, we have

$$2(3y - 1) - 3y = 4$$
$$6y - 2 - 3y = 4$$
$$3y - 2 = 4$$
$$3y = 6$$
$$y = 2$$

When
the equation
becomes
$$y = 2$$
$$x = 3y - 1$$
$$x = 3(2) - 1$$
$$x = 6 - 1$$
$$x = 5$$

The solution to our system is (5, 2).   ▲

Here are the steps to use in solving a system of equations by the substitution method.

Step 1: Solve either one of the equations for x or y. (This step is not necessary if one of the equations is already in the correct form, as in Examples 1 and 2.)

Step 2: Substitute the expression for the variable obtained in Step 1 into the other equation and solve it.

Step 3: Substitute the solution from Step 2 into any equation in the system that contains both variables and solve it.

Step 4: Check your results, if necessary.

Here are some more examples.

▼ **Example 4** Solve by substitution:

$$-2x + 4y = 14$$
$$-3x + \;\; y = 6$$

SOLUTION We can solve either equation for either variable. If we look at the system closely, it becomes apparent that solving the second equation for y is the easiest way to go. If we add $3x$ to both sides of the second equation, we have

$$y = 3x + 6$$

Substituting the expression $3x + 6$ back into the first equation in place of y yields the following results:

$$-2x + 4(3x + 6) = 14$$
$$-2x + 12x + 24 = 14$$
$$10x + 24 = 14$$
$$10x = -10$$
$$x = -1$$

Substituting $x = -1$ into the equation $y = 3x + 6$ leaves us with

$$y = 3(-1) + 6$$
$$y = -3 + 6$$
$$y = 3$$

The solution to our system is $(-1, 3)$. ▲

▼ **Example 5** Solve by substitution:

$$4x + 2y = 8$$
$$y = -2x + 4$$

SOLUTION Substituting the expression $-2x + 4$ for y from the second equation into the first equation, we have

$$4x + 2(-2x + 4) = 8$$
$$4x - 4x + 8 = 8$$
$$8 = 8 \quad \text{A true statement}$$

Both variables have been eliminated and we are left with a true statement. Recall from the last section that a true statement in this situation tells us the lines coincide. That is, the equations $4x + 2y = 8$ and $y = -2x + 4$ have exactly the same graph. Any point on that graph has coordinates that satisfy both equations and is a solution to the system. ▲

Solve the following systems by substitution. Substitute the expression in the second equation into the first equation and solve.

Problem Set 3.5

1. $x + y = 11$
$\quad\ y = 2x - 1$

2. $x - y = -3$
$\quad\ y = 3x + 5$

3. $x + y = 20$
    $y = 5x + 2$

4. $3x - y = -1$
    $y = -2x + 6$

5. $-2x + y = 0$
    $y = -3x - 5$

6. $4x - y = 6$
    $y = -x + 4$

7. $3x - 2y = -2$
    $y = -x + 6$

8. $2x - 3y = 17$
    $y = -x + 6$

9. $5x - 4y = -16$
    $y = 4x + 4$

10. $6x + 2y = 18$
    $y = 2x - 6$

11. $5x + 4y = 7$
    $y = -3x$

12. $10x + 2y = -6$
    $y = -5x$

Solve the following systems by solving the second equation for x and then using the substitution method.

13. $x + 3y = 4$
    $x - 2y = -1$

14. $x - y = 5$
    $x + 2y = -1$

15. $2x + y = 1$
    $x - 5y = 17$

16. $2x - 2y = 2$
    $x - 3y = -7$

17. $3x + 5y = 14$
    $x - 5y = 18$

18. $x + 2y = 5$
    $2x - 4y = 2$

19. $5x + 3y = 0$
    $2x - 6y = -36$

20. $x - 3y = -5$
    $3x - 6y = 0$

21. $-9x - 3y = 7$
    $3x + y = 2$

22. $x + 3y = 9$
    $2x + 6y = 18$

Solve the following systems using the substitution method.

23. $5x - 8y = 7$
    $y = 2x - 5$

24. $3x + 4y = 10$
    $y = 8x - 15$

25. $7x - y = 24$
    $x = 2y + 9$

26. $3x - y = -8$
    $x = 6y + 3$

27. $-3x + 2y = 6$
    $y = 3x$

28. $-2x - y = -3$
    $y = -3x$

29. $x - 6y = -20$
    $x = y$

30. $2x - 4y = 0$
    $x = y$

31. $2x - y = 12$
    $3x + 2y = 11$

32. $y = 2 + 4x$
    $2x = y - 4$

33. $x - y = 8$
    $x + 3y = 48$

34. $y - 3x = -2$
    $2x + y = -17$

35. $x - y = 2$
    $-4x + 4y = -8$

36. $2x - y = 5$
    $-4x + 2y = -10$

Review Problems   The problems below review material we covered in Section 2.5. Reviewing these problems will help you in the next section.

For each word problem, write an equation that describes the situation and then solve it.

37. The sum of a number and 4 is 12. Find the number.

38.   The product of a number and −3 is 15. Find the number.
39.   The sum of six times a number and 2 is 20. Find the number.
40.   The sum of five times a number and 8 is −2. Find the number.

I have often heard this student remark about the word problems in beginning algebra: "What does this have to do with real life?" Most of the word problems we will encounter don't have much to do with "real life." We are actually just practicing. Ultimately, all problems requiring the use of algebra are word problems. That is, they are stated in words first, then translated to symbols. The problem then is solved by some system of mathematics, like algebra. Most real applications involve calculus or higher levels of mathematics. So, if the problems we solve are upsetting or frustrating to you, then you are probably taking them too seriously.

The word problems in this section have two unknown quantities. We will write two equations in two variables (each of which represents one of the unknown quantities), which of course is a system of equations. We then solve the system by one of the methods developed in the previous sections of this chapter. Here are the steps to follow in solving these word problems.

Step 1:   Read the problem carefully (and more than once) and decide what is needed. Let *x* represent one of the unknown quantities and *y* the other.

Step 2:   Write two equations using the two variables from Step 1 that together describe the situation. (This is the step that tends to make people anxious. Read the problem as many times as it takes to understand the situation clearly.)

Step 3:   Solve the system obtained from Step 2.

Step 4:   Check the values of *x* and *y* from Step 3 to see that they make sense in the original problem.

Remember, the more problems you work, the more problems you will be able to work. If you have trouble getting started on the problem set, come back to the examples and work through them yourself. The examples are similar to the problems found in the problem set.

▼   Example 1   One number is two more than five times another number. Their sum is 20. Find the two numbers.

SOLUTION

Step 1:   Let *x* represent one of the numbers and *y* represent the other.

Step 2: "One number is two more than five times another" translates to

$$y = 5x + 2$$

"Their sum is 20" translates to

$$x + y = 20$$

The system that describes the situation must be

$$x + y = 20$$
$$y = 5x + 2$$

Step 3: We can solve this system by substituting the expression $5x + 2$ in the second equation for y into the first equation:

$$x + 5x + 2 = 20$$
$$6x + 2 = 20$$
$$6x = 18$$
$$x = 3$$

Using $x = 3$ in either of the first two equations and then solving for y we get $y = 17$.

Step 4: The number 17 is 2 more than 5 times 3, and the sum of 17 and 3 is 20.

So 17 and 3 are the numbers we are looking for.   ▲

▼   **Example 2**   Mr. Smith had $15,000 to invest. He invested part at 6% and the rest at 7%. If he earns $980 in interest, how much did he invest at each rate?

SOLUTION

Step 1: Let $x =$ the amount invested at 6% and $y =$ the amount invested at 7%.

Step 2: Since Mr. Smith invested a total of $15,000, we have

$$x + y = 15,000$$

The interest he earns comes from 6% of the amount invested at 6% (x) and 7% of the amount invested at 7% (y). To find 6% of x we simply multiply x by 0.06, which gives us $0.06x$ (6% is equivalent to 0.06). To find 7% of y, we multiply 0.07 times y and get $0.07y$.

Interest +	interest =	total
at 6%	at 7%	interest
$0.06x$ +	$0.07y$ =	980

The system is

$$x + \quad y = 15{,}000 \qquad \text{The number of dollars}$$
$$0.06x + 0.07y = 980 \qquad \text{The value of the dollars}$$

Step 3: We multiply the first equation by -6 and the second by 100 to eliminate x.

$$
\begin{array}{l}
x + \quad y = 15{,}000 \quad \xrightarrow{\text{multiply by } -6} \quad -6x - 6y = -90{,}000 \\
0.06x + 0.07y = 980 \quad \xrightarrow{\text{multiply by } 100} \quad \underline{6x + 7y = \quad 98{,}000} \\
\hphantom{0.06x + 0.07y = 980 \quad \xrightarrow{\text{multiply by } 100} \quad } y = \quad 8{,}000
\end{array}
$$

Substituting $y = 8000$ into the first equation and solving for x, we get $x = 7000$.

Step 4: The sum of 8000 and 7000 is 15,000. Six percent of 7000 is $(0.06)(7000) = 420$. Seven percent of 8000 is $(0.07)(8000) = 560$. The total interest then is $\$420 + \$560 = \$980$.

He invested $7000 at 6% and $8000 at 7%. ▲

▼ **Example 3** John has $1.70 all in dimes and nickels. He has a total of 22 coins. How many of each kind does he have?

SOLUTION

Step 1: Let $x =$ the number of nickels and $y =$ the number of dimes.

Step 2: The total number of coins is 22, so

$$x + y = 22$$

The total amount of money he has is 1.70, which comes from nickels and dimes.

$$
\begin{array}{ccccc}
\text{Amount of money} & + & \text{amount of money} & = & \text{total amount} \\
\text{in nickels} & & \text{in dimes} & & \text{of money} \\
0.05x & + & 0.10y & = & 1.70
\end{array}
$$

The system that represents the situation is

$$x + \quad y = \quad 22 \qquad \text{The number of coins}$$
$$0.05x + 0.10y = \quad 1.70 \qquad \text{The value of the coins}$$

Step 3: We multiply the first equation by -5 and the second by 100 to eliminate the variable x.

$$x + \quad y = 22 \qquad \xrightarrow{\text{multiply by } -5} \quad -5x - 5y = -110$$
$$0.05x + 0.10y = 1.70 \quad \xrightarrow{\text{multiply by } 100} \quad \underline{5x + 10y = 170}$$
$$5y = 60$$
$$y = 12$$

Substituting $y = 12$ into our first equation, we get $x = 10$.

Step 4: Twelve dimes and 10 nickels total 22 coins with a total value of $0.10(12) + 0.05(10) = 1.20 + 0.50 = \$1.70$.

John has 12 dimes and 10 nickels.   ▲

▼ **Example 4**   A grocer mixes coffee beans selling for $0.60 a pound with coffee beans selling for $0.90 a pound to form a blend that sells for $0.80 a pound. How much of each type should he mix if he wants to have 30 pounds of the blend?

SOLUTION

Step 1:   Let $x =$ the number of pounds of coffee that sells for $0.60 a pound and $y =$ the number of pounds of coffee that sells for $0.90 a pound.

Step 2:   He wants a total of 30 pounds of coffee, so

$$x + y = 30$$

The total value of the $0.60 coffee is $0.60x$; the total value of the $0.90 coffee is $0.90y$; the total value of the blend is $0.80(30)$.

$$0.60x + 0.90y = 0.80(30)$$

Step 3:   Our system is

$$x + \quad y = 30 \qquad \text{The number of pounds of beans}$$
$$0.60x + 0.90y = 0.80(30) \qquad \text{The value of the beans}$$

$$x + \quad y = 30 \qquad \xrightarrow{\text{multiply by } -60} \quad -60x - 60y = -1800$$
$$0.60x + 0.90y = 0.80(30) \quad \xrightarrow{\text{multiply by } 100} \quad \underline{60x + 90y = 2400}$$
$$30y = 600$$
$$y = 20$$

If $y = 20$, then $x = 10$.

Step 4:   (1)   20 lb + 10 lb = 30 lb
(2)   20 lb of coffee selling for $0.90 per pound has a total value of $0.90(20) = \$18.00$
(3)   10 lb of coffee selling for $0.60 per pound has a total value of $0.60(10) = \$6.00$

    (4)   30 lb of coffee selling for $0.80 per pound has a total value of 0.80(30) = $24.00

    (5) $18.00 + $6.00 = $24.00

He mixes 10 pounds of the $0.60 coffee with 20 pounds of the $0.90 coffee to get 30 pounds of the $0.80 blend.   ▲

Solve the following word problems. Be sure to show the equations used.   Problem Set 3.6

1. Two numbers have a sum of 25. One number is five more than the other. Find the numbers.

2. The difference of two numbers is 6. Their sum is 30. Find the two numbers.

3. The sum of two numbers is 15. One number is four times the other. Find the numbers.

4. The difference of two numbers is 28. One number is three times the other. Find the two numbers.

5. Two numbers have a difference of 5. The larger number is one more than twice the smaller. Find the two numbers.

6. One number is two more than three times another. Their sum is 26. Find the two numbers.

7. One number is five more than four times another. Their sum is 35. Find the two numbers.

8. The difference of two numbers is 8. The smaller is twice the larger decreased by 7. Find the two numbers.

9. Ron has 14 coins with a total value of $2.30. The coins are nickels and quarters. How many of each coin does he have?

10. Diane has $0.95 in dimes and nickels. She has a total of 11 coins. How many of each kind does she have?

11. Suppose Tom has 21 coins totaling $3.45. If he has only dimes and quarters, how many of each type does he have?

12. A coin collector has 31 dimes and nickels with a total face value of $2.40. (They are actually worth a lot more.) How many of each coin does she have?

13. Mr. Wilson invested money in two accounts. His total investment was $20,000. If one account pays 6% in interest and the other pays 8% in interest, how much does he have in each account if he earned a total of $1380 in interest in one year?

14. A total of $11,000 was invested. Part of the $11,000 was invested at 4% and the rest was invested at 7%. If the investment earns $680 per year, how much was invested at each rate?

15. A woman invested four times as much at 5% as she did at 6%. The total amount of interest she earns in one year from both accounts is $520. How much did she invest at each rate?

16. Ms. Hagan invested twice as much money in an account that pays 7% interest as she did in an account that pays 6% in interest. Her total investment pays her $1000 a year in interest. How much did she invest at each amount?

17. A grocer mixed two kinds of coffee to form a blend that he will sell for $0.88 per pound. He mixed coffee selling for $0.85 per pound with coffee selling for $0.90 per pound to get 10 pounds of the blend. How many pounds of each kind of coffee did he use?

18. A grocer mixed coffee worth $0.75 a pound with coffee worth $0.95 a pound to get 12 pounds of coffee he will sell for $0.80 per pound. How much of each type of coffee did he use?

19. Football tickets are $1.50 for adults and $1.00 for children. If 275 tickets were sold to one game, for a total of $350, how many adults and how many children attended the game?

20. Tickets for the flower show were $2.50 for adults and $1.50 for kids under 12. If 70 people paid a total of $155 to see the show, how many adults and how many kids bought tickets?

Review Problems   The following problems review material we covered in Section 2.6.

Solve each inequality.

21. $2x - 6 > 5x + 6$
22. $4x - 3 \geq 8x + 1$
23. $3(2x + 4) \leq -6$
24. $5(3x - 2) < 20$
25. $4(2 - x) \geq 12$
26. $6(1 - x) > -12$
27. $6 - 2(x + 3) < -2$
28. $7 - 3(x - 5) \leq -2$

3.7
Linear Inequalities in Two Variables

A linear inequality in two variables is any expression that can be put in the form

$$ax + by < c$$

where *a, b,* and *c* are real numbers (*a* and *b* not both 0). The inequality symbol can be any one of the following four: $<, \leq, >, \geq$.

Some examples of linear inequalities are

$$2x + 3y < 6 \qquad y \geq 2x + 1 \qquad x - y \leq 0$$

Although not all of the above have the form $ax + by < c$, each one can be put in that form.

The solution set for a linear inequality is a section of the coordinate plane. The boundary for the section is found by replacing the inequality symbol with an equal sign and graphing the resulting equation. The boundary is included in the solution set (and represented with a solid line) if the inequality symbol used originally is $\leq$ or $\geq$. The boundary is not included (and is represented with a dotted line) if the original symbol is $<$ or $>$.

Let's look at some examples.

▼ **Example 1** Graph the solution set for $x + y \leq 4$.

SOLUTION The boundary for the graph is the graph of $x + y = 4$; we graph $x + y = 4$ by using the method described in Section 3.2. The boundary is included in the solution set because the inequality symbol is $\leq$.

Here is the graph of the boundary:

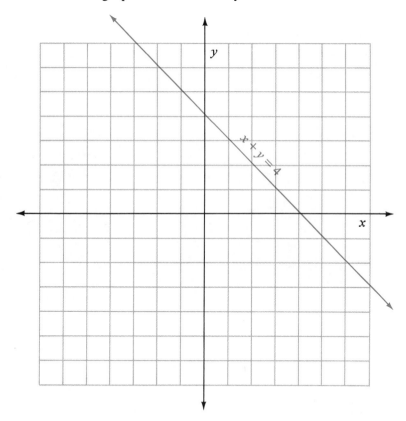

The boundary separates the coordinate plane into two sections, or regions—the region above the boundary and the region below the boundary. The solution set for $x + y \leq 4$ is one of these two regions along with the boundary. To find the correct region, we simply choose any convenient point that is *not* on the boundary.

We then substitute the coordinates of the point into the original inequality $x + y \leq 4$. If the point we choose satisfies the inequality, then it is a member of the solution set, and we can assume that all points on the same side of the boundary as the chosen point are also in the solution set. If the coordinates of our point do not satisfy the original inequality, then the solution set lies on the other side of the boundary.

In this example, a convenient point not on the boundary is the origin. Substituting $(0, 0)$ into $x + y \leq 4$ gives us

$$0 + 0 \leq 4$$
$$0 \leq 4 \qquad \text{A true statement}$$

Since the origin is a solution to the inequality $x + y \leq 4$, and the origin is below the boundary, all other points below the boundary are also solutions.

Here is the graph of $x + y \leq 4$:

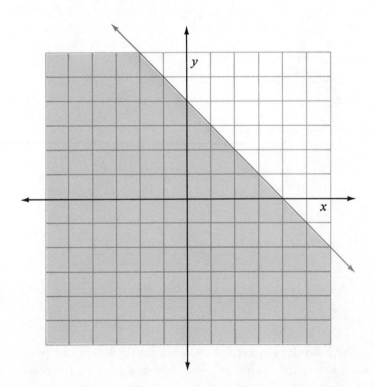

The region above the boundary is described by the inequality $x + y > 4$.

Here is a list of steps to follow when graphing the solution set for linear inequalities in two variables:

Step 1: Replace the inequality symbol with an equal sign. The resulting equation represents the boundary for the solution set.

Step 2: Graph the boundary found in Step 1, using a *solid line* if the boundary is included in the solution set (that is, if the original inequality symbol was either $\leq$ or $\geq$). Use a *broken line* to graph the boundary if it is *not* included in the solution set. (It is not included if the original inequality was either $<$ or $>$.)

Step 3: Choose any convenient point not on the boundary and substitute the coordinates into the *original* inequality. If the resulting statement is *true*, the graph lies on the *same* side of the boundary as the chosen point. If the resulting statement is *false*, the solution set lies on the *opposite* side of the boundary.

▼ **Example 2**   Graph the solution set for $y < 2x - 3$.

SOLUTION   The boundary is the graph of $y = 2x - 3$. The boundary is not included since the original inequality symbol is $<$. Therefore we use a broken line to represent the boundary:

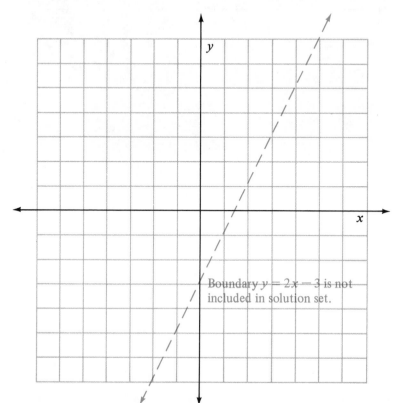

Boundary $y = 2x - 3$ is not included in solution set.

A convenient test point is again the origin. Using $(0, 0)$ in $y < 2x - 3$, we have

$$0 < 2(0) - 3$$
$$0 < -3 \qquad \text{A false statement}$$

Since our test point gives us a false statement and it lies above the boundary, the solution set must lie on the other side of the boundary, as shown here:

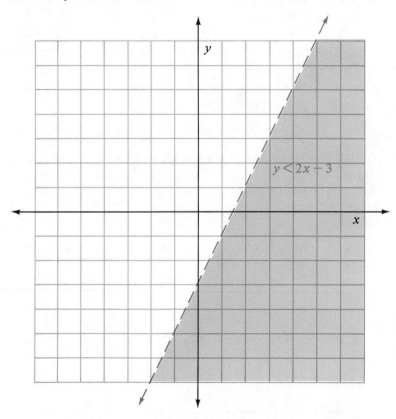

▼ **Example 3** Graph the solution set for $x \leq 5$.

SOLUTION The boundary is $x = 5$, which is a vertical line. All points to the left have x-coordinates less than 5 and all points to the right have x-coordinates greater than 5, as shown in the graph below:

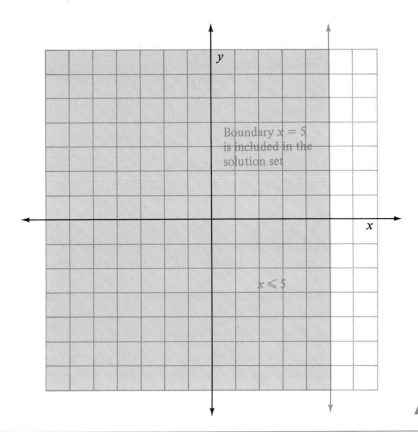

Boundary $x = 5$ is included in the solution set

$x \leq 5$

Graph the following linear inequalities.

1. $2x - 3y < 6$

2. $3x + 2y \geq 6$

3. $x - 2y \leq 4$

4. $2x + y > 4$

5. $x - y \leq 2$

6. $x - y \leq 1$

7. $3x - 4y \geq 12$

8. $4x + 3y < 12$

9. $5x - y \leq 5$

10. $4x + y > 4$

11. $2x + 6y \leq 12$

12. $x - 5y > 5$

13. $x \geq 1$

14. $x < 5$

15. $x \geq -3$

16. $y \leq -7$

17. $y < 2$

18. $3x - y > 1$

19. $2x + y > 3$

20. $5x + 2y < 2$

21. $y \leq 3x - 1$

22. $y \geq 3x + 2$

23. $y \geq x - 5$

24. $y \geq x + 3$

25. $y + \frac{1}{2}x \leq 2$

26. $y < \frac{1}{3}x + 3$

27. $y < -x + 4$

28. $y \geq -x - 3$

29. $y \geq 3x - 4$

30. $y < 5x + 1$

Review Problems   The problems below review material we covered in Section 2.7.

Graph each compound inequality.

31. $x < -2$ or $x > 3$ **32.** $x \leq -2$ or $x \geq 3$

33. $x \geq -2$ and $x \leq 3$ **34.** $x > -2$ and $x < 3$

35. $-3 < x < 1$ **36.** $0 \leq x \leq 2$

Chapter 3 Summary and Review

Examples

1. The solution to the system

$$x + 2y = 4$$
$$x - y = 1$$

is the ordered pair (2, 1). It is the only ordered pair that satisfies both equations.

DEFINITIONS [3.3]

1. A *system of linear equations*, as the term is used in this book, is two linear equations that each contain the same two variables.

2. The *solution* set for a system of equations is the set of all ordered pairs that satisfy *both* equations. The solution set to a system of linear equations will contain:

     (a)   One ordered pair when the graphs of the two equations intersect at only one point (this is the most common situation)

     (b)   No ordered pairs when the graphs of the two equations are parallel lines

     (c)   An infinite number of ordered pairs when the graphs of the two equations coincide (are the same line)

2. Solving the system in Example 1 by graphing looks like

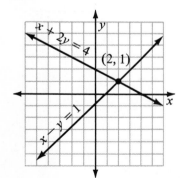

TO SOLVE A SYSTEM BY GRAPHING [3.3]

Step 1:   Graph the first equation.

Step 2:   Graph the second equation on the same set of axes.

Step 3:   The solution to the system consists of the coordinates of the point where the graphs cross each other (the coordinates of the point of intersection).

Step 4:   Check the solution to see that it satisfies *both* equations, if necessary.

TO SOLVE A SYSTEM BY THE ELIMINATION METHOD [3.4]

Step 1:   Look the system over to decide which variable will be easiest to eliminate.

Step 2: Use the multiplication property of equality on each equation separately to insure that the coefficients of the variable to be eliminated are opposites.

Step 3: Add the left and right sides of the system produced in Step 2 and solve the resulting equation.

Step 4: Substitute the solution from Step 3 back into any equation with both x and y variables and solve.

Step 5: Check your solutions in both equations, if necessary.

TO SOLVE A SYSTEM BY THE SUBSTITUTION METHOD [3.5]

Step 1: Solve either of the equations for one of the variables (this step is not necessary if one of the equations has the correct form already).

Step 2: Substitute the results of Step 1 into the other equation and solve.

Step 3: Substitute the results of Step 2 into an equation with both x and y variables and solve. (The equation produced in Step 1 is usually a good one to use.)

Step 4: Check your solution, if necessary.

SPECIAL CASES [3.3, 3.4, 3.5]

In some cases, using the elimination or substitution method eliminates both variables. The situation is interpreted as follows:

1. If the resulting statement is *false*, then the lines are parallel and there is no solution to the system.

2. If the resulting statement is *true*, then the equations represent the same line (the lines coincide). In this case, any ordered pair that satisfies either equation is a solution to the system.

TO GRAPH A LINEAR INEQUALITY IN TWO VARIABLES [3.7]

Step 1: Replace the inequality symbol with an equal sign. The resulting equation represents the boundary for the solution set.

Step 2: Graph the boundary found in Step 1, using a *solid line* if the original inequality symbol was either $\leq$ or $\geq$. Use a *broken line* otherwise.

3. We can eliminate the y variable from the system in Example 1 by multiplying both sides of the second equation by 2 and adding the result to the first equation:

$$
\begin{array}{rcl}
x + 2y = 4 & & x + 2y = 4 \\
x - y = 1 & \xrightarrow[\text{multiply by 2}]{} & 2x - 2y = 2 \\
\hline
& & 3x = 6 \\
& & x = 2
\end{array}
$$

Substituting $x = 2$ into either of the original two equations give $y = 1$. The solution is $(2, 1)$.

4. We can apply the substitution method to the system in Example 1 by first solving the second equation for x to get

$$x = y + 1$$

Substituting this expression of x into the first equation, we have

$$
\begin{aligned}
y + 1 + 2y &= 4 \\
3y + 1 &= 4 \\
3y &= 3 \\
y &= 1
\end{aligned}
$$

Using $y = 1$ in either of the original equations gives $x = 2$.

5. Graph $x - y > 3$.

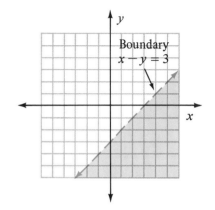

Step 3: Choose any convenient point not on the boundary and substitute the coordinates into the *original* inequality. If the resulting statement is *true*, the graph lies on the *same* side of the boundary as the chosen point. If the resulting statement is *false*, the solution set lies on the *opposite* side of the boundary.

 COMMON MISTAKES

The most common mistake encountered in solving linear systems is the failure to complete the problem. Here is an example:

$$\begin{array}{r} x + y = 8 \\ x - y = 4 \\ \hline 2x \phantom{{}+y} = 12 \\ x = 6 \end{array}$$

This is only half the solution. To find the other half, we must substitute the 6 back into one of the original equations and then solve for y.

Remember, solutions to systems of linear equations always consist of ordered pairs. We need an x-coordinate and a y-coordinate; $x = 6$ can never be a solution to a system of linear equations.

Chapter 3 Test

Solve the following systems by graphing.

1. $x + y = 3$
   $x - y = 11$

2. $2x + 3y = 6$
   $x - y = 3$

3. $x + 2y = 5$
   $y = 2x$

4. $3x - 7y = 21$
   $y = 2x - 3$

5. $-5x - 4y = 20$
   $x - 4y = -4$

6. $x - y = 5$
   $x = -3$

Solve the following systems by the substitution method.

7. $2x - 3y = 1$
   $x = y + 2$

8. $3x + 2y = 20$
   $y = 2x + 3$

9. $-y + 2x = 3$
   $y = 2x + 1$

10. $3x - 6y = -6$
    $x = y + 1$

11. $7x - 2y = -4$
    $y = 3x + 3$

12. $y = 2x + 1$
    $y = 3x - 4$

Solve the following systems. Use any convenient method.

13. $x - y = 1$
    $2x + y = -10$

14. $2x - 3y = -7$
    $x + 3y = -8$

15. $2x + y = 7$
    $3x + y = 12$

16. $6x - 10y = 6$
    $9x - 15y = -4$

17. $6x - 10y = 6$
    $9x - 15y = 9$

18. $3x + 2y = 10$
    $-2x + 2y = 0$

19. $7x + 8y = -2$
    $3x - 2y = 10$

20. $3x - 8y = 7$
    $5x + 2y = -19$

Solve the following word problems.

21. The sum of two numbers is 12. Their difference is 2. Find the numbers.

22. The sum of two numbers is 15. One number is six more than twice the other. Find the two numbers.

23. Mr. Stork has $10,000 to invest. He would like to earn $540 per year in interest. How much should he invest at 5% if the rest is to be invested at 6%.

24. Diane has 12 coins that total $1.60. If the coins are all nickels and quarters, how many of each type does she have?

25. Graph $5x - 2y < 10$.

Exponents and Polynomials

The main idea in this chapter is notation. We will define a shorthand notation to represent repeated multiplications. The new definition and notation is in Section 4.1. The rest of the chapter deals with developing properties associated with the new notation and then applying these properties to other situations.

To perform successfully in this chapter you should understand the following concepts. The concepts that are not familiar should be reviewed.

1. Subtraction is addition of the opposite.
2. Division is multiplication by the reciprocal.
3. Multiplication by positive integers is repeated addition.
4. Addition and multiplication are commutative operations.
5. Multiplication distributes over addition.

In Section 1.6, when we developed the rule for multiplication of real numbers, we used the fact that multiplication with whole numbers can be thought of as repeated addition. That is, $3 \cdot 5 = 5 + 5 + 5$. Multiplication by a whole number is shorthand notation for repeated addition. A similar relationship exists between what are called exponents and repeated multiplication.

An *exponent* is a number written just above and to the right of

4.1 Multiplication with Exponents

another number, which is called the *base*. In the expression 5^2, for example, the exponent is 2 and the base is 5. The expression 5^2 is read "5 to the second power," or "5 squared." The meaning of the expression is

$$5^2 = 5 \cdot 5 = 25$$

The exponent 2 indicates repeated multiplication.

In the expression 5^3, the exponent is 3 and the base is 5. The expression 5^3 is read "5 to the third power," or "5 cubed." The meaning of the expression is

$$5^3 = 5 \cdot 5 \cdot 5 = 125$$

The exponent 3 indicates repeated multiplication. The same is true of any exponent that is a positive integer.

Here are some further examples:

▼ **Examples**

1. $4^3 = 4 \cdot 4 \cdot 4 = 16 \cdot 4 = 64$ exponent 3, base 4

2. $(\frac{1}{3})^2 = \frac{1}{3} \cdot \frac{1}{3} = \frac{1}{9}$ exponent 2, base $\frac{1}{3}$

3. $(-2)^5 = (-2)(-2)(-2)(-2)(-2) = -32$ exponent 5, base -2

4. $(-\frac{3}{4})^2 = (-\frac{3}{4})(-\frac{3}{4}) = \frac{9}{16}$ exponent 2, base $-\frac{3}{4}$ ▲

QUESTION What is the difference between $(-5)^2$ and -5^2?
ANSWER In the first case the base is -5. In the second case the base is 5. The answer to the first is 25. The answer to the second is -25. Can you tell why? Would there be a difference in the answers if the exponent in each case were changed to 3?

We can simplify our work with exponents by developing some properties of exponents. We want to list the things we know are true about exponents and then use these properties to simplify expressions that contain exponents.

The first property of exponents applies to products with the same base. We can use the definition of exponents as indicating repeated multiplication to simplify expressions like $2^4 \cdot 2^3$ and $(-3)^2 (-3)^3$:

$$2^4 \cdot 2^3 = (2 \cdot 2 \cdot 2 \cdot 2)(2 \cdot 2 \cdot 2)$$
$$= (2 \cdot 2 \cdot 2 \cdot 2 \cdot 2 \cdot 2 \cdot 2)$$
$$= 2^7 \quad\quad\quad\quad \text{Notice: } 4 + 3 = 7$$

$$(-3)^2 \, (-3)^3 = [(-3)(-3)][(-3)(-3)(-3)]$$
$$= (-3)(-3)(-3)(-3)(-3)$$
$$= (-3)^5 \qquad\qquad \text{Notice: } 2 + 3 = 5$$

In each of these cases multiplication with the same base resulted in addition of exponents. We can summarize these results with the following property.

Property 1 for Exponents

If a is any real number and r and s are integers, then

$$a^r \cdot a^s = a^{r+s}$$

In words: To multiply two expressions with the same base, add exponents and use the common base.

Here are some examples using Property 1.

▼   **Examples**   Simplify the following expressions.

5. $5^3 \cdot 5^6 = 5^{3+6} = 5^9$

6. $x^7 \cdot x^8 = x^{7+8} = x^{15}$

7. $3^4 \cdot 3^8 \cdot 3^5 = 3^{4+8+5} = 3^{17}$ ▲

Note: In the preceding examples, notice that in each case the base in the original problem is the same base that appears in the answer and that it is written only once in the answer. A very common mistake that people make when they first begin to use Property 1 is to write a 2 in front of the base in the answer. For example, people making this mistake would get $2x^{15}$ or $(2x)^{15}$ as the result in Example 6. To avoid this mistake, you must be sure you understand the meaning of Property 1 exactly as it is written.

Another common type of expression involving exponents is one in which an expression containing an exponent is raised to another power. The expressions $(5^3)^2$ and $[(-3)^2]^4$ are examples:

$$(5^3)^2 = (5^3)(5^3)$$
$$= 5^{3+3}$$
$$= 5^6 \qquad\qquad \text{Notice: } 3 \cdot 2 = 6$$

$$[(-3)^2]^4 = (-3)^2(-3)^2(-3)^2(-3)^2$$
$$= (-3)^{2+2+2+2}$$
$$= (-3)^8 \qquad\qquad \text{Notice: } 2 \cdot 4 = 8$$

These results offer justification for the second property of exponents.

Property 2 for Exponents

If a is any real number and r and s are integers, then

$$(a^r)^s = a^{r \cdot s}$$

In words: A power raised to another power is the base raised to the product of the powers.

▼   **Examples**   Simplify the following expressions.

8. $(4^5)^6 = 4^{5 \cdot 6} = 4^{30}$

9. $(x^3)^5 = x^{3 \cdot 5} = x^{15}$ ▲

The third property of exponents applies to expressions in which the product of two or more numbers or variables is raised to a power. Let's look at how the expression $(2x)^3$ can be simplified:

$$\begin{aligned}
(2x)^3 &= (2x)(2x)(2x) \\
&= (2 \cdot 2 \cdot 2)(x \cdot x \cdot x) \\
&= 2^3 \cdot x^3 \qquad \text{Notice: The exponent 3 distributes} \\
&= 8x^3 \qquad \text{over the product } 2x
\end{aligned}$$

We can generalize this result into a third property of exponents.

Property 3 for Exponents

If a and b are any two real numbers and r is an integer, then

$$(ab)^r = a^r b^r$$

In words: The power of a product is the product of the powers.

Here are some examples using Property 3 to simplify expressions.

▼   **Examples**   Simplify the following expressions.

10. $(5x)^2 = 5^2 \cdot x^2$ Property 3
$= 25x^2$

11. $(2xy)^3 = 2^3 \cdot x^3 \cdot y^3$ Property 3
$= 8x^3y^3$

12. $(3x^2)^3 = 3^3(x^2)^3$ Property 3
$= 27x^6$ Property 2

13. $(2y)^3(3y^2) = 2^3y^3(3y^2)$ Property 3

         $= 8 \cdot 3(y^3 \cdot y^2)$ Commutative and associative properties

         $= 24y^5$ Property 1

14. $-3^3 = -3 \cdot 3 \cdot 3$ Definition of exponents

         $= -27$

15. $(-\frac{1}{4} x^2 y^3)^2 = (-\frac{1}{4})^2(x^2)^2(y^3)^2$ Property 3

         $= \frac{1}{16} x^4 y^6$ Property 2

Although all of the above problems can be done by just applying the definition of exponents, the amount of work necessary is shortened considerably by applying the properties directly.

Name the base and exponent in each of the following expressions. Then use the definition of exponents as repeated multiplication to simplify.

Problem Set 4.1

1. 4^2	**2.** 6^2	**3.** 7^2	**4.** 8^2
5. 4^3	**6.** 10^3	**7.** $(-5)^2$	**8.** -5^2
9. -2^3	**10.** $(-2)^3$	**11.** 3^4	**12.** $(-3)^4$
13. $(\frac{2}{3})^2$	**14.** $(\frac{2}{3})^3$	**15.** $(\frac{1}{2})^4$	**16.** $(\frac{4}{5})^2$

Use Property 1 for Exponents to simplify each expression. Leave all answers in terms of exponents.

17. $x^4 \cdot x^5$	**18.** $x^7 \cdot x^3$
19. $7^6 \cdot 7^1$	**20.** $8^9 \cdot 8^1$
21. $y^{10} \cdot y^{20}$	**22.** $y^{30} \cdot y^{30}$
23. $2^5 \cdot 2^4 \cdot 2^3$	**24.** $4^2 \cdot 4^3 \cdot 4^4$
25. $x^4 \cdot x^6 \cdot x^8 \cdot x^{10}$	**26.** $x^{20} \cdot x^{18} \cdot x^{16} \cdot x^{14}$

Use Property 2 for Exponents to write each of the problems below with a single exponent. (Assume all variables are positive numbers.)

27. $(x^2)^5$	**28.** $(x^5)^2$	**29.** $(5^4)^3$	**30.** $(5^3)^4$
31. $(y^3)^3$	**32.** $(y^2)^2$	**33.** $(2^5)^{10}$	**34.** $(10^5)^2$
35. $(a^3)^x$	**36.** $(a^5)^x$	**37.** $(b^x)^y$	**38.** $(b^r)^s$

Use Property 3 for Exponents to simplify each of the expressions below.

39. $(4x)^2$	**40.** $(2x)^4$
41. $(2y)^5$	**42.** $(5y)^2$
43. $(-3x)^4$	**44.** $(-3x)^3$
45. $(7ab)^2$	**46.** $(9ab)^2$
47. $(4xyz)^3$	**48.** $(5xyz)^3$

Simplify the expressions below by using the properties of exponents.

49. $(2x^4)^3$ **50.** $(3x^5)^2$

51. $(4a^3)^2$ **52.** $(5a^2)^2$

53. $(x^2)^3(x^4)^2$ **54.** $(x^5)^2(x^3)^5$

55. $(a^3)^1(a^2)^4$ **56.** $(a^4)^1(a^1)^3$

57. $(2x)^3(2x)^4$ **58.** $(3x)^2(3x)^3$

59. $(3x^2)^3 \cdot (2x)^4$ **60.** $(3x)^3 \cdot (2x^3)^2$

61. $(4x^2y^3)^2$ **62.** $(9x^3y^5)^2$

63. $(2a^4b^1)^3$ **64.** $(3a^1b^7)^3$

65. $(3x^2)(2x^3)(5x^4)$ **66.** $(x^3)(2x)(5x^5)$

67. $(4x^2y)^3 \cdot (2xy)^2$ **68.** $(3x^3y^2)^2 \cdot (2x^4y^5)^4$

69. If you were to use the definition of exponents as repeated multiplication to evaluate 10^2, 10^3, and 10^4, you would get $10^2 = 100$, $10^3 = 1000$, and $10^4 = 10,000$. Without actually multiplying, what is 10^5? What is the relationship between the exponent on each of these powers of 10 and the number of zeros in the result?

70. Use the results found in Problem 69 to evaluate 10^1 and 10^0.

71. In this section we evaluated expressions like $(2^3)^2$, but not expressions like 2^{3^2}. What is the difference in the meanings of the two expressions? Evaluate each of them.

72. Evaluate the expressions $(2^2)^3$ and 2^{2^3}. (The answers in both cases are a little larger than usual, so be careful when you multiply.)

73. Here is an example of an analogy. These kinds of questions are common on entrance exams and aptitude tests.

x^4 is to x^2 as 16 is to:

(a) 8 (b) 4 (c) 2 (d) 1

74. Here is an analogy problem that is a little more difficult than the first one:

x^2 is to x as Superman is to:

(a) Mr. White (b) Jimmy (c) Clark Kent (d) Lois Lane

Review Problems   The problems below review material on subtraction with negative numbers that we covered in Section 1.4. Reviewing these problems will help you understand the material in the next section.

Subtract.

75. $4 - 7$ **76.** $-4 - 7$

77. $4 - (-7)$ **78.** $-4 - (-7)$

79. $15 - 20$ **80.** $15 - (-20)$

81. $-15 - (-20)$ **82.** $-15 - 20$

In Section 4.1 we found that multiplication with the same base results in addition of exponents, that is, $a^r \cdot a^s = a^{r+s}$. Since division is the inverse operation of multiplication, we can expect division with the same base to result in subtraction of exponents.

To develop the properties for exponents under division we again apply the definition of exponents:

$$\frac{x^5}{x^3} = \frac{x \cdot x \cdot x \cdot x \cdot x}{x \cdot x \cdot x}$$

$$= \frac{x \cdot x \cdot x}{x \cdot x \cdot x} (x \cdot x)$$

$$= 1(x \cdot x)$$

$$= x^2 \qquad \qquad \text{Notice: } 5 - 3 = 2$$

$$\frac{2^4}{2^7} = \frac{2 \cdot 2 \cdot 2 \cdot 2}{2 \cdot 2 \cdot 2 \cdot 2 \cdot 2 \cdot 2 \cdot 2}$$

$$= \frac{2 \cdot 2 \cdot 2 \cdot 2}{2 \cdot 2 \cdot 2 \cdot 2} \cdot \frac{1}{2 \cdot 2 \cdot 2}$$

$$= \frac{1}{2 \cdot 2 \cdot 2}$$

$$= \frac{1}{2^3} \qquad \qquad \text{Notice: } 7 - 4 = 3$$

In both cases, division with the same base resulted in subtraction of the smaller exponent from the larger. The problem is deciding whether the answer is a fraction or not. The problem is resolved quite easily by the following definition.

DEFINITION If r is a positive integer, then $a^{-r} = \dfrac{1}{a^r} = \left(\dfrac{1}{a}\right)^r$.

The following examples illustrate how we use this definition to simplify expressions that contain negative exponents.

▼ **Examples** Write each expression with a positive exponent and then simplify.

1. $2^{-3} = \dfrac{1}{2^3} = \dfrac{1}{8}$ *Notice:* Negative exponents do not indicate negative numbers. They indicate reciprocals.

2. $5^{-2} = \dfrac{1}{5^2} = \dfrac{1}{25}$

3.　$3x^{-6} = 3 \cdot \dfrac{1}{x^6} = \dfrac{3}{x^6}$

▲

Now let us look back to our original problem and try to work it again with the help of a negative exponent. We know that $2^4/2^7 = 1/2^3$. Let us decide now that with division of the same base, we will always subtract the exponent in the denominator from the exponent in the numerator and see if this conflicts with what we know is true.

$\dfrac{2^4}{2^7} = 2^{4-7}$　　Subtracting the bottom exponent from the top exponent

$\quad = 2^{-3}$　　Subtraction

$\quad = \dfrac{1}{2^3}$　　Definition of negative exponents

Subtracting the exponent in the denominator from the exponent in the numerator and then using the definition of negative exponents gives us the same result we obtained previously. We can now continue the list of properties of exponents we started in Section 4.1.

Property 4 for Exponents

If *a* is any real number and *r* and *s* are integers, then

$$\frac{a^r}{a^s} = a^{r-s}$$

In words:　To divide with the same base, subtract the exponent in the denominator from the exponent in the numerator and raise the base to the exponent that results.

The examples below show how we use Property 4 and the definition for negative exponents to simplify expressions involving division.

▼ **Examples**　Simplify the following expressions.

4.　$\dfrac{x^9}{x^6} = x^{9-6} = x^3$

5.　$\dfrac{x^4}{x^{10}} = x^{4-10} = x^{-6} = \dfrac{1}{x^6}$

6.　$\dfrac{13^{15}}{13^{20}} = 13^{15-20} = 13^{-5} = \dfrac{1}{13^5}$

7. $\dfrac{(ab)^3}{(ab)^7} = (ab)^{3-7} = (ab)^{-4} = \dfrac{1}{(ab)^4}$ ▲

Our final property of exponents is similar to Property 3 from Section 4.1, but it involves division instead of multiplication. After we have stated the property, we will give a proof of it. The proof shows why this property is true.

Property 5 for Exponents

If a and b are any two real numbers ($b \neq 0$) and r is an integer, then

$$\left(\frac{a}{b}\right)^r = \frac{a^r}{b^r}$$

In words: A quotient raised to a power is the quotient of the powers.

PROOF $\left(\dfrac{a}{b}\right)^r = \left(a \cdot \dfrac{1}{b}\right)^r$ By the definition of division

$\qquad\qquad = a^r \cdot \left(\dfrac{1}{b}\right)^r$ By Property 3

$\qquad\qquad = a^r \cdot b^{-r}$ By the definition of negative exponents

$\qquad\qquad = a^r \cdot \dfrac{1}{b^r}$ By the definition of negative exponents

$\qquad\qquad = \dfrac{a^r}{b^r}$ By the definition of division

▼ **Examples**   Simplify these expressions.

8. $\left(\dfrac{x}{2}\right)^3 = \dfrac{x^3}{2^3} = \dfrac{x^3}{8}$

9. $\left(\dfrac{5}{y}\right)^2 = \dfrac{5^2}{y^2} = \dfrac{25}{y^2}$

10. $\left(\dfrac{2}{3}\right)^4 = \dfrac{2^4}{3^4} = \dfrac{16}{81}$ ▲

Zero and One as Exponents

We have two special exponents left to deal with before our rules for exponents are complete. We have learned to use positive and negative exponents, but we have not run across the exponents 0 and 1. To obtain an expression for x^1, we will solve a problem two different ways:

$$\left. \begin{array}{l} \dfrac{x^3}{x^2} = \dfrac{x \cdot x \cdot x}{x \cdot x} = x \\[2em] \dfrac{x^3}{x^2} = x^{3-2} = x^1 \end{array} \right\} \quad \text{Hence } x^1 = x$$

Stated generally, this rule says that $a^1 = a$. This seems reasonable and we will use it, since it is consistent with our property of division using the same base.

We use the same procedure to obtain an expression for x^0.

$$\left. \begin{array}{l} \dfrac{5^2}{5^2} = \dfrac{25}{25} = 1 \\[2em] \dfrac{5^2}{5^2} = 5^{2-2} = 5^0 \end{array} \right\} \quad \text{Hence } 5^0 = 1$$

It seems, therefore, that the best definition of x^0 is 1 for all x except $x = 0$. In the case of $x = 0$, we have 0^0, which we will not define. This definition will probably seem awkward at first. Most people would like to define x^0 as 0 when they first encounter it. Remember, the zero in this expression is an exponent. x^0 does not mean to multiply by zero. Thus, we can make the general statement that $a^0 = 1$ for all real numbers, except $a = 0$.

Here are some examples involving the exponents 0 and 1.

▼ **Examples**   Simplify the following expressions.

11. $8^0 = 1$

12. $8^1 = 8$

13. $4^0 + 4^1 = 1 + 4 = 5$

14. $(2x^2y)^0 = 1$

15. $\left(\frac{3}{4}\right)^0 = 1$

16. $\left(\frac{3}{4}\right)^1 = \frac{3}{4}$ ▲

Here is a summary of the definitions and properties of exponents we have developed so far. For each definition or property in the list, a and b are real numbers, and r and s are integers.

<table>
<tr><td align="center">DEFINITIONS</td><td align="center">PROPERTIES</td></tr>
</table>

$$a^{-r} = \frac{1}{a^r} = \left(\frac{1}{a}\right)^r \qquad a \neq 0$$

$$a^1 = a$$

$$a^0 = 1 \qquad a \neq 0$$

$$a^r \cdot a^s = a^{r+s}$$

$$(a^r)^s = a^{rs}$$

$$(ab)^r = a^r b^r$$

$$\frac{a^r}{a^s} = a^{r-s} \qquad a \neq 0$$

$$\left(\frac{a}{b}\right)^r = \frac{a^r}{b^r} \qquad b \neq 0$$

Here are some additional examples. These use a combination of the properties and definitions listed above.

▼   **Examples**   Simplify each expression. Write all answers with positive exponents only.

17. $\dfrac{(5x^3)^2}{x^4} = \dfrac{25x^6}{x^4}$ Properties 2 and 3

$\qquad\qquad = 25x^2$ Property 4

18. $\dfrac{x^{-8}}{(x^2)^3} = \dfrac{x^{-8}}{x^6}$ Property 2

$\qquad\qquad = x^{-8-6}$ Property 4

$\qquad\qquad = x^{-14}$ Subtraction

$\qquad\qquad = \dfrac{1}{x^{14}}$ Definition of negative exponents

19. $\left(\dfrac{y^5}{y^3}\right)^2 = \dfrac{(y^5)^2}{(y^3)^2}$ Property 5

$\qquad\qquad = \dfrac{y^{10}}{y^6}$ Property 2

$\qquad\qquad = y^4$ Property 4

Notice in Example 19 that we could have simplified inside the parentheses first and then raised the result to the second power:

$$\left(\frac{y^5}{y^3}\right)^2 = (y^2)^2 = y^4$$

20. $(3x^5)^{-2} = \dfrac{1}{(3x^5)^2}$ Definition of negative exponents

$= \dfrac{1}{9x^{10}}$ Properties 2 and 3

21. $x^{-8} \cdot x^5 = x^{-8+5}$ Property 1

$= x^{-3}$ Addition

$= \dfrac{1}{x^3}$ Definition of negative exponents   ▲

Problem Set 4.2

Write each of the following with positive exponents and then simplify, when possible.

1. 3^{-2}	**2.** 3^{-3}	**3.** 6^{-2}	**4.** 2^{-6}
5. 8^{-2}	**6.** 3^{-4}	**7.** 5^{-3}	**8.** 9^{-2}
9. $2x^{-3}$	**10.** $5x^{-1}$	**11.** $(2x)^{-3}$	**12.** $(5x)^{-1}$
13. $(5y)^{-2}$	**14.** $5y^{-2}$	**15.** 10^{-2}	**16.** 10^{-3}

Use Property 4 to simplify each of the expressions below. Write all answers that contain exponents with positive exponents only.

17. $\dfrac{5^3}{5^1}$	**18.** $\dfrac{7^8}{7^6}$	**19.** $\dfrac{5^1}{5^3}$	**20.** $\dfrac{7^6}{7^8}$
21. $\dfrac{x^{10}}{x^4}$	**22.** $\dfrac{x^4}{x^{10}}$	**23.** $\dfrac{4^3}{4^0}$	**24.** $\dfrac{4^0}{4^3}$
25. $\dfrac{(2x)^7}{(2x)^4}$	**26.** $\dfrac{(2x)^4}{(2x)^7}$	**27.** $\dfrac{6^{11}}{6}$	**28.** $\dfrac{8^7}{8}$
29. $\dfrac{6}{6^{11}}$	**30.** $\dfrac{8}{8^7}$	**31.** $\dfrac{2^{-5}}{2^3}$	**32.** $\dfrac{2^{-5}}{2^{-3}}$
33. $\dfrac{2^5}{2^{-3}}$	**34.** $\dfrac{2^{-3}}{2^{-5}}$	**35.** $\dfrac{(ab)^{-5}}{(ab)^{-8}}$	**36.** $\dfrac{(ab)^{10}}{(ab)^{-15}}$

Simplify the following expressions. Any answers that contain exponents should contain positive exponents only.

37. $(2x^{-3})^4$	**38.** $(3x^{-2})^3$	**39.** 10^0	**40.** 10^1
41. $(2a^2b)^1$	**42.** $(2a^2b)^0$	**43.** $(7y^3)^{-2}$	**44.** $(5y^4)^{-2}$
45. $x^{-3}x^{-5}$	**46.** $x^{-6} \cdot x^8$	**47.** $y^7 \cdot y^{-10}$	**48.** $y^{-4} \cdot y^{-6}$
49. $\dfrac{(x^2)^3}{x^4}$	**50.** $\dfrac{(x^5)^3}{x^{10}}$	**51.** $\dfrac{(a^4)^3}{(a^3)^2}$	**52.** $\dfrac{(a^5)^3}{(a^5)^2}$

53. $\dfrac{y^7}{(y^2)^8}$ **54.** $\dfrac{y^2}{(y^3)^4}$ **55.** $\dfrac{(x^0)^5}{x^3}$ **56.** $\dfrac{(x^5)^0}{x^3}$

57. $\dfrac{(x^{-2})^3}{x^{-5}}$ **58.** $\dfrac{(x^2)^{-3}}{x^{-5}}$

59. $\dfrac{(x^{-4})^{-2}}{x^{-10}}$ **60.** $\dfrac{(x^{-6})^{-5}}{x^{-20}}$

61. $\dfrac{(a^3)^2(a^4)^5}{(a^5)^2}$ **62.** $\dfrac{(a^4)^8(a^2)^5}{(a^3)^4}$

63. $\dfrac{(a^{-2})^3(a^4)^2}{(a^{-3})^{-2}}$ **64.** $\dfrac{(a^{-5})^{-3}(a^7)^{-1}}{(a^{-3})^5}$

65. $\dfrac{(x^{-7})^3(x^4)^5}{(x^3)^2(x^{-1})^8}$ **66.** $\dfrac{(x^{-9})^0(x^4)^{-3}}{(x^{-1})^5(x^5)^{-2}}$

Since division is defined to be equivalent to multiplication by the reciprocal, we can simplify expressions like $\dfrac{1}{\frac{1}{3}}$ as follows:

$$\frac{1}{\frac{1}{3}} = 1 \cdot \frac{3}{1} = 3$$

Use this idea to simplify the following.

67. $\dfrac{1}{3^{-1}}$ **68.** $\dfrac{1}{2^{-1}}$ **69.** $\dfrac{1}{3^{-2}}$ **70.** $\dfrac{1}{2^{-3}}$

71. $\dfrac{1}{7^{-2}}$ **72.** $\dfrac{1}{3^{-4}}$ **73.** $\dfrac{1}{5^{-2}}$ **74.** $\dfrac{1}{8^{-2}}$

Review Problems   The problems below review material we covered in Section 2.1. They will help you understand some of the material in the next section.

Simplify the following expressions.

75. $4x + 3x$ **76.** $9x + 7x$
77. $5a - 3a$ **78.** $10a - 2a$
79. $4y + 5y + y$ **80.** $6y - y + 2y$

We have developed all the tools necessary to perform the four basic operations on the simplest of polynomials: monomials.

DEFINITION   A *monomial* is a one-term expression that is either a con-

**4.3
Operations with
Monomials**

stant (number) or the product of a constant and one or more variables raised to whole number (positive integer) exponents.

EXAMPLES   The following are monomials:

$$-3 \qquad 15x \qquad -23x^2y \qquad 49x^4y^2z^4 \qquad \tfrac{3}{4}a^2b^3$$

The numerical part of each monomial is called the numerical coefficient, or just coefficient. Monomials are also called *terms*.

It is very important that you master the four basic operations with monomials. Many of the topics we will cover later will be extensions of the work we do with monomials.

Multiplication and Division of Monomials

There are two basic steps involved in the multiplication of monomials. First we rewrite the products using the commutative and associative properties. Then we simplify by multiplying coefficients and adding exponents of like bases.

▼   **Examples**   Multiply.

1. $(-3x^2)(4x^3) = (-3 \cdot 4)(x^2 \cdot x^3)$ Commutative and associative properties

$\qquad\qquad\qquad\quad = -12x^5$ Multiply coefficients, add exponents

2. $(\tfrac{4}{5}x^5 \cdot y^2)(10x^3 \cdot y) = (\tfrac{4}{5} \cdot 10)(x^5 \cdot x^3)(y^2 \cdot y)$ Commutative and associative properties

$\qquad\qquad\qquad\qquad\quad = 8x^8y^3$ Multiply coefficients, add exponents   ▲

You can see that in each case the work was the same—multiply coefficients and add exponents of the same base. We can expect division of monomials to proceed in a similar way. Since our properties are consistent, division of monomials will result in division of coefficients and subtraction of exponents of like bases.

▼   **Examples**   Divide.

3. $\dfrac{15x^3}{3x^2} = \dfrac{15}{3} \cdot \dfrac{x^3}{x^2}$ Write as separate fractions

$\qquad\quad = 5x$ Divide coefficients, subtract exponents

4. $\dfrac{39x^2y^3}{3xy^5} = \dfrac{39}{3} \cdot \dfrac{x^2}{x} \cdot \dfrac{y^3}{y^5}$ Write as separate fractions

$$= 13x \cdot \frac{1}{y^2} \quad \text{Divide coefficients, subtract exponents}$$

$$= \frac{13x}{y^2}$$ ▲

Addition and subtraction of monomials will be almost identical, since subtraction is defined as addition of the opposite. With multiplication and division of monomials, the key was rearranging the numbers and variables using the commutative and associative properties. With addition, the key is application of the distributive property. We sometimes use the phrase *combine monomials* to describe addition and subtraction of monomials.

Addition and Subtraction of Monomials

DEFINITION Two terms (monomials) with the same variable part (same variables raised to the same powers) are called *similar* (or *like*) terms.

You can add only *similar* terms. This is because the distributive property (which is the key to addition of monomials) cannot be applied to terms that are not similar.

▼ **Examples** Combine the following monomials.

5. $-3x^2 + 15x^2 = (-3 + 15)x^2$ Distributive property
           $= 12x^2$ Add coefficients

6. $9x^2y - 20x^2y = (9 - 20)x^2y$ Distributive property
             $= -11x^2y$ Add coefficients

7. $5x^2 + 8y^2$ In this case we cannot apply the distributive property, so we cannot add the monomials. ▲

A topic closely related to combining similar terms is adding fractions with the same denominators. You may recall from previous math classes that to add two fractions with the same denominator, you simply add their numerators and put the result over the common denominator:

$$\frac{3}{4} + \frac{2}{4} = \frac{3 + 2}{4} = \frac{5}{4}$$

The reason we add numerators but do not add denominators is that we must follow the distributive property. To see this, you first have to recall that $\frac{3}{4}$ can be written as $3 \cdot \frac{1}{4}$, and $\frac{2}{4}$ can be written as $2 \cdot \frac{1}{4}$ (dividing by 4 is equivalent to multiplying by $\frac{1}{4}$). Here is the addition problem again, this time showing the use of the distributive property:

$$\frac{3}{4} + \frac{2}{4} = 3 \cdot \frac{1}{4} + 2 \cdot \frac{1}{4}$$

$$= (3 + 2) \cdot \frac{1}{4} \qquad \text{Distributive property}$$

$$= 5 \cdot \frac{1}{4}$$

$$= \frac{5}{4}$$

Here are some further examples.

▼ **Examples** Combine

8. $\dfrac{9}{2} + \dfrac{15}{2} = 9 \cdot \dfrac{1}{2} + 15 \cdot \dfrac{1}{2}$

$$= (9 + 15) \cdot \frac{1}{2} \qquad \text{Distributive property}$$

$$= 24 \cdot \frac{1}{2}$$

$$= 12$$

9. $\dfrac{3}{8} - \dfrac{7}{8} = 3 \cdot \dfrac{1}{8} - 7 \cdot \dfrac{1}{8}$

$$= (3 - 7) \cdot \frac{1}{8} \qquad \text{Distributive property}$$

$$= -4 \cdot \frac{1}{8}$$

$$= -\frac{4}{8}$$

$$= -\frac{1}{2} \qquad\qquad\qquad\qquad\qquad\qquad ▲$$

The next examples show how we simplify expressions containing monomials when more than one operation is involved.

▼ **Example 10** Simplify

$$\frac{(6x^4y)(3x^7y^5)}{9x^5y^2}$$

SOLUTION We begin by multiplying the two monomials in the numerator:

$$\frac{(6x^4y)(3x^7y^5)}{9x^5y^2} = \frac{18x^{11}y^6}{9x^5y^2} \qquad \text{Simplify numerator}$$

$$= 2x^6y^4 \qquad \text{Divide} \qquad ▲$$

▼ **Example 11** Simplify

$$\frac{14x^5}{2x^2} + \frac{15x^8}{3x^5}$$

SOLUTION Simplifying each expression separately and then combining similar terms gives

$$\frac{14x^5}{2x^2} + \frac{15x^8}{3x^5} = 7x^3 + 5x^3 \qquad \text{Divide}$$

$$= 12x^3 \qquad \text{Add} \qquad ▲$$

We end this section with a list of the rules for working with monomials.

Multiplication:	*Multiply* coefficients and *add* exponents with common bases.
Division:	*Divide* coefficients and *subtract* exponents with common bases.
Addition:	*Add* coefficients of *similar* terms.
Subtraction:	*Subtract* coefficients of *similar* terms.

Problem Set 4.3

Multiply.

1. $(3x^4)(4x^3)$ **2.** $(6x^5)(-2x^2)$
3. $(-2y^4)(8y^7)$ **4.** $(5y^{10})(2y^5)$
5. $(8x)(4x)$ **6.** $(7x)(5x)$
7. $(10a^3)(10a)(2a^2)$ **8.** $(5a^4)(10a)(10a^4)$
9. $(6ab^2)(-4a^2b)$ **10.** $(-5a^3b)(4ab^4)$
11. $(2x^2y^2)(-3x^9y^8)$ **12.** $(4a^4b)(5a^7b^{10})$
13. $(4x^2y)(3x^3y^3)(2xy^4)$ **14.** $(5x^6)(-10xy^4)(-2x^2y^6)$

Divide. Write all answers with positive exponents only.

15. $\dfrac{15x^3}{5x^2}$ **16.** $\dfrac{25x^5}{5x^4}$

17. $\dfrac{18y^9}{3y^{12}}$ **18.** $\dfrac{24y^4}{-8y^7}$

19. $\dfrac{32a^3}{64a^4}$ **20.** $\dfrac{25a^5}{75a^6}$

21. $\dfrac{21a^2b^3}{-7ab^5}$ **22.** $\dfrac{32a^5b^6}{8ab^5}$

23. $\dfrac{27x^3y^2z}{3xy^2z^3}$ **24.** $\dfrac{30x^5y^4z}{5x^3yz^2}$

25. $\dfrac{144x^9y^2}{-12x^{10}y^8}$ **26.** $\dfrac{256x^9y^2}{32x^4}$

27. $\dfrac{30x^8}{5x^2y^2}$ **28.** $\dfrac{6x^5}{-3x^6y^4}$

Combine by adding or subtracting as indicated.

29. $3x^2 + 5x^2$ **30.** $4x^3 + 8x^3$
31. $8x^5 - 19x^5$ **32.** $75x^6 - 50x^6$
33. $45a + a$ **34.** $72a + a$
35. $2a + a - 3a$ **36.** $5a + a - 6a$
37. $19x^2 - 20x^2$ **38.** $37x^4 - 38x^4$
39. $10x^3 - 8x^3 + 2x^3$ **40.** $7x^5 + 8x^5 - 12x^5$
41. $20ab^2 - 19ab^2 + 30ab^2$ **42.** $18a^3b^2 - 20a^3b^2 + 10a^3b^2$
43. $-4abc - 9abc - abc$ **44.** $-7abc - abc - abc$

Combine the following using the method shown in Examples 8 and 9.

45. $\dfrac{3}{5} + \dfrac{4}{5}$ **46.** $\dfrac{3}{5} - \dfrac{4}{5}$

47. $\dfrac{9}{25} + \dfrac{6}{25}$

48. $\dfrac{3}{4} + \dfrac{5}{4}$

49. $\dfrac{3}{8} - \dfrac{11}{8}$

50. $\dfrac{1}{6} - \dfrac{7}{6}$

51. $\dfrac{5}{12} + \dfrac{3}{12}$

52. $\dfrac{6}{21} + \dfrac{8}{21}$

53. $\dfrac{12}{18} - \dfrac{4}{18}$

54. $\dfrac{15}{6} - \dfrac{5}{6}$

Simplify. Write all answers with positive exponents only.

55. $\dfrac{(3x^2)(8x^5)}{6x^4}$

56. $\dfrac{(7x^3)(6x^8)}{14x^5}$

57. $\dfrac{(9a^2b)(2a^3b^4)}{18a^5b^7}$

58. $\dfrac{(21a^5b)(2a^8b^4)}{14ab}$

59. $\dfrac{(4x^3y^2)(9x^4y^{10})}{(3x^5y)(2x^6y)}$

60. $\dfrac{(5x^4y^4)(10x^3y^3)}{(25xy^5)(2xy^7)}$

61. $\dfrac{18x^4}{3x} + \dfrac{21x^7}{7x^4}$

62. $\dfrac{24x^{10}}{6x^4} + \dfrac{32x^7}{8x}$

63. $\dfrac{45a^6}{9a^4} - \dfrac{50a^8}{2a^6}$

64. $\dfrac{16a^9}{4a} - \dfrac{28a^{12}}{4a^4}$

65. $\dfrac{6x^7y^4}{3x^2y^2} + \dfrac{8x^5y^8}{2y^6}$

66. $\dfrac{40x^{10}y^{10}}{8x^2y^5} + \dfrac{10x^8y^8}{5y^3}$

Use your knowledge of the properties and definitions of exponents to find x in each of the following.

67. $4^x \cdot 4^5 = 4^7$

68. $\dfrac{5^x}{5^3} = 5^4$

69. $(7^3)^x = 7^{12}$

70. $\dfrac{3^x}{3^4} = 9$

71. The statement $(a + b)^2 = a^2 + b^2$ looks similar to Property 3 of exponents. However, it is not a property of exponents because almost every time we replace a and b with numbers, the expression above becomes a false statement. Let $a = 4$ and $b = 5$ in the expressions $(a + b)^2$ and $a^2 + b^2$ and see what each simplifies to.

72. Find two numbers to use in place of a and b so that the expressions $(a + b)^2$ and $a^2 + b^2$ will be equal.

Review Problems   The problems below review material we covered in Section 2.1. Reviewing these problems will help you understand some of the material in the next section.

Find the value of each expression when x is -2.

73.	$4x$	**74.**	$-3x$
75.	$2x + 5$	**76.**	$-4x - 1$
77.	$3x + 2 + 5x + 1$	**78.**	$4x + 3 + 5x + 6$
79.	$7x - x - 3$	**80.**	$5x - x - 4$

4.4
Addition and Subtraction of Polynomials

In this section we will extend what we learned in Section 4.3 to expressions called polynomials. We begin this section with the definition of a polynomial.

DEFINITION   A *polynomial* is a finite sum of monomials (terms).

EXAMPLES   The following are polynomials:

$$3x^2 + 2x + 1 \qquad 15x^2y + 21xy^2 - y^2 \qquad 3a - 2b + 4c - 5d$$

Polynomials can be further classified by the number of terms they contain. A polynomial with two terms is called a binomial. If it has three terms, it is a trinomial. As stated before, a monomial has only one term.

DEFINITION   The *degree* of a polynomial in one variable is the highest power to which the variable is raised.

EXAMPLES

$3x^5 + 2x^3 + 1$	A trinomial of degree 5
$2x + 1$	A binomial of degree 1
$3x^2 + 2x + 1$	A trinomial of degree 2
$3x^5$	A monomial of degree 5
-9	A monomial of degree 0

There are no new rules for adding one or more polynomials. We rely only on our previous knowledge. Here are some examples.

▼   **Example 1**   Add: $(2x^2 - 5x + 3) + (4x^2 + 7x - 8)$.

SOLUTION   We use the commutative and associative properties to group similar terms together and then apply the distributive property to add:

$(2x^2 - 5x + 3) + (4x^2 + 7x - 8)$

$$= (2x^2 + 4x^2) + (-5x + 7x) + (3 - 8)$$
Commutative and associative properties
$$= (2 + 4)x^2 + (-5 + 7)x + (3 - 8)$$
Distributive property
$$= 6x^2 + 2x - 5$$
Addition

The results here indicate that to add two polynomials, we add coefficients of similar terms.　▲

▼　**Example 2**　Add: $x^2 + 3x + 2x + 6$.

SOLUTION　The only similar terms here are the two middle terms. We combine them as usual to get

$$x^2 + 3x + 2x + 6 = x^2 + 5x + 6$$　▲

You will recall from Chapter 1 the definition of subtraction: $a - b = a + (-b)$. To subtract one expression from another, we simply add its opposite. The letters a and b in the definition can each represent polynomials. The opposite of a polynomial is the opposite of each of its terms. When you subtract one polynomial from another you subtract each of its terms.

▼　**Example 3**　Subtract: $(3x^2 + x + 4) - (x^2 + 2x + 3)$.

SOLUTION　To subtract $x^2 + 2x + 3$, we change the sign of each of its terms and add. If you are having trouble remembering why we do this, remember that we can think of $-(x^2 + 2x + 3)$ as $-1(x^2 + 2x + 3)$. If we distribute the -1 across $x^2 + 2x + 3$, we get $-x^2 - 2x - 3$:

$(3x^2 + x + 4) - (x^2 + 2x + 3) = 3x^2 + x + 4 - x^2 - 2x - 3$
Drop the parentheses and take the opposite of each term in the second polynomial
$$= (3x^2 - x^2) + (x - 2x) + (4 - 3)$$
$$= 2x^2 - x + 1$$　▲

▼ **Example 4** Subtract $-4x^2 + 5x - 7$ from $x^2 - x - 1$.

SOLUTION The polynomial $x^2 - x - 1$ comes first, then the subtraction sign, and finally the polynomial $-4x^2 + 5x - 7$ in parentheses.

$$(x^2 - x - 1) - (-4x^2 + 5x - 7)$$
$$= x^2 - x - 1 + 4x^2 - 5x + 7 \qquad \text{We take the opposite of}$$
$$\qquad\qquad\qquad\qquad\qquad\qquad\qquad\text{each term in the}$$
$$\qquad\qquad\qquad\qquad\qquad\qquad\qquad\text{second polynomial}$$
$$= (x^2 + 4x^2) + (-x - 5x) + (-1 + 7)$$
$$= 5x^2 - 6x + 6 \qquad\qquad\qquad\qquad\qquad\qquad\qquad ▲$$

There are two important points to remember when adding or subtracting polynomials. First, to add or subtract two polynomials you always add or subtract *coefficients* of similar terms. Second, the exponents never increase in value when adding or subtracting similar terms.

The last topic we want to consider in this section is finding the value of a polynomial for a given value of the variable.

To find the value of the polynomial $3x^2 + 1$ when x is 5, we replace x with 5 and simplify the result:

When $\qquad\qquad\qquad\qquad\qquad x = 5$
the polynomial $\qquad\qquad\qquad 3x^2 + 1$
becomes $\qquad\qquad\qquad 3(5)^2 + 1 = 3(25) + 1$
$$\qquad\qquad\qquad\qquad\qquad\qquad\qquad = 75 + 1$$
$$\qquad\qquad\qquad\qquad\qquad\qquad\qquad = 76$$

Notice that when evaluating the expression $3 \cdot 5^2 + 1$, we find 5^2 before multiplying by 3. It would be incorrect to multiply $3 \cdot 5$ and then square the result. Here is another example:

▼ **Example 5** Find the value of $3x^2 - 5x + 4$ when $x = -2$.

SOLUTION When $\qquad\qquad\qquad x = -2$
the polynomial $\qquad\qquad\qquad 3x^2 - 5x + 4$
becomes $\qquad\qquad 3(-2)^2 - 5(-2) + 4 = 3(4) + 10 + 4$
$$\qquad\qquad\qquad\qquad\qquad\qquad\qquad = 12 + 10 + 4$$
$$\qquad\qquad\qquad\qquad\qquad\qquad\qquad = 26 \qquad\qquad ▲$$

Problem Set 4.4

Identify each of the following polynomials as a trinomial, binomial, or monomial, and give the degree in each case.

1. $2x^3 - 3x^2 + 1$ 2. $4x^2 - 4x + 1$
3. $5 + 8a - 9a^3$ 4. $6 + 12x^3 + x^4$
5. $2x - 1$ 6. $4 + 7x$
7. $45x^2 - 1$ 8. $3a^3 + 8$
9. $7a^2$ 10. $90x$
11. -4 12. 56

Perform the following additions and subtractions.

13. $(2x^2 + 3x + 4) + (3x^2 + 2x + 5)$
14. $(x^2 + 5x + 6) + (x^2 + 3x + 4)$
15. $(3a^2 - 4a + 1) + (2a^2 - 5a + 6)$
16. $(5a^2 - 2a + 7) + (4a^2 - 3a + 2)$
17. $x^2 + 4x + 2x + 8$
18. $x^2 + 5x - 3x - 15$
19. $6x^2 - 3x - 10x + 5$
20. $10x^2 + 30x - 2x - 6$
21. $x^2 - 3x + 3x - 9$
22. $x^2 - 5x + 5x - 25$
23. $3y^2 - 5y - 6y + 10$
24. $y^2 - 18y + 2y - 12$
25. $(6x^3 - 4x^2 + 2x) + (9x^2 - 6x + 3)$
26. $(5x^3 + 2x^2 + 3x) + (2x^2 + 5x + 1)$
27. $(3x^2 - 4x - 5) + (5x^2 + 4x + 3)$
28. $(22x^3 - 12x^2 + x) - (5x^3 - 3x^2 + 5)$
29. $(a^2 - a - 1) - (-a^2 + a + 1)$
30. $(2x^2 - 7x - 8) - (6x^2 + 6x - 8) + (4x^2 - 2x + 3)$
31. $(-8x^2 + 2x + 1) + (10x^2 - 33x - 5) - (3x^2 - 5x - 8)$
32. $(7x^4 - 5x^2 - 4) - (8x^4 + 9x^2 - 3)$
33. $(4y^2 - 3y + 2) + (5y^2 + 12y - 4) - (13y^2 - 6y + 20)$
34. $(9x^3 - 8x^2 + 7x) - (2x - 3x^2 - 4x^3)$
35. Subtract $10x^2 + 23x - 50$ from $11x^2 - 10x + 13$.
36. Subtract $2x^2 - 3x + 5$ from $4x^2 - 5x + 10$.
37. Subtract $3y^2 + 7y - 15$ from $11y^2 + 11y + 11$.
38. Subtract $15y^2 - 8y - 2$ from $3y^2 - 3y + 2$.
39. Add $50x^2 - 100x - 150$ to $25x^2 - 50x + 75$.
40. Add $7x^2 - 8x + 10$ to $-8x^2 + 2x - 12$.
41. Subtract $2x + 1$ from the sum of $3x - 2$ and $11x + 5$.
42. Subtract $3x - 5$ from the sum of $5x + 2$ and $9x - 1$.
43. Find the value of the polynomial $x^2 - 2x + 1$ when x is 3.
44. Find the value of the polynomial $(x - 1)^2$ when x is 3.
45. Find the value of the polynomial $(y - 5)^2$ when y is 10.
46. Find the value of the polynomial $y^2 - 10y + 25$ when y is 10.
47. Find the value of $a^2 + 4a + 4$ when a is 2.
48. Find the value of $(a + 2)^2$ when a is 2.

Review Problems   The problems below review material covered in Section 4.3. Reviewing these problems will help you in Section 4.5.

Multiply.

49. $2x(5x)$ 50. $2x(-2x)$
51. $3x(-5x)$ 52. $-3x(-7x)$
53. $2x(3x^2)$ 54. $x^2(3x)$
55. $3x^2(2x^2)$ 56. $4x^2(2x^2)$

**4.5
Multiplication with
Polynomials**

We begin our discussion of multiplication of polynomials by finding the product of a monomial and a trinomial.

▼   **Example 1**   Multiply: $3x^2(2x^2 + 4x + 5)$.

SOLUTION   Applying the distributive property gives us

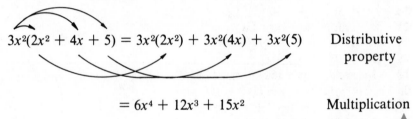

$$3x^2(2x^2 + 4x + 5) = 3x^2(2x^2) + 3x^2(4x) + 3x^2(5) \qquad \text{Distributive property}$$

$$= 6x^4 + 12x^3 + 15x^2 \qquad \text{Multiplication}$$
▲

The distributive property is the key to multiplication of polynomials. We can use it to find the product of any two polynomials. There are some shortcuts we can use in certain situations, however. Let's look at an example that involves the product of two binomials.

▼   **Example 2**   Multiply: $(3x - 5)(2x - 1)$.

SOLUTION
$$\begin{aligned}
(3x - 5)(2x - 1) &= 3x(2x - 1) - 5(2x - 1) \\
&= 3x(2x) + 3x(-1) + (-5)(2x) + (-5)(-1) \\
&= 6x^2 - 3x - 10x + 5 \\
&= 6x^2 - 13x + 5
\end{aligned}$$
▲

If we look closely at the second and third lines of work in this example, we can see that the terms in the answer come from all possible products of terms in the first binomial with terms in the second binomial. This result is generalized as follows.

RULE   To multiply any two polynomials, multiply each term in the first with each term in the second.

There are two ways we can put this rule to work. You should become familiar with each.

If we look at the original problem in Example 2 and then to the answer, we see that the first term in the answer came from multiplying the first terms in each binomial:

FOIL Method

$$3x \cdot 2x = 6x^2 \qquad \text{FIRST}$$

The middle term in the answer came from adding the products of the two outside terms with the two inside terms in each binomial:

$$
\begin{aligned}
3x(-1) &= -\ 3x \qquad \text{OUTSIDE}\\
-5(2x) &= \underline{-10x} \qquad \text{INSIDE}\\
&\ \ \ -13x
\end{aligned}
$$

The last term in the answer came from multiplying the two last terms:

$$-5(-1) = 5 \qquad \text{LAST}$$

To summarize the FOIL method we will multiply another two binomials.

▼ **Example 3**   Multiply: $(2x + 3)(5x - 4)$.

SOLUTION

$$(2x + 3)(5x - 4) = \underbrace{2x(5x)}_{\text{First}} + \underbrace{2x(-4)}_{\text{Outside}} + \underbrace{3(5x)}_{\text{Inside}} + \underbrace{3(-4)}_{\text{Last}}$$

$$
\begin{aligned}
&= 10x^2 - 8x + 15x - 12\\
&= 10x^2 + 7x - 12
\end{aligned}
$$

With practice $-8x + 15x = 7x$ can be done mentally.   ▲

The FOIL method can be applied only when multiplying two binomials. To find products of polynomials with more than two terms we use what is called the COLUMN method.

COLUMN Method

The COLUMN method of multiplying two polynomials is very similar to long multiplication with whole numbers. It is just another way of finding all possible products of terms in one polynomial with terms in another polynomial.

▼ **Example 4**   COLUMN method.

$$
\begin{array}{r}
3x - 5 \\
2x - 1 \\
\hline
\end{array}
$$

$$
\begin{array}{rl}
-3x + 5 & \quad \text{Multiply } 3x \text{ and } -5 \text{ by } -1 \\
6x^2 - 10x & \quad \text{Multiply } 3x \text{ and } -5 \text{ by } 2x \\
\hline
6x^2 - 13x + 5 & \quad \text{Add similar terms to get answer} \qquad ▲
\end{array}
$$

We can see by comparing our answer to that of Example 2 that the results are the same regardless of which method is used. The COLUMN method can be applied to polynomials with many terms.

▼ **Example 5**   Multiply: $(2x + 3)(3x^2 - 2x + 1)$.

SOLUTION

$$
\begin{array}{r}
3x^2 - 2x + 1 \\
2x + 3 \\
\hline
9x^2 - 6x + 3 \\
6x^3 - 4x^2 + 2x \\
\hline
6x^3 + 5x^2 - 4x + 3 \qquad ▲
\end{array}
$$

Again, each term in the first polynomial has been multiplied by each term in the second.

It will be to your advantage to become very fast and accurate at multiplying polynomials. You should be comfortable using either method. The following examples illustrate the three types of multiplication.

▼ **Examples**   Multiply.

6. $4a^2(2a^2 - 3a + 5) = 4a^2(2a^2) + 4a^2(-3a) + 4a^2(5)$
$$= 8a^4 - 12a^3 + 20a^2$$

7. $(5x - 1)(2x + 6) = 5x(2x) + 5x(6) + (-1)(2x) + (-1)(6)$
$$= 10x^2 + 30x + (-2x) + (-6)$$
$$ \text{F} \qquad \text{O} \qquad \text{I} \qquad \text{L}$$
$$= 10x^2 + 28x - 6$$

8. $(3x + 2)(x^2 - 5x + 6)$

$$
\begin{array}{r}
x^2 - 5x + 6 \\
3x + 2 \\
\hline
2x^2 - 10x + 12 \\
3x^3 - 15x^2 + 18x \\
\hline
3x^3 - 13x^2 + 8x + 12
\end{array}
$$

▲

Multiply the following by applying the distributive property.

1. $2x(3x + 1)$
2. $4x(2x - 3)$
3. $2x^2(3x^2 - 2x + 1)$
4. $5x(4x^3 - 5x^2 + x)$
5. $2ab(a^2 - ab + 1)$
6. $3a^2b(a^3 + a^2b^2 + b^3)$
7. $y^2(3y^2 + 9y + 12)$
8. $5y(2y^2 - 3y + 5)$
9. $4x^2y(2x^3y + 3x^2y^2 + 8y^3)$
10. $6xy^3(2x^2 + 5xy + 12y^2)$

Multiply the binomials below. You should do about half the problems using the FOIL method and the other half using the COLUMN method. Remember, you want to be comfortable using each method.

11. $(x + 3)(x + 4)$
12. $(x + 2)(x + 5)$
13. $(x + 6)(x + 1)$
14. $(x + 1)(x + 4)$
15. $(x + 1)(x - 3)$
16. $(x + 4)(x - 2)$
17. $(a + 5)(a - 3)$
18. $(a - 8)(a + 2)$
19. $(x + 4)(x + 4)$
20. $(x + 5)(x + 5)$
21. $(x + 6)(x - 6)$
22. $(x + 3)(x - 3)$
23. $(y - 2)(y + 2)$
24. $(y - 4)(y + 4)$
25. $(2x - 3)(x - 4)$
26. $(3x - 5)(x - 2)$
27. $(a + 2)(2a - 1)$
28. $(a - 6)(3a + 2)$
29. $(2x - 5)(3x - 2)$
30. $(3x + 6)(2x - 1)$
31. $(3x - 2)(3x - 2)$
32. $(2x - 5)(2x - 5)$
33. $(5x - 4)(5x + 4)$
34. $(6x + 5)(6x - 5)$
35. $(8x + 3)(x - 2)$
36. $(5x + 4)(3x - 2)$
37. $(1 - 2a)(3 - 4a)$
38. $(1 - 3a)(3 + 2a)$
39. $(7 - 6x)(8 - 5x)$
40. $(7 - 4x)(8 - 3x)$

Multiply the following

41. $(x + 1)(x^2 + 3x - 4)$
42. $(x - 2)(x^2 + 3x - 4)$
43. $(a - 3)(a^2 - 3a + 2)$
44. $(a + 5)(a^2 + 2a + 3)$
45. $(x + 2)(x^2 - 2x + 4)$
46. $(x + 3)(x^2 - 3x + 9)$
47. $(2x + 1)(x^2 + 8x + 9)$
48. $(3x - 2)(x^2 - 7x + 8)$
49. $(5x^2 + 2x + 1)(x^2 - 3x + 5)$
50. $(2x^2 + x + 1)(x^2 - 4x + 3)$

The commutative and associative properties, when used together, indicate that when we want to find the product of three numbers, we can multiply any two of them first and then multiply the result by the other number. To find the product of 3, 5, and 6, we can find $3 \cdot 5$ and then multiply the result by 6, or we can find $5 \cdot 6$ and multiply the result by 3, or we can use any other combination we arrive at by changing the order or the grouping of the three numbers. The same idea holds for polynomials. Use this idea to find the following products.

51. $2(x + 7)(x - 2)$ **52.** $3(x + 5)(x - 4)$

53. $3x(x + 3)(x - 2)$ **54.** $3x(x - 3)(x + 2)$

55. $(x + 1)(x + 2)(x + 3)$ **56.** $(x - 1)(x - 2)(x - 3)$

57. The diagram below shows how you can visualize the product $(x + 5)(x + 3)$. The length of the large rectangle is $x + 5$ and the width is $x + 3$. The area of each of the smaller rectangles has been filled in by multiplying length times width in each case.

	x	5
x	x^2	$5x$
3	$3x$	15

Since the area of the large rectangle is equal to the sum of the areas of the four smaller rectangles, and area is length times width, we have

$$(x + 5)(x + 3) = x^2 + 3x + 5x + 15$$
$$= x^2 + 8x + 15$$

What multiplication problem is illustrated by this diagram?

	x	4
x	x^2	$4x$
2	$2x$	8

58. What multiplication problem is illustrated by the following diagram?

	x	6
x		
2		

59. Draw a diagram like the diagrams in Problems 57 and 58 to illustrate the product $(x + 10)(x + 5)$.

60. Draw a diagram to illustrate the product $(x + 1)(x + 4)$.

Review Problems The problems below review material we covered in Section 3.1. In each case, complete the ordered pairs so that they will be solutions to the equation given.

61. $x + y = 8$ (6,), (−4,), (0,)
62. $2x − y = 6$ (, 6), (−2,), (, 0)
63. $2x + 3y = 12$ (0,), (, 0), (, 2)
64. $y = 2x + 1$ (1,), (−3,), (, 0)

In this section we will combine the results of the last section with our definition of exponents to find some special products.

**4.6
Binomial Squares
and Other
Special Products**

▼ **Example 1**   Find the square of $(3x − 2)$.

SOLUTION   To square $3x − 2$, we multiply it by itself.

$$
\begin{aligned}
(3x − 2)^2 &= (3x − 2)(3x − 2) &&\text{Definition of exponents} \\
&= 9x^2 − 6x − 6x + 4 &&\text{FOIL method} \\
&= 9x^2 − 12x + 4 &&\text{Combine similar terms} \quad ▲
\end{aligned}
$$

Notice that the first and last terms in the answer are the square of the first and last terms in the original problem and that the middle term is twice the product of the two terms in the original binomial.

▼ **Examples**

2. $(a + b)^2 = (a + b)(a + b)$
          $= a^2 + 2ab + b^2$

3. $(a − b)^2 = (a − b)(a − b)$
          $= a^2 − 2ab + b^2$ ▲

Binomial squares having the form of Examples 2 and 3 occur very frequently in algebra. It will be to your advantage to memorize the following rule for squaring a binomial.

RULE   The square of a binomial is the sum of the square of the first term, the square of the last term, and twice the product of the two original terms.

Here is what it looks like algebraically:

$$(x + y)^2 = \quad x^2 \quad + \quad 2xy \quad + \quad y^2$$

| | Square of first term | Twice product of the two terms | Square of last term |

▼ **Examples**

		First term squared		Twice their product		Last term squared		Answer
4.	$(x - 5)^2$	$=$	x^2	$+$	$2(x)(-5)$	$+$	25	$= x^2 - 10x + 25$
5.	$(x + 2)^2$	$=$	x^2	$+$	$2(x)(2)$	$+$	4	$= x^2 + 4x + 4$
6.	$(2x - 3)^2$	$=$	$4x^2$	$+$	$2(2x)(-3)$	$+$	9	$= 4x^2 - 12x + 9$
7.	$(5x - 4)^2$	$=$	$25x^2$	$+$	$2(5x)(-4)$	$+$	16	$= 25x^2 - 40x + 16$

▲

Another special product that occurs frequently has the form $(a + b)(a - b)$. The only difference in the two binomials is the sign between the two terms. The interesting thing about this type of product is that the middle term is always zero. Here are some examples.

▼ **Examples**

8. $(2x - 3)(2x + 3) = 4x^2 + 6x - 6x - 9$ FOIL method
$$= 4x^2 - 9$$

9. $(x - 5)(x + 5) = x^2 + 5x - 5x - 25$ FOIL method
$$= x^2 - 25$$

10. $(3x - 1)(3x + 1) = 9x^2 + 3x - 3x - 1$ FOIL method
$$= 9x^2 - 1$$

▲

Notice that in each case the middle term is zero and therefore doesn't appear in the answer. The answers all turn out to be the difference of two squares. Here is a rule to help you memorize the result.

RULE   When multiplying two binomials that differ only in the sign between their terms, subtract the square of the last term from the square of the first term. Or

$$(a - b)(a + b) = a^2 - b^2$$

Here are some problems that result in the difference of two squares.

▼ **Examples**

11. $(x - 3)(x + 3) = x^2 - 9$

12. $(a + 2)(a - 2) = a^2 - 4$

13. $(9a + 1)(9a - 1) = 81a^2 - 1$

14. $(2x - 5y)(2x + 5y) = 4x^2 - 25y^2$

15. $(3a - 7b)(3a + 7b) = 9a^2 - 49b^2$ ▲

Although all the problems in this section can be worked correctly using the methods in the last section, they can be done much faster if the two rules are *memorized.* Here is a summary of the two rules.

$$(a + b)^2 = (a + b)(a + b) = a^2 + 2ab + b^2$$
$$(a - b)^2 = (a - b)(a - b) = a^2 - 2ab + b^2$$
$$(a - b)(a + b) = a^2 - b^2$$

Note: A very common mistake when squaring binomials is to write

$$(a + b)^2 = a^2 + b^2$$

It just isn't true. Exponents do *not* distribute over addition or subtraction. If we try it with 2 and 3 it becomes obvious.

$$(2 + 3)^2 \neq 2^2 + 3^2$$
$$25 \neq 13$$

Perform the indicated operations. Problem Set 4.6

1. $(x - 2)^2$
2. $(x + 2)^2$
3. $(a + 3)^2$
4. $(a - 3)^2$
5. $(x - 5)^2$
6. $(x - 4)^2$
7. $(a - 7)^2$
8. $(a + 7)^2$
9. $(x + 10)^2$
10. $(x - 10)^2$
11. $(a + b)^2$
12. $(a - b)^2$
13. $(2x - 1)^2$
14. $(3x + 2)^2$
15. $(4a + 5)^2$
16. $(4a - 5)^2$
17. $(3x - 2)^2$
18. $(2x - 3)^2$
19. $(3a + 5b)^2$
20. $(5a - 3b)^2$
21. $(4x - 5y)^2$
22. $(5x + 4y)^2$
23. $(7m + 2n)^2$
24. $(2m - 7n)^2$
25. $(6x - 10y)^2$
26. $(10x + 6y)^2$
27. $(x^2 + 5)^2$
28. $(x^2 + 3)^2$
29. $(a^2 + 1)^2$
30. $(a^2 - 2)^2$
31. $(x^3 - 7)^2$
32. $(x^3 + 4)^2$
33. $(x - 3)(x + 3)$
34. $(x + 4)(x - 4)$
35. $(a + 5)(a - 5)$
36. $(a - 6)(a + 6)$
37. $(y - 1)(y + 1)$
38. $(y - 2)(y + 2)$
39. $(9 + x)(9 - x)$
40. $(10 - x)(10 + x)$
41. $(2x + 5)(2x - 5)$
42. $(3x + 5)(3x - 5)$

43. $(4x - 1)(4x + 1)$ **44.** $(6x + 1)(6x - 1)$
45. $(2a + 7)(2a - 7)$ **46.** $(3a + 10)(3a - 10)$
47. $(6 - 7x)(6 + 7x)$ **48.** $(7 - 6x)(7 + 6x)$
49. $(x^2 + 3)(x^2 - 3)$ **50.** $(x^2 + 2)(x^2 - 2)$
51. $(a^2 + 4)(a^2 - 4)$ **52.** $(a^2 + 9)(a^2 - 9)$
53. The formula for the difference of two squares can be used as a shortcut to multiplying certain whole numbers if they have the correct form. Use the difference of two squares formula to multiply 49(51) by first writing 49 as $(50 - 1)$, and 51 as $(50 + 1)$.
54. Use the difference of two squares formula to multiply 101(99) by first writing 101 as $(100 + 1)$ and 99 as $(100 - 1)$.
55. Evaluate the expression $(x + 3)^2$ and the expression $x^2 + 6x + 9$ for $x = 2$.
56. Evaluate the expression $x^2 - 25$ and the expression $(x - 5)(x + 5)$ for $x = 6$.
57. Simplify the expression $(x + 2)^2 + 3(x + 2) - 4$ by first squaring $(x + 2)$, then multiplying 3 times $(x + 2)$, and then combining similar terms. The result should be the same as the product $(x + 6)(x + 1)$ if it were multiplied out.
58. Simplify the expression $(x + 3)^2 + 2(x + 3) - 3$. The result should be the same as the product $(x + 6)(x + 2)$ if it were multiplied out.

Review Problems   The problems below review material we covered in Section 4.3. Reviewing this material will help you understand the next section.

Simplify each expression (divide).

59. $\dfrac{10x^3}{5x}$ **60.** $\dfrac{-15x^2}{5x}$

61. $\dfrac{15x^2y}{3xy}$ **62.** $\dfrac{21xy^2}{3xy}$

63. $\dfrac{35a^6b^8}{70a^2b^{10}}$ **64.** $\dfrac{75a^2b^6}{25a^4b^3}$

**4.7
Dividing a
Polynomial by
a Monomial**

To divide a polynomial by a monomial, we will use the definition of division and apply the distributive property. Follow the steps in this example closely.

▼ **Example 1**   Divide $10x^3 - 15x^2$ by $5x$.

SOLUTION

$$\frac{10x^3 - 15x^2}{5x} = (10x^3 - 15x^2)\,\frac{1}{5x}$$

Division by $5x$ is the same as multiplication by $1/5x$

$$= 10x^3\left(\frac{1}{5x}\right) - 15x^2\left(\frac{1}{5x}\right)$$

Distribute $1/5x$ to both terms

$$= \frac{10x^3}{5x} - \frac{15x^2}{5x}$$

Multiplication by $1/5x$ is the same as division by $5x$

$$= 2x^2 - 3x$$

Division of monomials as done in Section 4.3 ▲

If we were to leave out the first two steps, the problem would look like this:

$$\frac{10x^3 - 15x^2}{5x} = \frac{10x^3}{5x} - \frac{15x^2}{5x}$$
$$= 2x^2 - 3x$$

The problem is much shorter and clearer this way. You may leave out the first two steps from Example 1 when working problems in this section. They are part of Example 1 only to help show you why the following rule is true.

RULE To divide a polynomial by a monomial, simply divide each term in the polynomial by the monomial.

Here are some further examples using our rule for division of a polynomial by a monomial.

▼ **Examples** Divide.

2. $\dfrac{3x^2 - 6}{3} = \dfrac{3x^2}{3} - \dfrac{6}{3}$
$$= x^2 - 2$$

3. $\dfrac{4x^2 - 2}{2} = \dfrac{4x^2}{2} - \dfrac{2}{2}$
$$= 2x^2 - 1$$

4. $\dfrac{27x^3 - 9x^2}{3x} = \dfrac{27x^3}{3x} - \dfrac{9x^2}{3x}$

$= 9x^2 - 3x$

5. $\dfrac{15x^2y + 21xy^2}{3xy} = \dfrac{15x^2y}{3xy} + \dfrac{21xy^2}{3xy}$

$= 5x + 7y$

6. $\dfrac{10x^3 - 5x^2 + 20x}{10x^2} = \dfrac{10x^3}{10x^2} - \dfrac{5x^2}{10x^2} + \dfrac{20x}{10x^2}$

$= x - \dfrac{1}{2} + \dfrac{2}{x}$

7. $\dfrac{24x^3y^2 + 16x^2y^2 - 4x^2y^3}{8x^2y} = \dfrac{24x^3y^2}{8x^2y} + \dfrac{16x^2y^2}{8x^2y} - \dfrac{4x^2y^3}{8x^2y}$

$= 3xy + 2y - \dfrac{y^2}{2}$ ▲

Problem Set 4.7

Divide the following polynomials by $5x$.

1. $5x^2 - 10x$
2. $10x^3 - 15x$
3. $15x - 10x^3$
4. $50x^3 - 20x^2$
5. $25x^2y - 10xy$
6. $15xy^2 + 20x^2y$
7. $35x^5 - 30x^4 + 25x^3$
8. $40x^4 - 30x^3 + 20x^2$
9. $50x^5 - 25x^3 + 5x$
10. $75x^6 + 50x^3 - 25x$

Divide the following by $-2a$.

11. $8a^2 - 4a$
12. $a^3 - 6a^2$
13. $16a^5 + 24a^4$
14. $30a^6 + 20a^3$
15. $8ab + 10a^2$
16. $6a^2b - 10ab^2$
17. $12a^3b - 6a^2b^2 + 14ab^3$
18. $4ab^3 - 16a^2b^2 - 22a^3b$
19. $a^2 + 2ab + b^2$
20. $a^2b - 2ab^2 + b^3$

Perform the following divisions (find the following quotients).

21. $\dfrac{6x + 8y}{2}$
22. $\dfrac{9x - 3y}{3}$
23. $\dfrac{7y - 21}{-7}$
24. $\dfrac{14y - 12}{2}$
25. $\dfrac{10xy - 8x}{2x}$
26. $\dfrac{12xy^2 - 18x}{-6x}$

27. $\dfrac{x^2y - x^3y^2}{x}$

28. $\dfrac{x^2y - x^3y^2}{x^2}$

29. $\dfrac{x^2y - x^3y^2}{-x^2y}$

30. $\dfrac{ab + a^2b^2}{ab}$

31. $\dfrac{a^2b^2 - ab^2}{-ab^2}$

32. $\dfrac{a^2b^2c + ab^2c^2}{abc}$

33. $\dfrac{x^3 - 3x^2y + xy^2}{x}$

34. $\dfrac{x^2 - 3xy^2 + xy^3}{x}$

35. $\dfrac{10a^2 - 15a^2b + 25a^2b^2}{5a^2}$

36. $\dfrac{11a^2b^2 - 33ab}{-11ab}$

37. $\dfrac{26x^2y^2 - 13xy}{-13xy}$

38. $\dfrac{6x^2y^2 - 3xy}{6xy}$

39. $\dfrac{4x^2y^2 - 2xy}{4xy}$

40. $\dfrac{6x^2a + 12x^2b - 6x^2c}{36x^2}$

41. $\dfrac{5a^2x - 10ax^2 + 15a^2x^2}{20a^2x^2}$

42. $\dfrac{12ax - 9bx + 18cx}{6x^2}$

43. $\dfrac{16x^5 + 8x^2 + 12x}{12x^3}$

44. $\dfrac{27x^2 - 9x^3 - 18x^4}{-18x^3}$

45. Evaluate the expression $\dfrac{10x + 15}{5}$ and the expression $2x + 3$ when $x = 2$.

46. Evaluate the expression $\dfrac{6x^2 + 4x}{2x}$ and the expression $3x + 2$ when $x = 5$.

47. Show that the expression $\dfrac{3x + 8}{2}$ is not the same as the expression $3x + 4$ by replacing x with 10 in both expressions and simplifying the results.

48. Show that the expression $\dfrac{x + 10}{x}$ is not equal to 10 by replacing x with 5 and simplifying.

Review Problems   The following problems review material we covered in Section 3.4.

Solve each system of equations by the elimination method.

49. $x + y = 6$ **50.** $2x + y = 5$
         $x - y = 8$ $-x + y = -4$
51. $2x - 3y = -5$ **52.** $2x - 4y = 10$
         $x + \ y = 5$ $3x - 2y = -1$

Chapter 4 Summary and Review

Examples

1.

(a) $2^3 = 2 \cdot 2 \cdot 2 = 8$
(b) $x^5 \cdot x^3 = x^{5+3} = x^8$

(c) $\dfrac{x^5}{x^3} = x^{5-3} = x^2$

(d) $(3x)^2 = 3^2 \cdot x^2 = 9x^2$

(e) $\left(\dfrac{2}{3}\right)^3 = \dfrac{2^3}{3^3} = \dfrac{8}{27}$

(f) $(x^5)^3 = x^{5 \cdot 3} = x^{15}$

(g) $3^{-2} = \dfrac{1}{3^2} = \dfrac{1}{9}$

EXPONENTS: DEFINITION AND PROPERTIES [4.1, 4.2]

Integer exponents indicate repeated multiplications.

$a^r \cdot a^s = a^{r+s}$ To multiply with the same base you add exponents.

$\dfrac{a^r}{a^s} = a^{r-s}$ To divide with the same base you subtract exponents.

$(ab)^r = a^r \cdot b^r$ Exponents distribute over multiplication.

$\left(\dfrac{a}{b}\right)^r = \dfrac{a^r}{b^r}$ Exponents distribute over division.

$(a^r)^s = a^{r \cdot s}$ A power of a power is the product of the powers.

$a^{-r} = \dfrac{1}{a^r}$ Negative exponents imply reciprocals.

2. $(5x^2)(3x^4) = 15x^6$

MULTIPLICATION OF MONOMIALS [4.3]

To multiply two monomials, multiply coefficients and add exponents.

3. $\dfrac{12x^9}{4x^5} = 3x^4$

DIVISION OF MONOMIALS [4.3]

To divide two monomials, divide coefficients and subtract exponents.

ADDITION OF POLYNOMIALS [4.4]

To add two polynomials, add coefficients of similar terms.

4. $(3x^2 - 2x + 1) + (2x^2 + 7x - 3)$
$= 5x^2 + 5x - 2$

SUBTRACTION OF POLYNOMIALS [4.4]

To subtract one polynomial from another, add the opposite of the second to the first.

5. $(3x + 5) - (4x - 3)$
$= 3x + 5 - 4x + 3$
$= -x + 8$

MULTIPLICATION OF POLYNOMIALS [4.5]

To multiply two polynomials, multiply each term in the first by each term in the second. If both are binomials, the FOIL method will accomplish this. If not, the column method is used.

6. $(x + 2)(3x - 1)$
$= 3x^2 - x + 6x - 2$
$= 3x^2 + 5x - 2$

SPECIAL PRODUCTS [4.6]

$$\left.\begin{array}{l}(a + b)^2 = a^2 + 2ab + b^2 \\ (a - b)^2 = a^2 - 2ab + b^2\end{array}\right\}$$ Binomial squares

$(a + b)(a - b) = a^2 - b^2$ Difference of two squares

7. $(x + 3)^2 = x^2 + 6x + 9$
$(x - 3)^2 = x^2 - 6x + 9$
$(x + 3)(x - 3) = x^2 - 9$

DIVIDING A POLYNOMIAL BY A MONOMIAL [4.7]

To divide a polynomial by a monomial, divide each term in the polynomial by the monomial.

8. $\dfrac{12x^3 - 18x^2}{6x}$

$= 2x^2 - 3x$

COMMON MISTAKES

Trying to add nonsimilar terms: $2x + 3x^2 = 5x^3$. To avoid this, remember that exponents *never* increase or decrease when adding terms.

Exponents *do not* distribute over addition: $(a + b)^2 \neq a^2 + b^2$. Convince yourself this doesn't work by trying it when $a = 2$ and $b = 3$.

Use the properties of exponents to simplify the following problems.   Chapter 4 Test

1. $5^3 \cdot 5^2$ **2.** $10^2 \cdot 10^5$

3. $\dfrac{3^4}{3^7}$ **4.** $\dfrac{4^6}{4^8}$

5. 5^{-2} 6. 2^{-5}

7. $(\frac{3}{4})^2$ 8. $(-\frac{5}{6})^2$

9. 10^1 10. $(2x^2)^0$

11. $(3x^2)^3$ 12. $(2x^2y^3)^3$

13. $\dfrac{x^3 \cdot x^5}{x^{10}}$ 14. $x^{-5} \cdot x^2$

15. $\dfrac{a^{-3}}{a^{-7}}$ 16. $\dfrac{(2a^2)(6a^3)}{4a^{12}}$

17. $\dfrac{(5x^2y)(10x^3y^3)}{25xy^4}$ 18. $\dfrac{(6x^5y^6)(8x^2y)}{(2x^4y^{10})(3xy^3)}$

Perform the indicated operations.

19. $(3x^2 + 2x - 1) + (4x^2 - 8x - 2)$
20. $(x^2 - 5x - 2) + (7x^2 + 2x + 5)$
21. $(x^3 + 2x^2 - 3) - (2x^2 + 3x + 5)$
22. $(4a^3 + a - 5) - (2a^3 + a^2 - 7)$
23. $(3x - 5)(2x + 1)$ 24. $(3x + 2)(7x - 1)$
25. $(a + 1)(a - 3)(a + 2)$ 26. $(2a - 1)(a + 4)(a - 4)$
27. $(2x - 3)(3x^2 - 5x + 2)$ 28. $(3x - 1)(4x^2 - x + 1)$
29. $(a - 1)(a^2 + a + 1)$ 30. $(a - 2)(a^2 + 2a + 4)$
31. $(2x - 5)^2$ 32. $(3x - 4)^2$
33. $(2a + 3)^2$ 34. $(3a - 2)^2$
35. $(4x^3 - 8x^2 + 2x) \div (2x)$ 36. $(15x^4 - 10x^3 + 25x^2) \div (-5x)$
37. $(20x^5 + 15x^3 - 10x) \div (10x^3)$ 38. $(16x^6 - 8x^4 + 12x^2) \div (-8x^4)$

Factoring

In this chapter we will be doing the reverse of some of the things we did in the last chapter. We are going to do what is called factoring. We will then use factoring to help solve second-degree equations—equations in which the variable is raised to the second power. Factoring is a tool that allows us to solve some problems and manipulate some expressions that we would otherwise be unable to work with.

The two concepts from our previous work that are necessary for success in this chapter are (1) the distributive property and (2) multiplication of binomials. The more familiar you are with the idea of the distributive property and the process of multiplying binomials, the quicker you will understand factoring.

The following diagram shows the relationship between multiplication and factoring:

5.1
Factoring Integers

$$\text{Multiplication}$$

$$\overset{\frown}{3 \cdot 4} \; = \; 12$$

Factors   Product

Factoring

When we read the problem from left to right, we say the product of 3 and 4 is 12. Or we multiply 3 and 4 to get 12. When we read the problem in the other direction, from right to left, we say we have *factored* 12 into 3 times 4, or 3 and 4 are *factors* of 12.

The number 12 can be factored still further:

$$12 = 4 \cdot 3$$
$$= 2 \cdot 2 \cdot 3$$
$$= 2^2 \cdot 3$$

The numbers 2 and 3 are called *prime factors* of 12, because neither of them can be factored any further.

To generalize the above illustration, we can make the following definitions.

DEFINITION   If *a* and *b* represent integers, then *a* is said to be a *factor* (or divisor) of *b* if *a* divides *b* evenly—that is, if *a* divides *b* with no remainder.

DEFINITION   A *prime* number is any positive integer larger than 1 whose only positive factors (divisors) are itself and 1.

EXAMPLES   The number 15 is not a prime number since it has factors of 3 and 5

$$15 = 3 \cdot 5$$

The number 31 is a prime number since its only divisors (factors) are itself and 1.

Here is a list of the first few prime numbers.

Prime numbers = {2, 3, 5, 7, 11, 13, 17, 19, 23, 29, 31, 37, 41, . . .}

When a number is not prime, we can factor it into the product of prime numbers. To factor a number into the product of primes, we simply factor it until it cannot be factored further.

▼   **Example 1**   Factor the number 60 into the product of prime numbers.

SOLUTION   We begin by writing 60 as the product of any two positive numbers whose product is 60, like 6 and 10:

$$60 = 6 \cdot 10$$

We then factor these numbers:

$$60 = 6 \cdot 10$$
$$= (2 \cdot 3) \cdot (2 \cdot 5)$$

Each of these factors is prime. It is customary to write the prime factors in order from smallest to largest.

$$60 = 2 \cdot 2 \cdot 3 \cdot 5$$
$$= 2^2 \cdot 3 \cdot 5$$ ▲

▼ **Example 2** Factor 630 into the product of primes.

SOLUTION

$$630 = 63 \cdot 10$$
$$= (7 \cdot 9) \cdot (2 \cdot 5)$$
$$= 7 \cdot 3 \cdot 3 \cdot 2 \cdot 5$$
$$= 2 \cdot 3^2 \cdot 5 \cdot 7$$

It makes no difference which two numbers we start with, as long as their product is 630. We will always get the same result because a number has only one set of prime factors.

$$630 = 18 \cdot 35$$
$$= 3 \cdot 6 \cdot 5 \cdot 7$$
$$= 3 \cdot 2 \cdot 3 \cdot 5 \cdot 7$$
$$= 2 \cdot 3^2 \cdot 5 \cdot 7$$ ▲

Every positive integer factors into the product of primes one and only one way.

▼ **Example 3** Factor 525 into the product of primes

SOLUTION

$$525 = 5 \cdot 105$$
$$= 5 \cdot 5 \cdot 21$$
$$= 5 \cdot 5 \cdot 3 \cdot 7$$
$$= 3 \cdot 5^2 \cdot 7$$ ▲

We can see from the above examples the relationship between multiplication and factoring. Factoring is the reverse of multiplication.

Label each of the following numbers as *prime* or *not prime*. If a number is not prime, then list at least two of its factors besides itself and 1.

1. 48	**2.** 72	**3.** 37
4. 23	**5.** 33	**6.** 39
7. 221	**8.** 169	**9.** 29
10. 598	**11.** 156	**12.** 171
13. 53	**14.** 420	**15.** 3150
16. 1023	**17.** 1024	**18.** 43
19. 21	**20.** 543	

Factor the following into the product of primes. When the number has been factored completely, write its prime factors from smallest to largest.

21. 35	**22.** 70	**23.** 128
24. 256	**25.** 144	**26.** 288
27. 38	**28.** 63	**29.** 105
30. 210	**31.** 180	**32.** 900
33. 385	**34.** 1925	**35.** 121
36. 546	**37.** 420	**38.** 598
39. 620	**40.** 2310	

41. Write the number 6^2 without an exponent and then factor the result into the product of primes.

42. Write the expression $4^3 \cdot 9^2$ as a single number and then factor it into the product of primes.

43. Write $9^4 \cdot 16^2$ as the product of primes without actually multiplying it out first.

44. Write $10^2 \cdot 12^3$ as the product of primes without actually multiplying it out first.

Review Problems　The problems below review material we covered in Section 4.5. Reviewing these problems will help you understand the next section.

Multiply.

45. $3(x - 5)$	**46.** $4(x + 1)$
47. $5x^2(x - 3)$	**48.** $5x^2(x + 3)$
49. $4x^2y^3(2x - 3y)$	**50.** $2x^3y^2(3x + 2y)$
51. $6x^2(x^2 - 2x + 1)$	**52.** $3x^4(x^2 - x - 1)$

**5.2
The Greatest
Common Factor**

In this section we will apply the distributive property to polynomials to factor from them what is called the greatest common factor.

DEFINITION   The *greatest common factor* for a polynomial is the largest monomial that divides (is a factor of) each term of the polynomial.

We use the term *largest monomial* to mean the monomial with the greatest coefficient and highest power of the variable.

▼ **Example 1**   Find the greatest common factor for the polynomial $3x^5 + 12x^2$.

SOLUTION   The terms of the polynomial are $3x^5$ and $12x^2$. The largest number that divides the coefficients is 3, and the highest power of x that is a factor of x^5 and x^2 is x^2. The greatest common factor for $3x^5 + 12x^2$ is $3x^2$. That is, $3x^2$ is the largest monomial that divides each term of $3x^5 + 12x^2$.   ▲

▼ **Example 2**   Find the greatest common factor for $8a^3b^2 + 16a^2b^3 + 20a^3b^3$.

SOLUTION   The largest number that divides each of the coefficients is 4. The highest power of the variable that is a factor of a^3b^2, a^2b^3, and a^3b^3 is a^2b^2. The greatest common factor for $8a^3b^2 + 16a^2b^3 + 20a^3b^3$ is $4a^2b^2$. It is the largest monomial that is a factor of each term.   ▲

Once we have recognized the greatest common factor of a polynomial, we can apply the distributive property and factor it out of each term. We rewrite the polynomial as the product of its greatest common factor with the polynomial that remains after the greatest common factor has been factored from each term in the original polynomial.

▼ **Example 3**   Factor out the greatest common factor from $5x^3 - 15x^2$.

SOLUTION   The greatest common factor is $5x^2$. We rewrite the polynomial as

$$5x^3 - 15x^2 = 5x^2 \cdot x - 5x^2 \cdot 3$$

Then we apply the distributive property to get

$$5x^2 \cdot x - 5x^2 \cdot 3 = 5x^2(x - 3)$$

To check our work, we simply multiply $5x^2$ and $(x - 3)$ to get $5x^3 - 15x^2$, which is our original polynomial.   ▲

▼ **Example 4** Factor out the greatest common factor from $16x^5 - 20x^4 + 8x^3$.

SOLUTION The greatest common factor is $4x^3$. We rewrite the polynomial so we can see the greatest common factor $4x^3$ in each term; then we apply the distributive property to factor it out:

$$16x^5 - 20x^4 + 8x^3 = 4x^3 \cdot 4x^2 - 4x^3 \cdot 5x + 4x^3 \cdot 2$$
$$= 4x^3(4x^2 - 5x + 2)$$

Again, we could check our work by simply multiplying. ▲

▼ **Example 5** Factor out the greatest common factor for $6x^3y - 18x^2y^2 + 12xy^3$.

SOLUTION The greatest common factor is $6xy$. We rewrite the polynomial in terms of $6xy$ and then apply the distributive property as follows:

$$6x^3y - 18x^2y^2 + 12xy^3 = 6xy \cdot x^2 - 6xy \cdot 3xy + 6xy \cdot 2y^2$$
$$= 6xy(x^2 - 3xy + 2y^2)$$ ▲

▼ **Example 6** Factor out the greatest common factor from $3a^2b - 6a^3b^2 + 9a^3b^3$.

SOLUTION The greatest common factor is $3a^2b$.

$$3a^2b - 6a^3b^2 + 9a^3b^3 = 3a^2b(1) - 3a^2b(2ab) + 3a^2b(3ab^2)$$
$$= 3a^2b(1 - 2ab + 3ab^2)$$ ▲

Notice in Example 6 that we write the first term, $3a^2b$, as $3a^2b(1)$ before we apply the distributive property. It is a very common mistake when first factoring this type of problem to forget the 1 and write

$$3a^2b - 6a^3b^2 + 9a^3b^3 = 3a^2b(-2ab + 3ab^2)$$

The mistake is obvious when we multiply the right side and notice we get something different from what we started with.

We never need to make a mistake with factoring, since we can always multiply our results and compare them with our original polynomial. They must be identical.

Problem Set 5.2

Factor the following by taking out the greatest common factor.

1. $15x + 25$ **2.** $14x + 21$
3. $6a + 9$ **4.** $8a + 10$
5. $4x - 8y$ **6.** $9x - 12y$

7. $3x^2 - 6x - 9$

8. $2x^2 + 6x + 4$

9. $3a^2 - 3a - 60$

10. $2a^2 - 18a + 28$

11. $24y^2 - 52y + 24$

12. $18y^2 + 48y + 32$

13. $9x^2 - 8x^3$

14. $7x^3 - 4x^2$

15. $13a^2 - 26a^3$

16. $5a^2 - 10a^3$

17. $21x^2y - 28xy^2$

18. $30xy^2 - 25x^2y$

19. $22a^2b^2 - 11ab^2$

20. $15x^3 - 25x^2 + 30x$

21. $7x^3 + 21x^2 - 28x$

22. $16x^4 - 20x^2 - 16x$

23. $121y^4 - 11x^4$

24. $25a^4 - 5b^4$

25. $100x^4 - 50x^3 + 25x^2$

26. $36x^5 + 72x^3 - 81x^2$

27. $8a^2 + 16b^2 + 32c^2$

28. $-9a^2 - 18b^2 - 27c^2$

29. $-31x^5 - 28x^4 - 22x^3$

30. $-21x^6 - 13x^5 - 22x^3$

31. $4a^2b - 16ab^2 + 32a^2b^2$

32. $5ab^2 + 10a^2b^2 + 15a^2b$

33. $121a^3b^2 - 22a^2b^3 + 33a^3b^3$

34. $20a^4b^3 - 18a^3b^4 + 22a^4b^4$

35. $27rs^2t + 9r^2s^2t - 18rst^2$

36. $-38r^2st^2 + 28r^2s^2t^2 - 18r^2s^2t^3$

37. $12x^2y^3 - 72x^5y^3 - 36x^4y^4$

38. $49xy - 21x^2y^2 + 35x^3y^3$

39. The greatest common factor of the binomial $3x + 6$ is 3. The greatest common factor of the binomial $2x + 4$ is 2. What is the greatest common factor of their product, $(3x + 6)(2x + 4)$, when it has been multiplied out?

40. The greatest common factors of the binomials $4x + 2$ and $5x + 10$ are 2 and 5, respectively. What is the greatest common factor of their product, $(4x + 2)(5x + 10)$, when it has been multiplied out?

41. The following factorization is incorrect. Find the mistake and correct the right-hand side:

$$12x^2 + 6x + 3 = 3(4x^2 + 2x)$$

42. Find the mistake in the following factorization and then rewrite the right-hand side correctly.

$$10x^2 + 2x + 6 = 2(5x^2 + 3)$$

Review Problems The problems below review material we covered in Section 4.5. Reviewing these problems will help you with the next section.

Multiply using the FOIL method.

43. $(x + 3)(x + 4)$

44. $(x - 5)(x + 2)$

45. $(x + 7)(x - 2)$

46. $(x + 7)(x + 2)$

47. $(x - 7)(x + 2)$

48. $(x - 7)(x - 2)$

49. $(x - 3)(x + 2)$

50. $(x + 3)(x - 2)$

In this section we will limit the kind of trinomials we factor to those in which the coefficient of the squared term is 1. The more familiar we are with multiplication of binomials the easier factoring trinomials will be.

5.3 Factoring Trinomials

Recall multiplication of binomials from Chapter 4:

$$(x + 3)(x + 4) = x^2 + 7x + 12$$
$$(x - 5)(x + 2) = x^2 - 3x - 10$$

The first term in the answer is the product of the first terms in each binomial. The last term in the answer is the product of the last terms in each binomial. The middle term in the answer comes from adding the product of the outside terms with the product of the inside terms.

Lets let a and b represent real numbers and look at the product $(x + a)(x + b)$:

$$(x + a)(x + b) = x^2 + ax + bx + ab$$
$$= x^2 + (a + b)x + ab$$

The coefficient of the middle term is the sum of a and b. The last term is the product of a and b. Writing this as a factoring problem, we have

$$x^2 + \underset{\text{Sum}}{(a + b)}x + \underset{\text{Product}}{ab} = (x + a)(x + b)$$

To factor a trinomial in which the coefficient of x^2 is 1, we need only find the numbers a and b whose sum is the coefficient of the middle term and whose product is the constant term (last term).

▼ **Example 1**   Factor $x^2 + 8x + 12$.

SOLUTION   The coefficient of x^2 is 1. We need two numbers whose sum is 8 and whose product is 12. The numbers are 6 and 2:

$$x^2 + 8x + 12 = (x + 6)(x + 2)$$

We can easily check our work by multiplying $(x + 6)$ and $(x + 2)$:

$$(x + 6)(x + 2) = x^2 + 6x + 2x + 12$$
$$= x^2 + 8x + 12 \qquad ▲$$

▼ **Example 2**   Factor $x^2 - 2x - 15$.

SOLUTION   The coefficient of x^2 is again 1. We need to find a pair of numbers whose sum is -2 and whose product is -15. Here are ll the possibilities for products that are -15:

Products	Sums
$-1(15) = -15$	$-1 + 15 = 14$
$1(-15) = -15$	$1 + (-15) = -14$
$-5(3) = -15$	$-5 + 3 = -2$
$5(-3) = -15$	$5 + (-3) = 2$

The third line gives us what we want. The factors of $x^2 - 2x - 15$ are $(x - 5)$ and $(x + 3)$.

$$x^2 - 2x - 15 = (x - 5)(x + 3)$$

Again, we never need to make a mistake when factoring because we can always check our results by multiplying our factors to see if their product is the original polynomial:

$$(x - 5)(x + 3) = x^2 + 3x - 5x - 15$$
$$= x^2 - 2x - 15 \qquad \blacktriangle$$

▼ **Example 3**   Factor $2x^2 + 10x - 28$.

SOLUTION   The coefficient of x^2 is 2. We begin by factoring out the greatest common factor, which is 2:

$$2x^2 + 10x - 28 = 2(x^2 + 5x - 14)$$

Now we factor the remaining trinomial by finding a pair of numbers whose sum is 5 and whose product is -14. Here are the possibilities:

Products	Sums
$-1(14) = -14$	$-1 + 14 = 13$
$1(-14) = -14$	$1 + (-14) = -13$
$-7(2) = -14$	$-7 + 2 = -5$
$7(-2) = -14$	$7 + (-2) = 5$

The last line gives us the desired result. Our factors of $x^2 + 5x - 14$ are $(x + 7)$ and $(x - 2)$. Here is the complete problem:

$$2x^2 + 10x - 28 = 2(x^2 + 5x - 14)$$
$$= 2(x + 7)(x - 2) \qquad \blacktriangle$$

Note: In Example 3 we began by factoring out the greatest common factor. The first step in factoring any trinomial is to look for the greatest common factor. If the trinomial in question has as the greatest common factor a factor other than 1, we factor it out first and then try to factor the trinomial that remains.

▼ **Example 4**   Factor $3x^3 - 3x^2 - 18x$.

SOLUTION   We begin by factoring out the greatest common factor $3x$:

$$3x^3 - 3x^2 - 18x = 3x(x^2 - x - 6)$$

We now find a pair of numbers whose product is -6 and whose sum is -1. [The coefficient of the middle term is -1, since $-x = -1(x)$.]

Products	Sums
$-1(6) = -6$	$-1 + 6 = 5$
$1(-6) = -6$	$1 + (-6) = -5$
$-2(3) = -6$	$-2 + 3 = 1$
$2(-3) = -6$	$2 + (-3) = -1$

The last line gives us the pair we are looking for. The factors of $x^2 - x - 6$ are $(x + 2)$ and $(x - 3)$:

$$3x^3 - 3x^2 - 18x = 3x(x^2 - x - 6)$$
$$= 3x(x - 3)(x + 2) \qquad \blacktriangle$$

▼ **Example 5**   Factor $x^2 + 8xy + 12y^2$

SOLUTION   This time we need two expressions whose product is $12y^2$ and whose sum is $8y$. The two expresions are $6y$ and $2y$ (see Example 1 in this section):

$$x^2 + 8xy + 12y^2 = (x + 6y)(x + 2y)$$

You should convince yourself that these factors are correct by finding their product.   ▲

   Trinomials in which the coefficient of the second-degree term is 1 are the easiest to factor. Success in factoring any type of polynomial is directly related to the amount of time spent working problems. The more we practice, the more accomplished we become at factoring.

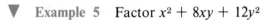

Problem Set 5.3

Factor the following trinomials.

1. $x^2 + 7x + 12$ 2. $x^2 + 7x + 10$
3. $x^2 + 3x + 2$ 4. $x^2 + 7x + 6$
5. $a^2 + 10a + 21$ 6. $a^2 - 7a + 12$
7. $x^2 - 7x + 10$ 8. $x^2 - 3x + 2$
9. $y^2 - 10y + 21$ 10. $y^2 - 7y + 6$
11. $x^2 - x - 12$ 12. $x^2 - 4x - 5$
13. $y^2 + y - 12$ 14. $y^2 + 3y - 18$
15. $x^2 + 5x - 14$ 16. $x^2 - 5x - 24$
17. $r^2 - 8r - 9$ 18. $r^2 - r - 2$

19. $x^2 - x - 30$
20. $x^2 + 8x + 12$
21. $a^2 + 15a + 56$
22. $a^2 - 9a + 20$
23. $y^2 - y - 42$
24. $y^2 + y - 42$
25. $x^2 + 13x + 42$
26. $x^2 - 13x + 42$

Factor the following completely. First factor out the greatest common factor, then factor the remaining trinomial.

27. $2x^2 + 6x + 4$
28. $3x^2 - 6x - 9$
29. $3a^2 - 3a - 60$
30. $2a^2 - 18a + 28$
31. $4x^3 + 16x^2 - 48x$
32. $2x^3 + 12x^2 + 18x$
33. $a^3 - 5a^2 + 6a$
34. $a^3 - 5a^2 - 6a$
35. $x^4 - x^3 - 12x^2$
36. $x^4 - 11x^3 + 24x^2$
37. $2r^3 + 4r^2 - 30r$
38. $5r^3 + 45r^2 + 100r$
39. $2y^4 - 6y^3 - 8y^2$
40. $3r^3 - 3r^2 - 6r$
41. $x^5 + 4x^4 + 4x^3$
42. $x^5 + 13x^4 + 42x^3$
43. $3y^4 - 12y^3 - 15y^2$
44. $5y^4 - 10y^3 + 5y^2$
45. $4x^4 - 52x^3 + 144x^2$
46. $3x^3 - 3x^2 - 18x$

Factor the following trinomials.

47. $x^2 + 5xy + 6y^2$
48. $x^2 - 5xy + 6y^2$
49. $x^2 - 9xy + 20y^2$
50. $x^2 + 9xy + 20y^2$
51. $a^2 + 2ab - 8b^2$
52. $a^2 - 2ab - 8b^2$
53. $a^2 - 10ab + 25b^2$
54. $a^2 + 6ab + 9b^2$
55. $a^2 + 10ab + 25b^2$
56. $a^2 - 6ab + 9b^2$
57. $x^2 + 2xa - 48a^2$
58. $x^2 - 3xa - 10a^2$
59. $x^2 - 5xb - 36b^2$
60. $x^2 - 13xb + 36b^2$
61. If one of the factors of $x^2 + 24x + 128$ is $x + 8$, what is the other factor?
62. If one factor of $x^2 + 260x + 2500$ is $x + 10$, what is the other factor?
63. What polynomial, when factored, gives $(4x + 3)(x - 1)$?
64. What polynomial factors to $(4x - 3)(x + 1)$?

Review Problems The problems below review material we covered in Section 4.5. Reviewing these problems will help you in the next section.

Multiply using the FOIL method.

65. $(6a + 1)(q + 2)$
66. $(6a - 1)(a - 2)$
67. $(3a + 2)(2a + 1)$
68. $(3a - 2)(2a - 1)$
69. $(6a + 2)(a + 1)$
70. $(3a + 1)(2a + 2)$

We will now consider trinomials whose greatest common factor is 1 and whose leading coefficient (the coefficient of the squared term) is a number other than one. The methods of the last section no longer apply. The factors are, in many cases, not as easy to find.

**5.4
More Trinomials
to Factor**

Suppose we want to factor the trinomial $2x^2 - 5x - 3$. We know the factors will be a pair of binomials. The product of their first terms is $2x^2$ and the product of their last terms is -3. Let us list all the possible factors along with the trinomial that would result if we were to multiply them together. Remember, the middle term comes from the product of the inside terms plus the product of the outside terms.

Binomial Factors	First Term	Middle Term	Last Term
$(2x - 3)(x + 1)$	$2x^2$	$-x$	-3
$(2x + 3)(x - 1)$	$2x^2$	$+x$	-3
$(2x - 1)(x + 3)$	$2x^2$	$+5x$	-3
$(2x + 1)(x - 3)$	$2x^2$	$-5x$	-3

We can see from the last line that the factors of $2x^2 - 5x - 3$ are $(2x + 1)(x - 3)$. There is no straightforward way, as there was in the last section, in which to find the factors other than by trial and error or by simply listing all the possibilities. The idea is to be sure that the first and last terms work out and then to look for combinations that will make the middle term work out. We look for possible factors that, when multiplied, will give the correct first and last terms, and then we see if we can adjust them to give the correct middle term.

▼   **Example 1**   Factor $6a^2 + 7a + 2$.

SOLUTION   We list all the possible pairs of factors that, when multiplied together, give a trinomial whose first term is $6a^2$ and whose last term is $+2$:

Binomial Factors	First Term	Middle Term	Last Term
$(6a + 1)(a + 2)$	$6a^2$	$+13a$	$+2$
$(6a - 1)(a - 2)$	$6a^2$	$-13a$	$+2$
$(3a + 2)(2a + 1)$	$6a^2$	$+7a$	$+2$
$(3a - 2)(2a - 1)$	$6a^2$	$-7a$	$+2$

The factors of $6a^2 + 7a + 2$ are $(3a + 2)$ and $(2a + 1)$:

$$6a^2 + 7a + 2 = (3a + 2)(2a + 1)$$ ▲

Notice in the above list that we did not include the factors $(6a + 2)$ and $(a + 1)$. We do not need to try these, since the first factor has a 2 common to each term and so could be factored again, giving $2(3a + 1)(a + 1)$. Since our original trinomial, $6a^2 + 7a + 2$, did *not* have a greatest common factor of 2, neither of its factors will. We need to consider only factors that cannot themselves be factored further.

▼ **Example 2** Factor $4x^2 - x - 3$.

SOLUTION We list all the possible factors that, when multiplied, give a trinomial whose first term is $4x^2$ and whose last term is -3:

Binomial Factors	First Term	Middle Term	Last Term
$(4x + 1)(x - 3)$	$4x^2$	$-11x$	-3
$(4x - 1)(x + 3)$	$4x^2$	$+11x$	-3
$(4x + 3)(x - 1)$	$4x^2$	$-x$	-3
$(4x - 3)(x + 1)$	$4x^2$	$+x$	-3
$(2x + 1)(2x - 3)$	$4x^2$	$-4x$	-3
$(2x - 1)(2x + 3)$	$4x^2$	$+4x$	-3

The third line shows that the factors of $4x^2 - x - 3$ are $(4x + 3)$ and $(x - 1)$.

$$4x^2 - x - 3 = (4x + 3)(x - 1)$$ ▲

You will find that the more practice you have at factoring this type of trinomial, the faster you will get the correct factors. You will pick up some shortcuts along the way or maybe come across a system of eliminating some factors as possibilities. Whatever works best for you is the method you should use. Factoring is a very important tool and you must be good at it.

▼ **Example 3** Factor $12y^3 + 10y^2 - 12y$.

SOLUTION We begin by factoring out the greatest common factor, $2y$.

$$12y^3 + 10y^2 - 12y = 2y(6y^2 + 5y - 6)$$

We now list all possible factors of a trinomial with the first term $6y^2$ and last term -6, along with the associated middle terms:

Possible Factors	Middle Term when Multiplied
$(3y + 2)(2y - 3)$	$-5y$
$(3y - 2)(2y + 3)$	$+5y$
$(6y + 1)(y - 6)$	$-35y$
$(6y - 1)(y + 6)$	$+35y$

The second line gives the correct factors. The complete problem is

$$12y^3 + 10y^2 - 12y = 2y(6y^2 + 5y - 6)$$
$$= 2y(3y - 2)(2y + 3) \quad \blacktriangle$$

▼ **Example 4** Factor $30x^2y - 5xy^2 - 10y^3$.

SOLUTION The greatest common factor is $5y$:

$$30x^2y - 5xy^2 - 10y^3 = 5y(6x^2 - xy - 2y^2)$$

A list of possible factors with the first term $6x^2$ and last term $-2y^2$ and the associated middle terms looks like this:

Possible Factors	Middle Term when Multiplied
$(2x - y)(3x + 2y)$	$+xy$
$(2x + y)(3x - 2y)$	$-xy$
$(6x + y)(x - 2y)$	$-11xy$
$(6x - y)(x + 2y)$	$+11xy$

The factors of $6x^2 - xy - 2y^2$ are $(2x + y)$ and $(3x - 2y)$:

$$30x^2y - 5xy^2 - 10y^3 = 5y(6x^2 - xy - 2y^2)$$
$$= 5y(2x + y)(3x - 2y) \quad \blacktriangle$$

Remember, we never need to make a mistake in factoring, for we can always check our results by multiplying the factors and comparing the results with our original polynomial.

Problem Set 5.4

Factor the following trinomials.

1.	$2x^2 + 7x + 3$	**2.**	$2x^2 + 5x + 3$
3.	$2a^2 - a - 3$	**4.**	$2a^2 + a - 3$
5.	$3x^2 + 2x - 5$	**6.**	$3x^2 - 2x - 5$
7.	$3y^2 - 14y - 5$	**8.**	$3y^2 + 14y - 5$
9.	$6x^2 + 13x + 6$	**10.**	$6x^2 - 13x + 6$
11.	$4x^2 - 12xy + 9y^2$	**12.**	$4x^2 + 12xy + 9y^2$
13.	$4y^2 - 11y - 3$	**14.**	$4y^2 + y - 3$
15.	$20x^2 - 41x + 20$	**16.**	$20x^2 + 9x - 20$
17.	$20a^2 + 48ab - 5b^2$	**18.**	$20a^2 + 29ab + 5b^2$
19.	$20x^2 - 21x - 5$	**20.**	$20x^2 - 48x - 5$
21.	$12m^2 + 16m - 3$	**22.**	$12m^2 + 20m + 3$
23.	$20x^2 + 37x + 15$	**24.**	$20x^2 + 13x - 15$
25.	$12a^2 - 25ab + 12b^2$	**26.**	$12a^2 + 7ab - 12b^2$

27. $3x^2 - xy - 14y^2$
28. $3x^2 + 19xy - 14y^2$
29. $14x^2 + 29x - 15$
30. $14x^2 + 11x - 15$

Factor each of the following completely. Look first for the greatest common factor.

31. $4x^2 + 2x - 6$
32. $6x^2 - 51x + 63$
33. $24a^2 - 50a + 24$
34. $18a^2 + 48a + 32$
35. $10x^3 - 23x^2 + 12x$
36. $10x^4 + 7x^3 - 12x^2$
37. $6x^4 - 11x^3 - 10x^2$
38. $6x^3 + 19x^2 + 10x$
39. $10a^3 - 6a^2 - 4a$
40. $6a^3 + 15a^2 + 9a$
41. $15x^3 - 102x^2 - 21x$
42. $2x^4 - 24x^3 + 14x^2$
43. $35y^3 - 60y^2 - 20y$
44. $14y^4 - 32y^3 + 8y^2$
45. $15a^4 - 2a^3 - a^2$
46. $10a^5 - 17a^4 + 3a^3$
47. $24x^2y - 6xy - 45y$
48. $8x^2y^2 + 26xy^2 + 15y^2$
49. $12x^2y - 34xy^2 + 14y^3$
50. $12x^2y - 46xy^2 + 14y^3$
51. Evaluate the expression $2x^2 + 7x + 3$ and the expression $(2x + 1)(x + 3)$ for $x = 2$.
52. Evaluate the expression $2a^2 - a - 3$ and the expression $(2a - 3)(a + 1)$ for $a = 5$.
53. What polynomial factors to $(2x + 3)(2x - 3)$?
54. What polynomial factors to $(x + 2)(x - 2)(x^2 + 4)$?

Review Problems   The following problems review material we covered in Section 4.6. Reviewing these problems will help you in the next section.

Multiply the following expressions.

55. $(x + 3)(x - 3)$
56. $(x + 5)(x - 5)$
57. $(6x + 1)(6x - 1)$
58. $(4a + 5)(4a - 5)$
59. $(x + 4)^2$
60. $(x - 5)^2$
61. $(2x + 3)^2$
62. $(2x - 3)^2$

In Chapter 4 we listed the following three special products:

$$(a + b)^2 = (a + b)(a + b) = a^2 + 2ab + b^2$$
$$(a - b)^2 = (a - b)(a - b) = a^2 - 2ab + b^2$$
$$(a + b)(a - b) = a^2 - b^2$$

**5.5
The Difference of
Two Squares**

Since factoring is the reverse of multiplication, we can also consider the three special products as three special factorizations:

$$a^2 + 2ab + b^2 = (a + b)^2$$
$$a^2 - 2ab + b^2 = (a - b)^2$$
$$a^2 - b^2 = (a + b)(a - b)$$

Any trinomial of the form $a^2 + 2ab + b^2$ or $a^2 - 2ab + b^2$ can be factored by the methods of Section 5.4. The last line is the factorization for the difference of two squares. The difference of two squares always factors in this way. Again, these are patterns to be recognized on sight.

▼ **Example 1** Factor $16x^2 - 25$.

SOLUTION We can see that the first term is a perfect square and the last term is also. This fact becomes even more obvious if we rewrite the problem as

$$16x^2 - 25 = (4x)^2 - (5)^2$$

The first term is the square of the quantity $4x$ and the last term is the square of 5. The completed problem looks like this:

$$16x^2 - 25 = (4x)^2 - (5)^2$$
$$= (4x + 5)(4x - 5)$$

To check our results we multiply:

$$(4x + 5)(4x - 5) = 16x^2 + 20x - 20x - 25$$
$$= 16x^2 - 25 \qquad ▲$$

▼ **Example 2** Factor $36a^2 - 1$.

SOLUTION We rewrite the two terms to show they are perfect squares and then factor. Remember, 1 is its own square, $1^2 = 1$.

$$36a^2 - 1 = (6a)^2 - (1)^2$$
$$= (6a + 1)(6a - 1)$$

To check our results we multiply:

$$(6a + 1)(6a - 1) = 36a^2 + 6a - 6a - 1$$
$$= 36a^2 - 1 \qquad ▲$$

▼ **Example 3** Factor $x^4 - y^4$.

SOLUTION x^4 is the perfect square $(x^2)^2$, and y^4 is $(y^2)^2$.

$$x^4 - y^4 = (x^2)^2 - (y^2)^2$$
$$= (x^2 - y^2)(x^2 + y^2)$$

The factor $(x^2 - y^2)$ is itself the difference of two squares and therefore can be factored again. The factor $(x^2 + y^2)$ is the *sum* of two squares and cannot be factored again. The complete problem is this:

$$x^4 - y^4 = (x^2)^2 - (y^2)^2$$
$$= (x^2 - y^2)(x^2 + y^2)$$
$$= (x + y)(x - y)(x^2 + y^2) \qquad \blacktriangle$$

Note: If you think the sum of two squares, $x^2 + y^2$, factors, you should try it. Write down the factors you think it has and them multiply them using the FOIL method. You won't get $x^2 + y^2$.

▼ **Example 4** Factor $25x^2 - 60x + 36$.

SOLUTION Although this trinomial can be factored by the method we used in Section 5.4, we notice that the first and last terms are the perfect squares $(5x)^2$ and $(6)^2$. Before going through the method for factoring trinomials by listing all possible factors, we can check to see if $25x^2 - 60x + 36$ factors to $(5x - 6)^2$. We need only multiply to check:

$$(5x - 6)^2 = (5x - 6)(5x - 6)$$
$$= 25x^2 - 30x - 30x + 36$$
$$= 25x^2 - 60x + 36$$

The trinomial $25x^2 - 60x + 36$ factors to $(5x - 6)(5x - 6) = (5x - 6)^2$. $\qquad \blacktriangle$

▼ **Example 5** Factor $m^2 + 14m + 49$.

SOLUTION Since the first and last terms are perfect squares, we can try the factors $(m + 7)(m + 7)$:

$$(m + 7)^2 = (m + 7)(m + 7)$$
$$= m^2 + 7m + 7m + 49$$
$$= m^2 + 14m + 49$$

The factors of $m^2 + 14m + 49$ are $(m + 7)(m + 7) = (m + 7)^2$. $\qquad \blacktriangle$

Factor the following. Problem Set 5.5

1. $x^2 - 9$
2. $x^2 - 25$
3. $a^2 - 36$
4. $a^2 - 64$
5. $x^2 - 49$
6. $x^2 - 121$
7. $4a^2 - 16$
8. $4a^2 + 16$
9. $9x^2 + 25$
10. $16x^2 - 36$
11. $25x^2 - 169$
12. $x^2 - y^2$
13. $9a^2 - 16b^2$
14. $49a^2 - 25b^2$

15. $9 - m^2$

16. $16 - m^2$

17. $25 - 4x^2$

18. $36 - 49y^2$

19. $2x^2 - 18$

20. $3x^2 - 27$

21. $32a^2 - 128$

22. $3a^3 - 48a$

23. $8x^2y - 18y$

24. $50a^2b - 72b$

25. $a^4 - b^4$

26. $a^4 - 16$

27. $16m^4 - 81$

28. $81 - m^4$

29. $3x^3y - 75xy^3$

30. $2xy^3 - 8x^3y$

Factor the following.

31. $x^2 - 2x + 1$

32. $x^2 - 6x + 9$

33. $x^2 + 2x + 1$

34. $x^2 + 6x + 9$

35. $a^2 - 10a + 25$

36. $a^2 + 10a + 25$

37. $y^2 + 4y + 4$

38. $y^2 - 8y + 16$

39. $x^2 - 4x + 4$

40. $x^2 + 8x + 16$

41. $m^2 - 12m + 36$

42. $m^2 + 12m + 36$

43. $4a^2 + 12a + 9$

44. $9a^2 - 12a + 4$

45. $49x^2 - 14x + 1$

46. $64x^2 - 16x + 1$

47. $9y^2 - 30y + 25$

48. $25y^2 + 30y + 9$

49. $x^2 + 10xy + 25y^2$

50. $25x^2 + 10xy + y^2$

51. $9a^2 + 6ab + b^2$

52. $9a^2 - 6ab + b^2$

Factor the following by first factoring out the greatest common factor.

53. $3a^2 + 18a + 27$

54. $4a^2 - 16a + 16$

55. $2x^2 + 20xy + 50y^2$

56. $3x^2 + 30xy + 75y^2$

57. $5x^3 + 30x^2y + 45xy^2$

58. $12x^2y - 36xy^2 + 27y^3$

59. Find a value for b so that the polynomial $x^2 + bx + 49$ factors to $(x + 7)^2$.

60. Find a value of b so that the polynomial $x^2 + bx + 81$ factors to $(x + 9)^2$.

61. Find the value of c for which the polynomial $x^2 + 10x + c$ factors to $(x + 5)^2$.

62. Find the value of a for which the polynomial $ax^2 + 12x + 9$ factors to $(2x + 3)^2$.

Review Problems The following problems review material we covered in Sections 2.2 and 2.3. Reviewing these problems will help you understand the next section.

Solve each equation.

63. $x + 5 = 8$

64. $x - 6 = -3$

65. $x + 2 = 0$

66. $x - 4 = 0$

67. $3x - 6 = 9$

68. $5x - 1 = 14$

69. $2x + 3 = 0$

70. $4x - 5 = 0$

In this section we will use the methods of factoring developed in the last three sections, along with a special property of 0, to solve quadratic equations.

DEFINITION   Any equation that can be put in the form $ax^2 + bx + c = 0$, where a, b, and c are real numbers ($a \neq 0$), is called a *quadratic equation*. $ax^2 + bx + c = 0$ is called *standard form* for a quadratic equation:

$$\text{an } x^2 \text{ term} \quad \text{an } x \text{ term} \quad \text{and a constant term}$$
$$a(\text{variable})^2 + b(\text{variable}) + (\text{absence of the variable}) = 0$$

The number zero has a special property. If we multiply two numbers and the product is 0, then one or both of the original two numbers must be 0. In symbols, this property looks like this:

Special Property of 0

Let a and b represent real numbers. If $a \cdot b = 0$, then
$a = 0$ or $b = 0$

Suppose we want to solve the quadratic equation $x^2 + 5x + 6 = 0$. We can factor the left side into $(x + 2)(x + 3)$. Then we have

$$x^2 + 5x + 6 = 0$$
$$(x + 2)(x + 3) = 0$$

Now $(x + 2)$ and $(x + 3)$ both represent real numbers. Their product is 0; therefore, either $(x + 3)$ is 0 or $(x + 2)$ is 0. Either way we have a solution to our equation. We use the property of zero stated above to finish the problem:

$$x^2 + 5x + 6 = 0$$
$$(x + 2)(x + 3) = 0$$
$$x + 2 = 0 \quad \text{or} \quad x + 3 = 0$$
$$x = -2 \quad \text{or} \quad x = -3$$

Our solution set is $\{-2, -3\}$. Our equation has two solutions. To check our solutions we have to check each one separately to see that they both produce a true statement when used in place of the variable:

When $x = -3$
the equation $x^2 + 5x + 6 = 0$
becomes $(-3)^2 + 5(-3) + 6 = 0$
                      $9 + (-15) + 6 = 0$
                                $0 = 0$ True statement

When $x = -2$
the equation $x^2 + 5x + 6 = 0$
becomes $(-2)^2 + 5(-2) + 6 = 0$
                              $4 + (-10) + 6 = 0$
                                              $0 = 0$ True statement

We have solved a quadratic equation by replacing it with two linear equations in one variable. Here are some steps to follow when solving quadratic equations.

Step 1: Put the equation in standard form, that is, 0 on one side and decreasing powers of the variable on the other.

Step 2: Factor completely.

Step 3: Use the special property of 0 to set each factor from Step 2 to 0; i.e., $a \cdot b = 0$ implies $a = 0$ or $b = 0$.

Step 4: Solve each equation produced in Step 3.

Step 5: Check each solution, if necessary.

▼ **Example 1** Solve the equation $2x^2 - 5x = 12$.

SOLUTION

Step 1: We begin by adding -12 to both sides, so the equation is in standard form:

$$2x^2 - 5x = 12$$
$$2x^2 - 5x - 12 = 0$$

Step 2: We factor the left side completely:

$$(2x + 3)(x - 4) = 0$$

Step 3: We set each factor to 0:

$$2x + 3 = 0 \quad \text{or} \quad x - 4 = 0$$

Step 4: Solve each of the equations from Step 3:

$$2x + 3 = 0 \qquad x - 4 = 0$$
$$2x = -3 \qquad\quad x = 4$$
$$x = -\tfrac{3}{2}$$

The solution set is $\{-\tfrac{3}{2}, 4\}$.

Step 5: Substitute each solution into $2x^2 - 5x = 12$ to check:

Check $-\tfrac{3}{2}$ Check 4

$2(-\tfrac{3}{2})^2 - 5(-\tfrac{3}{2}) = 12$ $2(4)^2 - 5(4) = 12$
$2(\tfrac{9}{4}) + 5(\tfrac{3}{2}) = 12$ $2(16) - 20 = 12$

$$\tfrac{9}{2} + \tfrac{15}{2} = 12 \qquad\qquad 32 - 20 = 12$$
$$\tfrac{24}{2} = 12 \qquad\qquad 12 = 12$$
$$12 = 12 \qquad\qquad\qquad \blacktriangle$$

▼ **Example 2**   Find the solution set for $16a^2 - 25 = 0$.

SOLUTION   The equation is already in standard form, so we begin by factoring:

$$16a^2 - 25 = 0$$

$(4a - 5)(4a + 5) = 0$	Factor left side
$4a - 5 = 0 \quad \text{or} \quad 4a + 5 = 0$	Set each factor to 0
$4a = 5 \qquad\qquad 4a = -5$	Solve the resulting equations
$a = \tfrac{5}{4} \qquad\qquad a = -\tfrac{5}{4}$	

The solution set is $\{\tfrac{5}{4}, -\tfrac{5}{4}\}$. We can check the solutions by substituting $\tfrac{5}{4}$ and $-\tfrac{5}{4}$ for a in the original equation.   ▲

▼ **Example 3**   Solve $24x^3 = -10x^2 + 6x$ for x.

SOLUTION

$24x^3 + 10x^2 - 6x = 0$	Standard form
$2x(12x^2 + 5x - 3) = 0$	Factor out $2x$
$2x(3x - 1)(4x + 3) = 0$	Factor remaining trinomial
$2x = 0 \quad \text{or} \quad 3x - 1 = 0 \quad \text{or} \quad 4x + 3 = 0$	Set factors to 0
$x = 0 \quad \text{or} \qquad x = \tfrac{1}{3} \quad \text{or} \qquad x = -\tfrac{3}{4}$	Solutions

The solution set is $\{0, \tfrac{1}{3}, -\tfrac{3}{4}\}$.   ▲

Notice that in order to solve a quadratic equation by this method, it must be possible to factor it. We will learn how to solve quadratic equations that do not factor in Chapter 8.

The following quadratic equations are already in factored form. Use the special property with zero to set the factors to zero and solve.

1. $(x + 2)(x - 1) = 0$
2. $(x + 3)(x + 2) = 0$
3. $(a - 4)(a - 5) = 0$
4. $(a + 6)(a - 1) = 0$
5. $x(x + 1)(x - 3) = 0$
6. $x(2x + 1)(x - 5) = 0$
7. $(3x + 2)(2x + 3) = 0$
8. $(4x - 5)(x - 6) = 0$
9. $m(3m + 4)(3m - 4) = 0$
10. $m(2m - 5)(3m - 1) = 0$
11. $2y(3y + 1)(5y + 3) = 0$
12. $3y(2y - 3)(3y - 4) = 0$

Solve the following quadratic equations.

13. $x^2 + 3x + 2 = 0$ 14. $x^2 - x - 6 = 0$
15. $x^2 - 9x + 20 = 0$ 16. $x^2 + 2x - 3 = 0$
17. $a^2 - 2a - 24 = 0$ 18. $a^2 - 11a + 30 = 0$
19. $m^2 - 10m = -9$ 20. $m^2 + m = 42$
21. $x^2 = -6x - 9$ 22. $x^2 = 10x - 25$
23. $a^2 - 16 = 0$ 24. $a^2 - 36 = 0$
25. $2x^2 + 5x - 12 = 0$ 26. $3x^2 + 14x - 5 = 0$
27. $9x^2 + 12x + 4 = 0$ 28. $12x^2 - 24x + 9 = 0$
29. $2x^2 = 3x + 20$ 30. $6x^2 = x + 2$
31. $3m^2 = 20 - 7m$ 32. $2m^2 = -18 + 15m$
33. $4x^2 - 49 = 0$ 34. $16x^2 - 25 = 0$
35. $x^2 - 3x = 0$ 36. $x^2 + 5x = 0$
37. $18r^2 - 50 = 0$ 38. $27r^2 - 75 = 0$
39. $6x^2 = -5x + 4$ 40. $9x^2 = 12x - 4$
41. $4x^2 = 8x$ 42. $3x^2 = -9x$
43. $4x^2 - 2x - 30 = 0$ 44. $9x^2 + 6x - 24 = 0$
45. $5x^4 + 34x^3 = 7x^2$ 46. $6x^3 = -x^2 + 5x$
47. $8y^2 + 16y = 10$ 48. $24y^2 - 22y = -4$
49. $20x^3 = -18x^2 + 18x$ 50. $60x^4 = 122x^3 - 60x^2$

Review Problems   The following problems review material we covered in Section 2.5. They are taken from the book *Academic Algebra,* written by William J. Milne and published by the American Book Company in 1901. The hint in parentheses is mine.

51. A bicycle and a suit cost $90. How much did each cost, if the bicycle cost five times as much as the suit. (Hint: Let $x =$ the cost of the suit; then the cost of the bicycle is $5x$.)
52. A man bought a cow and a calf for $36, paying eight times as much for the cow as for the calf. What was the cost of each?
53. A house and a lot cost $3000. If the house cost four times as much as the lot, what was the cost of each?
54. A plumber and two helpers together earned $7.50 per day. How much did each earn per day, if the plumber earned four times as much as each helper.

5.7
Word Problems

We will use the same four steps in solving word problems in this section that we have used in the past. The equation that describes the situation will turn out to be a quadratic equation.

▼ **Example 1**   The product of two consecutive odd integers is 63. Find the integers.

SOLUTION

Step 1:   Let x = the first odd integer, then $x + 2$ = the second odd integer.

Step 2:   An equation that describes the situation is

$$x(x + 2) = 63 \qquad \text{(Their product is 63.)}$$

Step 3:   We solve the equation by the method developed in Section 5.6.

$$x(x + 2) = 63$$
$$x^2 + 2x = 63$$
$$x^2 + 2x - 63 = 0$$
$$(x - 7)(x + 9) = 0$$
$$x - 7 = 0 \quad \text{or} \quad x + 9 = 0$$
$$x = 7 \quad \text{or} \qquad x = -9$$

If the first odd integer is 7, the next odd integer is $7 + 2 = 9$. If the first odd integer is -9, the next consecutive odd integer is $-9 + 2 = -7$. We have two pairs of consecutive odd integers that are solutions. They are 7, 9 and -9, -7.

Step 4:   We check to see that their products are 63:

$$7(9) = 63$$
$$-7(-9) = 63 \qquad\qquad ▲$$

Many word problems dealing with area can best be described algebraically by quadratic equations.

The area of a rectangle is length times width:

Area = length · width
$A = L \cdot W$

The area of a triangle is one-half the product of the base and height:

Area = $\frac{1}{2}$(base · height)
$A = \frac{1}{2}(b \cdot h)$

▼   **Example 2**   The length of a rectangle is three more than twice the width. The area is 44 square inches (sq. in.). Find the dimensions (find the length and width.)

SOLUTION   Let x = the width of the rectangle. Then $2x + 3 =$ the length of the rectangle, because the length is three more than twice the width.

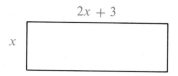

Since the area is 44 sq. in., an equation that describes the situation is

$$x(2x + 3) = 44 \qquad \text{(Length} \cdot \text{width} = \text{area)}$$

We now solve the equation:

$$x(2x + 3) = 44$$
$$2x^2 + 3x = 44$$
$$2x^2 + 3x - 44 = 0$$
$$(2x + 11)(x - 4) = 0$$
$$2x + 11 = 0 \quad \text{or} \quad x - 4 = 0$$
$$x = -\tfrac{11}{2} \quad \text{or} \qquad x = 4$$

The solution $x = -\tfrac{11}{2}$ cannot be used, since length and width are always given in positive units. The width is 4. The length is three more than twice the width or $2(4) + 3 = 11$.

$$\text{Width} = 4 \text{ inches}$$
$$\text{Length} = 11 \text{ inches}$$

The solutions check in the original problem, since $4(11) = 44$. ▲

▼   **Example 3**   The area of a square is twice its perimeter. What is the length of its side?

SOLUTION   Let x = the length of its side. Then x^2 = the area of the square and $4x$ = the perimeter of the square:

An equation that describes the situation is

$$x^2 = 2(4x) \qquad \text{The area is 2 times}$$
$$x^2 = 8x \qquad\qquad \text{the perimeter}$$

$$x^2 - 8x = 0$$
$$x(x - 8) = 0$$
$$x = 0 \quad \text{or} \quad x = 8$$

Since $x = 0$ does not make sense in our original problem, we use $x = 8$. If the side has length 8, then the perimeter is $4(8) = 32$ and the area is $8^2 = 64$. Since 64 is twice 32, our solution is correct. ▲

Another application of quadratic equations involves the Pythagorean theorem, an important theorem from geometry. This theorem gives the relationship between the sides of any right triangle. A right triangle is a triangle in which one of the angles is 90°. We will state the Pythagorean theorem here, but we will not prove it.

Pythagorean Theorem

In any right triangle, the square of the longest side (called the hypotenuse) is equal to the sum of the squares of the other two sides (called legs).

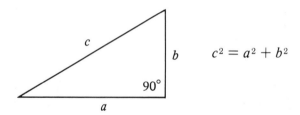

$$c^2 = a^2 + b^2$$

▼ **Example 4** The three sides of a right triangle are three consecutive integers. Find the lengths of the three sides.

SOLUTION Let $x =$ the first integer (shortest side)
then $\quad x + 1 =$ the next consecutive integer
and $\quad x + 2 =$ the last consecutive integer (longest side)

A diagram of the triangle looks like this:

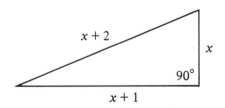

The Pythagorean theorem tells us that the square of the longest side, $(x + 2)^2$, is equal to the sum of the squares of the two shorter sides, $(x + 1)^2 + x^2$. Here is the equation:

$$
\begin{aligned}
(x + 2)^2 &= (x + 1)^2 + x^2 & \\
x^2 + 4x + 4 &= x^2 + 2x + 1 + x^2 & \text{Expand squares} \\
x^2 - 2x - 3 &= 0 & \text{Standard form} \\
(x - 3)(x + 1) &= 0 & \text{Factor} \\
x - 3 = 0 \quad &\text{or} \quad x + 1 = 0 & \text{Set factors to 0} \\
x = 3 \quad &\text{or} \quad\quad x = -1 &
\end{aligned}
$$

Since a triangle cannot have a side with a negative number for its length, we must not use -1 for a solution to our original problem. Therefore the shortest side is 3. The other two sides are the next two consecutive integers, 4 and 5.   ▲

Problem Set 5.7

Solve the following word problems. Be sure to show the equation used.

1. The product of two consecutive even integers is 80. Find the two integers.
2. The product of two consecutive integers is 72. Find the two integers.
3. The product of two consecutive odd integers is 99. Find the two integers.
4. The product of two consecutive integers is 132. Find the integers.
5. The product of two consecutive even integers is ten less than five times their sum. Find the integers.
6. The product of two consecutive odd integers is one less than four times their sum. Find the two integers.
7. The sum of two numbers is -14. Their product is 48. Find the two numbers.
8. The sum of two numbers is 12. Their product is 32. Find the numbers.
9. One number is two more than five times another. Their product is 24. Find the numbers.
10. One number is one more than twice another. Their product is 55. Find the numbers.
11. One number is four times another. Their product is four times their sum. Find the numbers.
12. One number is two more than twice another. Their product is two more than twice their sum. Find the numbers.
13. The length of a rectangle is one more than the width. The area is 12 sq. in. Find the dimensions.
14. The length of a rectangle is three more than twice the width. The area is 44 sq. in. Find the dimensions.
15. The numerical value of the area of a rectangle is 11 more than the numerical value of the perimeter. Find the width, if the length is two more than the width.

16. The numerical value of the area of a rectangle is 19 more than two times the perimeter. Find the width, if the length is two more than the width.
17. The height of a triangle is twice the base. The area is 9 sq. in. Find the base.
18. The height of a triangle is two more than twice the base. The area is 20 sq. ft. Find the base.
19. The numerical value of the area of a rectangle is twice the length. The length is one more than three times the width. Find the dimensions.
20. The area of a rectangle is three times the length. The length is one more than twice the width. Find the dimensions.
21. The three sides of a right triangle are given by three consecutive even integers. Find the three sides.
22. One leg of a right triangle is three more than the other leg. The hypotenuse is 15 inches. Find the lengths of the two legs.
23. The shorter leg of a right triangle is 5 meters. The hypotenuse is 1 meter longer than the longer leg. Find the length of the longer leg.
24. The shorter leg of a right triangle is 12 yards. If the hypotenuse is 20 yards, how long is the other leg?

Review Problems   The problems below review material we covered in Section 3.5.

Solve each system by the substitution method.

25. $x + y = 5$
$\quad\quad y = x + 1$

26. $x - y = 7$
$\quad\quad x = 2y + 10$

27. $2x + y = 6$
$\quad\quad y = 3x + 1$

28. $2x - 3y = 6$
$\quad\quad y = 2x - 2$

29. $4x - 3y = -1$
$\quad -x + y = 2$

30. $5x - 2y = 2$
$\quad x + y = 6$

Chapter 5   Summary and Review

FACTORING [5.1]

Factoring is the reverse of multiplication.

Multiplication

$$3 \cdot 5 \;=\; 15$$

Factors   Product

Factoring

Examples

1. The number 150 can be factored into the product of prime numbers:

$$150 = 15 \cdot 10$$
$$= (3 \cdot 5)(2 \cdot 5)$$
$$= 2 \cdot 3 \cdot 5^2$$

2. $8x^4 - 10x^3 + 6x^2$
$= 2x^2 \cdot 4x^2 - 2x^2 \cdot 5x + 2x^2 \cdot 3$
$= 2x^2(4x^2 - 5x + 3)$

GREATEST COMMON FACTOR [5.2]

The largest monomial that divides each term of a polynomial is called the greatest common factor for that polynomial. We begin all factorizations by factoring out the greatest common factor.

3. $x^2 + 5x + 6 = (x + 2)(x + 3)$
$x^2 - 5x + 6 = (x - 2)(x - 3)$
$6x^2 - x - 2 = (2x + 1)(3x - 2)$
$6x^2 + 7x + 2 = (2x + 1)(3x + 2)$

FACTORING TRINOMIALS [5.3, 5.4]

One method of factoring a trinomial is to list all pairs of binomials the product of whose first terms gives the first term of the trinomial and the product of whose last terms gives the last term of the trinomial. We then choose the pair that gives the correct middle term for the original trinomial.

4. $x^2 + 10x + 25 = (x + 5)^2$
$x^2 - 10x + 25 = (x - 5)^2$
$x^2 - 25 = (x + 5)(x - 5)$

SPECIAL FACTORIZATIONS [5.5]

$$a^2 + 2ab + b^2 = (a + b)^2$$
$$a^2 - 2ab + b^2 = (a - b)^2$$
$$a^2 - b^2 = (a + b)(a - b)$$

5. Solve $x^2 - 6x = -8$:

$$x^2 - 6x + 8 = 0$$
$$(x - 4)(x - 2) = 0$$
$$x - 4 = 0 \quad \text{or} \quad x - 2 = 0$$
$$x = 4 \quad \text{or} \quad x = 2$$

Both solutions check.

TO SOLVE A QUADRATIC EQUATION [5.6]

Step 1: Write the equation in standard form: $ax^2 + bx + c = 0$.
Step 2: Factor completely.
Step 3: Set each factor equal to zero.
Step 4: Solve the equations found in Step 3.
Step 5: Check solutions, if necessary.

COMMON MISTAKES

Trying to apply the special property of 0 to other numbers. For example, consider the equation $(x - 3)(x + 4) = 18$. A fairly common mistake is to attempt to solve it with the following steps:

$$(x - 3)(x + 4) = 18$$
$$x - 3 = 18 \quad \text{or} \quad x + 4 = 18$$
$$x = 21 \quad \text{or} \quad x = 14$$

These are obviously not solutions, as a quick check will verify:

Check $x = 21$ Check $x = 14$

$(21 - 3)(21 + 4) = 18$ $(14 - 3)(14 + 4) = 18$

$18 \cdot 25 = 18$ false statements $11 \cdot 18 = 18$

$450 = 18$ ←———————→ $198 = 18$

The mistake is in setting each factor equal to 18. It is not necessarily true that since the product of two numbers is 18 that either one of them is itself 18.

The correct solution looks like this:

$$(x - 3)(x + 4) = 18$$
$$x^2 + x - 12 = 18$$
$$x^2 + x - 30 = 0$$
$$(x + 6)(x - 5) = 0$$
$$x + 6 = 0 \quad \text{or} \quad x - 5 = 0$$
$$x = -6 \quad \text{or} \quad x = 5$$

To avoid this mistake, remember that before you factor a quadratic equation, you must write it in standard form. It is only in standard form when 0 is on one side and decreasing powers of the variable are on the other.

Write each term as the product of prime numbers.   Chapter 5 Test

1. 35   **2.** 77   **3.** 41

4. 588   **5.** 128   **6.** 324

Factor the following completely.

7. $x^2 - 5x + 6$ **8.** $x^2 - x - 6$

9. $a^2 - 16$ **10.** $x^2 + 25$

11. $x^4 - 81$ **12.** $27x^2 - 75y^2$

13. $4a^2 + 22a + 10$ **14.** $3m^2 - 3m - 18$

15. $6y^2 + 7y - 5$ **16.** $12x^3 - 14x^2 - 10x$

Solve the following quadratic equations.

17. $x^2 + 7x + 12 = 0$ **18.** $x^2 - 4x + 4 = 0$

19. $x^2 - 36 = 0$ **20.** $x^2 = x + 20$

21. $x^2 - 11x = -30$ **22.** $y^3 = 16y$

23. $2a^2 = a + 15$ **24.** $30x^3 - 20x^2 = 10x$

Solve the following word problems. Be sure to show the equation used.

25. Two numbers have a sum of 20. Their product is 64. Find the numbers.

26. The product of two consecutive odd integers is seven more than their sum. Find the integers.

27. The length of a rectangle is five more than three times the width. The area is 42 sq. ft. Find the dimensions.

Rational Expressions

Rational expressions are to polynomials what fractions are to integers. As you will see, the problems you will work with fractions are directly related to the problems you will work with rational expressions. The methods used in both cases are the same.

Although there are many properties and principles used in this chapter, the three most essential are

1. The ability to factor polynomials.
2. Division as multiplication by the reciprocal.
3. The multiplication property of equality.

Most important is your ability to factor polynomials. There are many problems in this chapter that are impossible to solve without factoring.

In Chapter 1 we defined the set of rational numbers to be the set of all numbers that could be put in the form $\frac{a}{b}$, where a and b are integers ($b \neq 0$):

$$\text{Rational numbers} = \left\{ \frac{a}{b} \;\middle|\; a \text{ and } b \text{ are integers, } b \neq 0 \right\}$$

**6.1
Reducing Rational
Expressions to
Lowest Terms**

Since rational numbers are commonly called fractions, we will refer to them as such throughout this chapter. A rational expression is any expression that can be put in the form $\dfrac{P}{Q}$, where P and Q are polynomials and $Q \neq 0$:

Rational expressions $= \left\{ \dfrac{P}{Q} \,\middle|\, P \text{ and } Q \text{ are polynomials, } Q \neq 0 \right\}$

Each of the following is an example of a rational expression:

$$\frac{2x + 3}{x} \qquad \frac{x^2 - 6x + 9}{x^2 - 4} \qquad \frac{5}{x^2 + 6} \qquad \frac{2x^2 + 3x + 4}{2}$$

For the rational expression

$$\frac{x^2 - 6x + 9}{x^2 - 4}$$

the polynomial on top, $x^2 - 6x + 9$, is called the numerator, and the polynomial on the bottom, $x^2 - 4$, is called the denominator. The same is true of the other rational expressions.

DEFINITION   The *domain* of a rational expression is all replacements of the variable for which the rational expression is defined.

We must be careful that we do not use a value of the variable that will give us a denominator of zero. Remember, division by zero is not defined. Actually, to specify the domain for a rational expression, we need list only those values of the variable that give a denominator of zero. We list them as restrictions on the variable.

▼   **Examples**   Give the domain for each of the following rational expressions:

1. $\dfrac{x + 2}{x - 3}$

   SOLUTION   The domain is all real numbers except $x = 3$, since, when $x = 3$, the denominator is $3 - 3 = 0$.

2. $\dfrac{5}{x^2 - x - 6}$

   SOLUTION   If we factor the denominator, we have $x^2 - x - 6 = (x - 3)(x + 2)$. If either of the factors is zero, the whole denominator is zero. Our restrictions are $x = 3$ and $x = -2$, since either one makes $x^2 - x + 6 = 0$. ▲

We will not always list each restriction on a rational expression, but we should be aware of them and keep in mind that no rational expression can have a denominator of zero.

The two fundamental properties of rational expressions are listed below. We will use these two properties many times in this chapter.

PROPERTY 1 *Multiplying* the numerator and denominator of a rational expression by the same nonzero quantity will not change the value of the rational expression.

PROPERTY 2 *Dividing* the numerator and denominator of a rational expression by the same nonzero quantity will not change the value of the rational expression.

We can use Property 2 to reduce rational expressions to lowest terms. Since this process is almost identical to the process of reducing fractions to lowest terms, let's recall how the fraction $\frac{6}{15}$ is reduced to lowest terms:

$$\frac{6}{15} = \frac{2 \cdot 3}{5 \cdot 3} \qquad \text{Factor numerator and denominator}$$

$$= \frac{2 \cdot \cancel{3}}{5 \cdot \cancel{3}} \qquad \text{Divide out the common factor, 3}$$

$$= \frac{2}{5} \qquad \text{Reduce to lowest terms}$$

The same procedure applies to reducing rational expressions to lowest terms. The process is summarized in the following rule.

RULE To reduce a rational expression to lowest terms, first factor each expression completely and then divide both the numerator and denominator by any factors they have in common.

▼ **Example 3** Reduce $\dfrac{x^2 - 9}{x^2 + 5x + 6}$ to lowest terms.

SOLUTION We begin by factoring:

$$\frac{x^2 - 9}{x^2 + 5x + 6} = \frac{(x - 3)(x + 3)}{(x + 2)(x + 3)}$$

Notice that both polynomials contain the factor $(x + 3)$. If we divide the numerator by $(x + 3)$, we are left with $(x - 3)$. If we divide the denominator by $(x + 3)$, we are left with $(x + 2)$. The complete problem looks like this:

$$\frac{x^2 - 9}{x^2 + 5x + 6} = \frac{(x - 3)(x + 3)}{(x + 2)(x + 3)} \qquad \text{Factor the numerator and}$$
$$\text{denominator completely}$$

$$= \frac{x - 3}{x + 2} \qquad \text{Divide out the common}$$
$$\text{factor, } x + 3 \qquad \blacktriangle$$

It is convenient to draw a line through the factors as we divide them out. It is especially helpful when the problems become longer.

▼ **Example 4**   Reduce $\dfrac{2x^3 + 2x^2 - 24x}{x^3 + 2x^2 - 8x}$ to lowest terms.

SOLUTION   We begin by factoring the numerator and denominator completely. Then we divide out all factors common to the numerator and denominator. Here is what it looks like:

$$\frac{2x^3 + 2x^2 - 24x}{x^3 + 2x^2 - 8x} = \frac{2x(x^2 + x - 12)}{x(x^2 + 2x - 8)} \qquad \text{Factor out the greatest}$$
$$\text{common factor first}$$

$$= \frac{2x(x - 3)(x + 4)}{x(x - 2)(x + 4)} \qquad \text{Factor the remaining}$$
$$\text{trinomials}$$

$$= \frac{2(x - 3)}{(x - 2)} \qquad \text{Divide out the factors}$$
$$\text{common to the}$$
$$\text{numerator and}$$
$$\text{denominator} \qquad \blacktriangle$$

▼ **Example 5**   Reduce $\dfrac{x - 5}{x^2 - 25}$ to lowest terms.

SOLUTION

$$\frac{x - 5}{x^2 - 25} = \frac{(x - 5)}{(x - 5)(x + 5)} \qquad \text{Factor numerator and denominator}$$
$$\text{completely}$$

$$= \frac{1}{x + 5} \qquad \text{Divide out the common factor,}$$
$$(x - 5) \qquad \blacktriangle$$

You should look over the above example closely and see that you understand each step. There are two common mistakes with this type of problem. The first arises in dividing the numerator $(x - 5)$ by $(x - 5)$. Many times first year algebra students will give the result as 0. It is not 0 but 1. That is, $(x - 5)$ goes into $(x - 5)$ once.

   The second mistake is usually made by trying to do what some people mistake for canceling and looks like this:

$$\frac{x - 5}{x^2 - 25} = \frac{\overset{1}{\cancel{x}} - \overset{5}{\cancel{5}}}{\underset{x}{\cancel{x^2}} - \underset{5}{\cancel{25}}} \quad \text{Mistake}$$

$$= \frac{1}{x - 5}$$

But if we replaced x by 10, we would have

$$\frac{10 - 5}{100 - 25} = \frac{1}{10 - 5}$$

$$\frac{5}{75} = \frac{1}{5}$$

$$\frac{1}{15} = \frac{1}{5}$$

which is not true.

   To avoid this mistake, remember that we can divide out (sometimes called canceling) *factors* common to the numerator and denominator of a rational expression. Neither x nor 5 is a factor of either the numerator or denominator in the above problem and therefore cannot be divided out.

Reduce the following rational expressions to lowest terms, if possible. Specify any restrictions on the variable in reduced expression.

1. $\dfrac{5}{5x - 10}$ **2.** $\dfrac{-4}{2x - 8}$

3. $\dfrac{-8}{8x + 16}$ **4.** $\dfrac{3}{6x - 12}$

5. $\dfrac{a - 3}{a^2 - 9}$ **6.** $\dfrac{a + 4}{a^2 - 16}$

7. $\dfrac{x + 5}{x^2 - 25}$ **8.** $\dfrac{x - 2}{x^2 - 4}$

9. $\dfrac{2a}{10a^2}$ **10.** $\dfrac{22a^3}{11a}$

11. $\dfrac{2x^2 - 8}{4}$

12. $\dfrac{5x - 10}{x - 2}$

13. $\dfrac{35m^2}{7m^4}$

14. $\dfrac{27m^5n^2}{9m^2n^3}$

15. $\dfrac{2x - 10}{3x - 6}$

16. $\dfrac{4x - 8}{x - 2}$

17. $\dfrac{10a + 20}{5a + 10}$

18. $\dfrac{11a + 33}{6a + 18}$

19. $\dfrac{5x^2 - 5}{4x + 4}$

20. $\dfrac{7x^2 - 28}{2x + 4}$

21. $\dfrac{x - 3}{x^2 - 6x + 9}$

22. $\dfrac{x^2 - 10x + 25}{x - 5}$

23. $\dfrac{y + 2}{y^2 - y - 6}$

24. $\dfrac{y + 1}{y^2 - 2y - 3}$

25. $\dfrac{x + 5}{x^2 + 8x + 15}$

26. $\dfrac{x + 3}{x^2 + 8x + 15}$

27. $\dfrac{a - 3}{a^2 - 8a + 15}$

28. $\dfrac{a + 3}{a^2 - 2a - 15}$

29. $\dfrac{3x - 2}{9x^2 - 4}$

30. $\dfrac{2x - 3}{4x^2 - 9}$

31. $\dfrac{x^2 + 8x + 15}{x^2 + 5x + 6}$

32. $\dfrac{x^2 - 8x + 15}{x^2 - x - 6}$

33. $\dfrac{m^2 - m - 6}{m^2 - 5m + 6}$

34. $\dfrac{m^2 + 2m - 3}{m^2 - m - 12}$

35. $\dfrac{3x^2 - 3x}{2(x - 1)}$

36. $\dfrac{16y^2 - 4}{4y - 2}$

37. $\dfrac{x^2 + 3x - 4}{x^2 - 16}$

38. $\dfrac{4x^2 - 10x + 6}{2x^2 + x - 3}$

39. $\dfrac{3a^2 - 8a + 4}{9a^2 - 4}$

40. $\dfrac{3a^2 - 8a + 5}{4a^2 - 5a + 1}$

41. $\dfrac{4x^2 - 12x + 9}{4x^2 - 9}$

42. $\dfrac{5x^2 + 18x - 8}{5x^2 + 13x - 6}$

43. $\dfrac{x + 3}{x^4 - 81}$

44. $\dfrac{x^2 + 9}{x^4 - 81}$

45. $\dfrac{x-2}{x^4-16}$ **46.** $\dfrac{x^2-4}{x^4-16}$

47. Replace x with 5 and y with 4 in the expression

$$\frac{x^2-y^2}{x-y}$$

and simplify the result. Is the result equal to $5-4$ or $5+4$?

48. Replace x with 2 in the expression

$$\frac{x^3-1}{x-1}$$

and simplify the result. Your answer should be equal to what you would get if you replaced x with 2 in x^2+x+1.

Review Problems   The problems below review material we covered in Section 4.4. Reviewing these problems will help you in the next section.

Subtract.

49. $(4x+3)-(2x+7)$
50. $(5x+2)-(3x+8)$
51. $(x^2+3x-4)-(4x^2+x-1)$
52. $(7x^2+x-3)-(2x^2+3x-5)$
53. Subtract x^2-3x from x^2-5x.
54. Subtract $-2x+6$ from $-2x+8$.
55. Subtract $2x^3-10x^2$ from $2x^3$.
56. Subtract $10x^2-50x$ from $10x^2-3x$.

In this section we will learn how to divide a polynomial by a polynomial. Part of what we will do here is very similar to what we did in the previous section, and the rest is similar to long division with whole numbers.

6.2
Dividing a Polynomial by a Polynomial

 Example 1   Divide x^2-5x+6 by $x-3$.

SOLUTION   Written in symbols, the problem looks like this:

$$\frac{x^2-5x+6}{x-3}$$

Factoring the numerator and then dividing out the common factor, we have

$$\frac{x^2 - 5x + 6}{x - 3} = \frac{(x - 3)(x - 2)}{x - 3}$$

$$= x - 2$$

The binomial $x - 3$ divides the trinomial $x^2 - 5x + 6$ exactly. The quotient is $x - 2$. ▲

For this type of division, the denominator must be a factor of the numerator. If it is, we simply factor the numerator and divide out the common factor.

The next question is, "What do we do if we can't factor the numerator?" The answer is we use a form of long division for polynomials that is similar to long division with whole numbers.

Since long division for polynomials is very similar to long division with whole numbers, we will begin by reviewing a division problem with whole numbers. You may realize when looking at Example 2 that you don't have a very good idea why you proceed as you do with long division. What you do know is that the process always works. We are going to approach the explanations in this section in much the same manner. That is, we won't always be sure why the steps we will use are important, only that they always produce the correct result.

▼ **Example 2** $27\overline{)3962}$

   SOLUTION

$$\begin{array}{r} 1 \\ 27\overline{)3962} \\ \underline{27} \\ 12 \end{array}$$

⟵ Estimate 27 into 39

⟵ Multiply $1 \times 27 = 27$

⟵ Subtract $39 - 27 = 12$

$$\begin{array}{r} 1 \\ 27\overline{)3962} \\ \underline{27\downarrow} \\ 126 \end{array}$$

⟵ Bring down the 6

These are the four basic steps in long division. Estimate, multiply, subtract, and bring down the next term. To finish the problem, we simply perform the same four steps again:

$$\begin{array}{r} 14 \\ 27\overline{)3962} \\ \underline{27} \\ 126 \\ \underline{108} \\ 182 \end{array}$$

⟵ 4 is the estimate

⟵ Multiply to get 108

⟵ Subtract to get 18,
then bring down the 2

One more time.

$$\begin{array}{r} 146 \longleftarrow 6 \text{ is the estimate} \\ 27\overline{)3962} \\ \underline{27} \\ 126 \\ \underline{108} \\ 182 \\ \underline{162} \longleftarrow \text{Multiply to get } 162 \\ 20 \longleftarrow \text{Subtract to get } 20 \end{array}$$

Since there is nothing left to bring down we have our answer.

$$\frac{3962}{27} = 146 + \frac{20}{27} \quad \text{or} \quad 146\frac{20}{27} \qquad \blacktriangle$$

Here is how it works with polynomials.

▼ **Example 3** Divide $\dfrac{x^2 - 5x + 8}{x - 3}$.

SOLUTION

$$\begin{array}{r} x \quad\longleftarrow \text{ Estimate } x^2 \div x = x \\ x - 3\overline{)\,x^2 - 5x + 8} \\ -\quad + \\ \underline{\cancel{+}x^2 \cancel{-} 3x} \quad\longleftarrow \text{ Multiply } x(x-3) = x^2 - 3x \\ -2x \quad\longleftarrow \text{ Subtract } (x^2 - 5x) - (x^2 - 3x) = -2x \end{array}$$

$$\begin{array}{r} x \\ x - 3\overline{)\,x^2 - 5x + 8} \\ -\quad + \quad \downarrow \\ \underline{\cancel{+}x^2 \cancel{-} 3x \quad \downarrow} \\ -2x + 8 \quad\longleftarrow \text{ Bring down the } 8 \end{array}$$

Notice that to subtract one polynomial from another, we add its opposite. That is why we change the signs on $x^2 - 3x$ and add what we get to $x^2 - 5x$. (To subtract the second polynomial, simply change the signs and add.)

We perform the same four steps again:

$$\begin{array}{r} x - 2 \quad\longleftarrow -2 \text{ is the estimate } (-2x \div x = -2) \\ x - 3\overline{)\,x^2 - 5x + 8} \\ -\quad + \quad \downarrow \\ \underline{\cancel{+}x^2 \cancel{-} 3x \quad \downarrow} \\ -2x + 8 \\ +\quad - \\ \underline{\cancel{-} 2x \cancel{+} 6} \quad\longleftarrow \text{ Multiply } -2(x-3) = -2x + 6 \\ 2 \quad\longleftarrow \text{ Subtract } (-2x + 8) - (-2x + 6) = 2 \end{array}$$

Since there is nothing left to bring down, we have our answer:

$$\frac{x^2 - 5x + 8}{x - 3} = x - 2 + \frac{2}{x - 3}$$

To check our answer, we multiply $(x - 3)(x - 2)$ to get $x^2 - 5x + 6$. Then, adding on the remainder, 2, we have $x^2 - 5x + 8$.

▲

▼ **Example 4**   Divide $\dfrac{6x^2 - 11x - 14}{2x - 5}$.

SOLUTION

$$\begin{array}{r}
3x + 2 \\
2x - 5\overline{)\ 6x^2 - 11x - 14} \\
-\quad + \\
\cancel{6x^2} \cancel{+} 15x \quad\downarrow \\
\hline
+\ \ 4x - 14 \\
-\quad + \\
\cancel{+}\ \ 4x \cancel{+} 10 \\
\hline
-\ 4
\end{array}$$

$$\frac{6x^2 - 11x - 14}{2x - 5} = 3x + 2 + \frac{-4}{2x - 5}$$ ▲

One last step is sometimes necessary. The two polynomials in a division problem must both be in descending powers of the variable and cannot skip any powers from the highest power down to the constant term.

▼ **Example 5**   Divide $\dfrac{2x^3 - 3x + 2}{x - 5}$.

SOLUTION   The problem will be much less confusing if we write $2x^3 - 3x + 2$ as $2x^3 + 0x^2 - 3x + 2$. Adding $0x^2$ does not change our original problem.

$$\begin{array}{r}
2x^2 \\
x - 5\overline{)\ 2x^3 +\ \ 0x^2 - 3x + 2} \\
-\quad + \\
\cancel{2x^3} \cancel{+} 10x^2 \quad\ \ \big| \\
\hline
+\ 10x^2 - 3x
\end{array}$$

⟵ Estimate $2x^3 \div x = 2x^2$

⟵ Multiply $2x^2(x - 5) = 2x^3 - 10x^2$

⟵ Subtract $(2x^3 + 0x^2) - (2x^3 - 10x^2) = 10x^2$
Bring down the next term.

Adding the term $0x^2$ gives us a column in which to write $10x^2$. (Remember, you can add and subtract only similar terms.)

Here is the completed problem:

$$
\begin{array}{r}
2x^2 + 10x + 47 \\
x - 5 \overline{)\, 2x^3 + 0x^2 - 3x + 2} \\
\underline{-+} \\
\cancel{+}2x^3 \cancel{-} 10x^2 \downarrow \\
+ 10x^2 - 3x \\
\underline{-+} \\
\cancel{+}\, 10x^2 \cancel{-} 50x \downarrow \\
+ 47x + 2 \\
\underline{-+} \\
\cancel{+}\, 47x \cancel{-}\, 235 \\
237
\end{array}
$$

Our answer is $\dfrac{2x^3 - 3x + 2}{x - 5} = 2x^2 + 10x + 47 + \dfrac{237}{x - 5}$. ▲

As you can see, long division with polynomials is a mechanical process. Once you have done it correctly a couple of times, it becomes very easy to produce the correct answer.

Divide by factoring the numerator and then dividing out the common factor (see Example 1).

Problem Set 6.2

1. $\dfrac{x^2 - 5x + 6}{x - 3}$

2. $\dfrac{x^2 - 5x + 6}{x - 2}$

3. $\dfrac{x^2 - 5x - 6}{x - 6}$

4. $\dfrac{x^2 - 5x - 6}{x + 1}$

5. $\dfrac{a^2 + 9a + 20}{a + 5}$

6. $\dfrac{a^2 + 9a + 20}{a + 4}$

7. $\dfrac{x^2 - 6x + 9}{x - 3}$

8. $\dfrac{x^2 + 10x + 25}{x + 5}$

9. $\dfrac{a^2 - 12a + 36}{a - 6}$

10. $\dfrac{a^2 + 4a + 4}{a + 2}$

11. $\dfrac{2x^2 + 5x - 3}{2x - 1}$

12. $\dfrac{4x^2 + 4x - 3}{2x - 1}$

13. $\dfrac{2a^2 - 9a - 5}{2a + 1}$

14. $\dfrac{4a^2 - 8a - 5}{2a + 1}$

15. $\dfrac{x^2 - 49}{x - 7}$

16. $\dfrac{x^2 - 36}{x + 6}$

17. $\dfrac{x^2 - 64}{x + 8}$

18. $\dfrac{x^2 - 9}{x + 3}$

Divide using long division (see Examples 3 and 4).

19. $\dfrac{x^2 + 5x + 8}{x + 3}$

20. $\dfrac{x^2 + 5x + 4}{x + 3}$

21. $\dfrac{a^2 + 3a + 2}{a + 5}$

22. $\dfrac{a^2 + 4a + 3}{a + 5}$

23. $\dfrac{x^2 + 2x + 1}{x - 2}$

24. $\dfrac{x^2 + 6x + 9}{x - 3}$

25. $\dfrac{x^2 + 5x - 6}{x + 1}$

26. $\dfrac{x^2 - x - 6}{x + 1}$

27. $\dfrac{a^2 + 3a + 1}{a + 2}$

28. $\dfrac{a^2 - a + 3}{a + 1}$

29. $\dfrac{2x^2 - 2x + 5}{2x + 4}$

30. $\dfrac{15x^2 + 19x - 4}{3x + 8}$

31. $\dfrac{6a^2 + 5a + 1}{2a + 3}$

32. $\dfrac{4a^2 + 4a + 3}{2a + 1}$

33. $\dfrac{6a^3 - 13a^2 - 4a + 15}{3a - 5}$

34. $\dfrac{2a^3 - a^2 + 3a + 2}{2a + 1}$

Fill in the missing terms in the numerator and then use long division to find the quotients (see Example 5).

35. $\dfrac{x^3 + 4x + 5}{x + 1}$

36. $\dfrac{x^3 + 4x^2 - 8}{x + 2}$

37. $\dfrac{x^3 - 1}{x - 1}$

38. $\dfrac{x^3 + 1}{x + 1}$

39. $\dfrac{x^3 - 8}{x - 2}$

40. $\dfrac{x^3 + 27}{x + 3}$

Review Problems   The problems below review material we covered in Section 1.6 and Problem Set 1.7. Reviewing these problems will help you understand the next section.

Multiply or divide as indicated.

41. $\dfrac{3}{8} \cdot \dfrac{5}{2}$

42. $\dfrac{4}{5} \cdot \dfrac{2}{3}$

43. $\dfrac{3}{10} \cdot \dfrac{5}{9}$

44. $\dfrac{2}{7} \cdot \dfrac{21}{30}$

45. $\dfrac{2}{3} \div \dfrac{1}{3}$

46. $\dfrac{3}{4} \div \dfrac{1}{4}$

47. $\dfrac{4}{5} \div \dfrac{9}{10}$

48. $\dfrac{1}{6} \div \dfrac{3}{12}$

6.3 Multiplication and Division of Rational Expressions

Recall that to multiply two fractions, we simply multiply numerators and multiply denominators and then reduce to lowest terms, if possible:

$$\dfrac{3}{4} \cdot \dfrac{10}{21} = \dfrac{30}{84} \quad \longleftarrow \text{ Multiply numerators}$$
$$\phantom{\dfrac{3}{4} \cdot \dfrac{10}{21} = \dfrac{30}{84}} \quad \longleftarrow \text{ Multiply denominators}$$

$$= \dfrac{5}{14} \qquad \text{Reduce to lowest terms}$$

The same problem can also be done by factoring numerators and denominators first and then dividing out the factors they have in common:

$$\dfrac{3}{4} \cdot \dfrac{10}{21} = \dfrac{3}{2 \cdot 2} \cdot \dfrac{2 \cdot 5}{3 \cdot 7} \qquad \text{Factor}$$

$$= \dfrac{\cancel{3} \cdot \cancel{2} \cdot 5}{2 \cdot 2 \cdot \cancel{3} \cdot 7} \qquad \begin{array}{l} \text{Multiply numerators} \\ \text{Multiply denominators} \end{array}$$

$$= \dfrac{5}{14} \qquad \text{Divide out common factors}$$

We can apply the second process to the product of two rational expressions, as the following example illustrates.

▼ **Example 1** Multiply $\dfrac{x - 2}{x + 3} \cdot \dfrac{x^2 - 9}{2x - 4}$.

SOLUTION We begin by factoring numerators and denominators as much as possible. Then we multiply the numerators and denominators. The last step consists of dividing out all factors common to the numerator and denominator:

$$\dfrac{x - 2}{x + 3} \cdot \dfrac{x^2 - 9}{2x - 4} = \dfrac{x - 2}{x + 3} \cdot \dfrac{(x - 3)(x + 3)}{2(x - 2)} \qquad \text{Factor completely}$$

$$= \frac{(x-2)(x-3)(x+3)}{(x+3)(2)(x-2)}$$ Multiply numerators and denominators

$$= \frac{x-3}{2}$$ Divide out common factors

▲

In Chapter 1 we defined division as the equivalent of multiplication by the reciprocal. This is how it looks with fractions:

$$\frac{4}{5} \div \frac{8}{9} = \frac{4}{5} \cdot \frac{9}{8}$$ Division as multiplication by the reciprocal

$$\left. \begin{array}{c} = \dfrac{2 \cdot 2 \cdot 3 \cdot 3}{5 \cdot 2 \cdot 2 \cdot 2} \\[2mm] = \dfrac{9}{10} \end{array} \right\}$$ Factor and divide out common factors

The same idea holds for division with rational expressions. The rational expression that follows the division symbol is called the *divisor*; to divide, we multiply by the reciprocal of the divisor.

▼ **Example 2** Divide $\dfrac{3x - 9}{x^2 - x - 20} \div \dfrac{x^2 + 2x - 15}{x^2 - 25}$.

SOLUTION We begin by taking the reciprocal of the divisor and writing the problem again in terms of multiplication. We then factor, multiply, and, finally, divide out all factors common to the numerator and denominator of the resulting expression. The complete problem looks like this:

$$\frac{3x - 9}{x^2 - x - 20} \div \frac{x^2 + 2x - 15}{x^2 - 25}$$ Multiply by the reciprocal of the divisor

$$= \frac{3x - 9}{x^2 - x - 20} \cdot \frac{x^2 - 25}{x^2 + 2x - 15}$$

$$= \frac{3(x - 3)}{(x + 4)(x - 5)} \cdot \frac{(x - 5)(x + 5)}{(x + 5)(x - 3)}$$ Factor

$$= \frac{3(x-3)(x-5)(x+5)}{(x + 4)(x-5)(x+5)(x-3)}$$ Divide out common factors

$$= \frac{3}{x + 4}$$ ▲

As you can see, factoring is the single most important tool we use in working with rational expressions. Most of the work we have done or will do with rational expressions is most easily accomplished if the rational expressions are in factored form. Here are some more examples of multiplication and division with rational expressions.

▼ **Examples**

3. Multiply $\dfrac{3a + 6}{a^2} \cdot \dfrac{a}{2a + 4}$.

SOLUTION

$$\dfrac{3a + 6}{a^2} \cdot \dfrac{a}{2a + 4}$$

$$= \dfrac{3(a + 2)}{a^2} \cdot \dfrac{a}{2(a + 2)} \qquad \text{Factor completely}$$

$$= \dfrac{3(a + 2)a}{a^2(2)(a + 2)} \qquad \text{Multiply}$$

$$= \dfrac{3}{2a} \qquad \begin{array}{l}\text{Divide numerator and} \\ \text{denominator by} \\ a(a + 2) \qquad \blacktriangle\end{array}$$

4. Divide $\dfrac{x^2 + 7x + 12}{x^2 - 16} \div \dfrac{x^2 + 6x + 9}{2x - 8}$.

SOLUTION

$$\dfrac{x^2 + 7x + 12}{x^2 - 16} \div \dfrac{x^2 + 6x + 9}{2x - 8} \qquad \begin{array}{l}\text{Division is multiplication} \\ \text{by the reciprocal}\end{array}$$

$$= \dfrac{x^2 + 7x + 12}{x^2 - 16} \cdot \dfrac{2x - 8}{x^2 + 6x + 9}$$

$$= \dfrac{(x + 3)(x + 4)(2)(x - 4)}{(x - 4)(x + 4)(x + 3)(x + 3)} \qquad \text{Factor and multiply}$$

$$= \dfrac{2}{x + 3} \qquad \begin{array}{l}\text{Divide out} \\ \text{common factors} \qquad \blacktriangle\end{array}$$

In Example 4 we factored and multiplied the two expressions in a single step. This saves writing the problem one extra time.

▼ **Example 5** Multiply $(x^2 - 49) \left(\dfrac{x + 4}{x + 7} \right)$

SOLUTION   We can think of the polynomial $x^2 - 49$ as having a denominator of 1. Thinking of $x^2 - 49$ in this way allows us to proceed as we did in previous examples:

$$(x^2 - 49)\left(\frac{x + 4}{x + 7}\right) = \frac{x^2 - 49}{1} \cdot \frac{x + 4}{x + 7} \qquad \text{Write } x^2 - 49 \text{ with denomi-nator 1}$$

$$= \frac{(x + 7)(x - 7)(x + 4)}{x + 7} \qquad \text{Factor}$$

$$= (x - 7)(x + 4) \qquad \text{Divide out common factors}$$

We can leave the answer in this form or multiply to get $x^2 - 3x - 28$. In this section let's agree to leave our answers in factored form. ▲

▼   **Example 6**   Multiply $a(a + 5)(a - 5)\left(\dfrac{a + 4}{a^2 + 5a}\right)$.

SOLUTION   We can think of the expression $a(a + 5)(a - 5)$ as having a denominator of 1:

$$a(a + 5)(a - 5)\left(\frac{a + 4}{a^2 + 5a}\right)$$

$$= \frac{a(a + 5)(a - 5)}{1} \cdot \frac{a + 4}{a^2 + 5a}$$

$$= \frac{a(a + 5)(a - 5)(a + 4)}{a(a + 5)} \qquad \text{Factor}$$

$$= (a - 5)(a + 4) \qquad \text{Divide out common factors}$$ ▲

Problem Set 6.3

Multiply or divide as indicated. Be sure to reduce all answers to lowest terms. (That is, the numerator and denominator of the answer should not have any factors in common.)

1. $\dfrac{x + y}{3} \cdot \dfrac{6}{x + y}$

2. $\dfrac{x - 1}{x + 1} \cdot \dfrac{5}{x - 1}$

3. $\dfrac{2x + 10}{3} \cdot \dfrac{6}{4x + 20}$

4. $\dfrac{5}{3x - 6} \cdot \dfrac{x - 2}{10}$

5. $\dfrac{9}{2a - 8} \div \dfrac{3}{a - 4}$

6. $\dfrac{8}{a^2 - 25} \div \dfrac{16}{a + 5}$

7. $\dfrac{x + 1}{x^2 - 9} \div \dfrac{2x + 2}{x + 3}$

8. $\dfrac{11}{x - 2} \div \dfrac{22}{2x^2 - 8}$

9. $\dfrac{a^2 + 5a}{7} \cdot \dfrac{4}{a^2 + 4a}$

10. $\dfrac{4a + 4}{a^2 - 25} \cdot \dfrac{a^2 - 5a}{8}$

11. $\dfrac{y^2 - 5y + 6}{2y + 4} \div \dfrac{2y - 6}{y + 2}$

12. $\dfrac{y^2 - 7y}{3y^2 - 48} \div \dfrac{y^2 - 9}{y^2 - 7y + 12}$

13. $\dfrac{2x - 8}{x^2 - 4} \cdot \dfrac{x^2 + 6x + 8}{x - 4}$

14. $\dfrac{x^2 + 5x + 1}{7x - 7} \cdot \dfrac{x - 1}{x^2 + 5x + 1}$

15. $\dfrac{x - 1}{x^2 - x - 6} \cdot \dfrac{x^2 + 5x + 6}{x^2 - 1}$

16. $\dfrac{x^2 - 3x - 10}{x^2 - 4x + 3} \cdot \dfrac{x^2 - 5x + 6}{x^2 - 3x - 10}$

17. $\dfrac{a^2 + 10a + 25}{a + 5} \div \dfrac{a^2 - 25}{a - 5}$

18. $\dfrac{a^2 + a - 2}{a^2 + 5a + 6} \div \dfrac{a - 1}{a}$

19. $\dfrac{y - 5}{y^2 + 3y + 2} \div \dfrac{y^2 - 5y + 6}{y^2 - 2y - 3}$

20. $\dfrac{y - 5}{y^2 + 7y + 12} \div \dfrac{y^2 - 7y + 10}{y^2 + 9y + 18}$

21. $\dfrac{x^2 + 10x + 21}{x^2 + 2x - 35} \cdot \dfrac{x^2 - 25}{x^2 - 2x - 15}$

22. $\dfrac{x^2 - 13x + 42}{x^2 + 10x + 21} \cdot \dfrac{x^2 + x - 6}{x^2 - 4}$

23. $\dfrac{x^2 + 5x + 6}{x^2 + 6x + 8} \cdot \dfrac{x^2 + 9x + 20}{x^2 + 8x + 15}$

24. $\dfrac{x^2 - 1}{x^2 + 3x + 2} \cdot \dfrac{x^2 - 4}{x^2 - 3x + 2}$

25. $\dfrac{2a^2 + 7a + 3}{a^2 - 16} \div \dfrac{4a^2 + 8a + 3}{2a^2 - 5a - 12}$

26. $\dfrac{3a^2 + 7a - 20}{a^2 + 3a - 4} \div \dfrac{3a^2 - 2a - 5}{a^2 - 2a + 1}$

27. $\dfrac{4y^2 - 12y + 9}{y^2 - 36} \div \dfrac{2y^2 - 5y + 3}{y^2 + 5y - 6}$

28. $\dfrac{5y^2 - 6y + 1}{y^2 - 1} \div \dfrac{16y^2 - 9}{4y^2 + 7y + 3}$

29. $\dfrac{x^2 - 1}{x^2 + 7x + 10} \cdot \dfrac{x^2 + 5x + 6}{x + 1} \cdot \dfrac{x + 5}{x^2 + 2x - 3}$

30. $\dfrac{4x^2 - 1}{x - 5} \cdot \dfrac{x^2 - 2x - 15}{2x^2 - x - 1} \cdot \dfrac{x - 1}{x^2 - 9}$

Multiply the following expressions using the method shown in Examples 5 and 6 in this section.

31. $(5x - 5)\left(\dfrac{3}{x - 1}\right)$

32. $(4x - 8)\left(\dfrac{x}{x - 2}\right)$

33. $(x^2 - 9)\left(\dfrac{2}{x + 3}\right)$

34. $(x^2 - 9)\left(\dfrac{-3}{x - 3}\right)$

35. $a(a + 5)(a - 5) \left(\dfrac{2}{a^2 - 25} \right)$ **36.** $a(a^2 - 4) \left(\dfrac{a}{a + 2} \right)$

37. $(x^2 - x - 6) \left(\dfrac{x + 1}{x - 3} \right)$ **38.** $(x^2 - 2x - 8) \left(\dfrac{x + 3}{x - 4} \right)$

39. $(x^2 - 4x - 5) \left(\dfrac{-2x}{x + 1} \right)$ **40.** $(x^2 - 6x + 8) \left(\dfrac{4x}{x - 2} \right)$

Review Problems   The problems below review material we covered in Section 4.3. Reviewing these problems will help you in the next section.

Add the following fractions.

41. $\dfrac{4}{2} + \dfrac{5}{2}$ **42.** $\dfrac{2}{3} + \dfrac{8}{3}$

43. $\dfrac{6}{8} + \dfrac{2}{8}$ **44.** $\dfrac{3}{7} + \dfrac{4}{7}$

45. $\dfrac{4}{15} + \dfrac{6}{15}$ **46.** $\dfrac{5}{14} + \dfrac{2}{14}$

47. $\dfrac{1}{5} + \dfrac{3}{5}$ **48.** $\dfrac{2}{7} + \dfrac{3}{7}$

**6.4
Addition and
Subtraction
of Rational
Expressions**

In Section 4.3 we combined fractions having the same denominator by combining their numerators and putting the result over the common denominator. We use the same process to add two rational expressions with the same denominator.

▼ **Examples**

1.   Add $\dfrac{5}{x} + \dfrac{3}{x}$.

   SOLUTION   Adding numerators, we have

$$\frac{5}{x} + \frac{3}{x} = \frac{8}{x}$$

2.   Add $\dfrac{x}{x^2 - 9} + \dfrac{3}{x^2 - 9}$.

   SOLUTION   Since both expressions have the same denominator, we add numerators and reduce to lowest terms:

$$\frac{x}{x^2 - 9} + \frac{3}{x^2 - 9} = \frac{x + 3}{x^2 - 9}$$

$$= \frac{\cancel{x + 3}}{(\cancel{x + 3})(x - 3)} \left.\begin{array}{l} \\ \\ \\ \\ \\ \end{array}\right\} \begin{array}{l} \text{Reduce to lowest} \\ \text{terms by factoring} \\ \text{the denominator and} \\ \text{then dividing out} \\ \text{common factor } x + 3 \end{array}$$

$$= \frac{1}{x - 3}$$

▲

It is the distributive property that allows us to add rational expressions by simply adding numerators. Here is Example 1 again, this time showing the use of the distributive property:

$$\frac{5}{x} + \frac{3}{x} = 5 \cdot \frac{1}{x} + 3 \cdot \frac{1}{x} \quad \begin{array}{l} \text{Dividing by } x \text{ is the same as multiplying} \\ \text{by its reciprocal, } 1/x \end{array}$$

$$= (5 + 3) \cdot \frac{1}{x} \quad \text{Distributive property}$$

$$= 8 \cdot \frac{1}{x}$$

$$= \frac{8}{x}$$

It is not necessary to show the distributive property when adding rational expressions. The only reason for showing it here is so you can see that rational expressions cannot be combined by addition unless they have the same denominator. Because of this property, we must begin all addition problems involving rational expressions by first making sure all the expressions have the same denominator.

DEFINITION   The *least common denominator* (LCD) for a set of denominators is the simplest quantity that is exactly divisible by all the denominators.

The least common denominator for the fractions $\frac{1}{4}$ and $\frac{2}{3}$ is 12, because 12 is the smallest positive number that is divisible by both 4 and 3. Likewise, the LCD for the rational expressions $1/x$ and $\frac{2}{3}$ is the expression $3x$, because $3x$ is the simplest expression that both 3 and x divide evenly. Not all LCDs are so easy to find. Sometimes we have to factor each denominator and then build the LCD from the factors.

▼   **Example 3**   Add $\dfrac{1}{10} + \dfrac{3}{14}$.

SOLUTION

Step 1: Find the LCD for 10 and 14. To do so, we factor each denominator and build the LCD from the factors:

$$\left.\begin{array}{l} 10 = 2 \cdot 5 \\[1em] 14 = 2 \cdot 7 \end{array}\right\} \quad \text{LCD} = 2 \cdot 5 \cdot 7 = 70$$

We know the LCD is divisible by 10 because it contains the factors 2 and 5. It is also divisible by 14 because it contains the factors 2 and 7.

Step 2: Change to equivalent fractions that each have denominator 70. To accomplish this task, we multiply the numerator and denominator of each fraction by the factor of the LCD that is not also a factor of its denominator:

Original Fractions		*Denominators in Factored Form*		*Multiply by Factor Needed to Obtain LCD*		*These have the Same Value as the Original Fractions*
$\dfrac{1}{10}$	$=$	$\dfrac{1}{2 \cdot 5}$	$=$	$\dfrac{1}{2 \cdot 5} \cdot \dfrac{7}{7}$	$=$	$\dfrac{7}{70}$
$\dfrac{3}{14}$	$=$	$\dfrac{3}{2 \cdot 7}$	$=$	$\dfrac{3}{2 \cdot 7} \cdot \dfrac{5}{5}$	$=$	$\dfrac{15}{70}$

The fraction $\frac{7}{70}$ has the same value as the fraction $\frac{1}{10}$. Likewise, the fractions $\frac{15}{70}$ and $\frac{3}{14}$ are equivalent; they have the same value.

Step 3: Add numerators and put the result over the LCD:

$$\frac{7}{70} + \frac{15}{70} = \frac{7 + 15}{70} = \frac{22}{70}$$

Step 4: Reduce to lowest terms:

$$\frac{22}{70} = \frac{11}{35} \qquad \text{Divide numerator and denominator by 2} \qquad \blacktriangle$$

The main idea in adding fractions is to write each fraction again with the LCD for a denominator. Once we have done that, we simply add numerators. The same process can be used to add rational expressions, as the next example illustrates.

▼ **Example 4** Subtract $\dfrac{3}{x} - \dfrac{1}{2}$.

SOLUTION

Step 1: The LCD for x and 2 is $2x$. It is the smallest expression divisible by x and by 2.

Step 2: To change to equivalent expressions with the denominator $2x$, we multiply the first fraction by $2/2$ and the second by x/x.

$$\frac{3}{x} \cdot \frac{2}{2} = \frac{6}{2x}$$

$$\frac{1}{2} \cdot \frac{x}{x} = \frac{x}{2x}$$

Step 3: Subtracting numerators of the rational expressions in Step 2, we have

$$\frac{6}{2x} - \frac{x}{2x} = \frac{6-x}{2x}$$

Step 4: Since $6-x$ and $2x$ do not have any factors in common, we cannot reduce any further.

This is how the complete problem looks:

$$\frac{3}{x} - \frac{1}{2} = \frac{3}{x} \cdot \frac{2}{2} - \frac{1}{2} \cdot \frac{x}{x}$$

$$= \frac{6}{2x} - \frac{x}{2x}$$

$$= \frac{6-x}{2x} \qquad \blacktriangle$$

▼ **Example 5** Add $\dfrac{5}{2x-6} + \dfrac{x}{x-3}$.

SOLUTION If we factor $2x - 6$, we have $2x - 6 = 2(x - 3)$. We need only multiply the second rational expression in our problem by $\frac{2}{2}$ to have two expressions with the same denominator:

$$\frac{5}{2x-6} + \frac{x}{x-3} = \frac{5}{2(x-3)} + \frac{x}{x-3}$$

$$= \frac{5}{2(x-3)} + \frac{2}{2}\left(\frac{x}{x-3}\right)$$

$$= \frac{5}{2(x - 3)} + \frac{2x}{2(x - 3)}$$

$$= \frac{2x + 5}{2(x - 3)}$$ ▲

▼ **Example 6** Add $\dfrac{2}{x^2 + 5x + 6} + \dfrac{x}{x^2 - 9}$.

SOLUTION

Step 1: We factor each denominator and build the LCD from the factors:

$$\left. \begin{array}{l} x^2 + 5x + 6 = (x + 2)(x + 3) \\ x^2 - 9 = (x + 3)(x - 3) \end{array} \right\} \ \text{LCD} = (x + 2)(x + 3)(x - 3)$$

Step 2: Change to equivalent rational expressions:

$$\frac{2}{x^2 + 5x + 6} = \frac{2}{(x + 2)(x + 3)} \cdot \frac{(x - 3)}{(x - 3)} = \frac{2x - 6}{(x + 2)(x + 3)(x - 3)}$$

$$\frac{x}{x^2 - 9} = \frac{x}{(x + 3)(x - 3)} \cdot \frac{(x + 2)}{(x + 2)} = \frac{x^2 + 2x}{(x + 2)(x + 3)(x - 3)}$$

Step 3: Add numerators of the rational expressions produced in Step 2:

$$\frac{2x - 6}{(x + 2)(x + 3)(x - 3)} + \frac{x^2 + 2x}{(x + 2)(x + 3)(x - 3)}$$

$$= \frac{x^2 + 4x - 6}{(x + 2)(x + 3)(x - 3)}$$

The numerator and denominator do not have any factors in common.   ▲

▼ **Example 7** Subtract $\dfrac{x}{x^2 - 25} - \dfrac{2}{3x + 15}$

SOLUTION We begin by factoring each denominator:

$$\frac{x}{x^2 - 25} - \frac{2}{3x + 15} = \frac{x}{(x + 5)(x - 5)} + \frac{-2}{3(x + 5)}$$

The LCD is $3(x + 5)(x - 5)$. Completing the problem, we have

$$= \frac{3}{3} \cdot \frac{x}{(x + 5)(x - 5)} + \frac{-2}{3(x + 5)} \cdot \frac{(x - 5)}{(x - 5)}$$

$$= \frac{3x}{3(x+5)(x-5)} + \frac{-2x+10}{3(x+5)(x-5)}$$

$$= \frac{x+10}{3(x+5)(x-5)} \qquad \blacktriangle$$

Notice in the first step that we replaced subtraction by addition of the opposite. There seems to be less chance for error when this is done on longer problems.

Find the following sums and differences.   Problem Set 6.4

1. $\dfrac{3}{x} + \dfrac{4}{x}$

2. $\dfrac{5}{x} + \dfrac{3}{x}$

3. $\dfrac{9}{a} - \dfrac{5}{a}$

4. $\dfrac{8}{a} - \dfrac{7}{a}$

5. $\dfrac{1}{x+1} + \dfrac{x}{x+1}$

6. $\dfrac{3}{x-3} - \dfrac{x}{x-3}$

7. $\dfrac{y^2}{y-1} - \dfrac{1}{y-1}$

8. $\dfrac{y^2}{y+3} - \dfrac{9}{y+3}$

9. $\dfrac{x^2}{x+2} + \dfrac{4x+4}{x+2}$

10. $\dfrac{x^2-6x}{x-3} + \dfrac{9}{x-3}$

11. $\dfrac{5}{x} + \dfrac{1}{3}$

12. $\dfrac{2}{x} + \dfrac{1}{2}$

13. $\dfrac{3}{a} - \dfrac{5}{6}$

14. $\dfrac{5}{a} - \dfrac{3}{4}$

15. $\dfrac{y}{2} - \dfrac{2}{y}$

16. $\dfrac{3}{y} + \dfrac{y}{3}$

17. $\dfrac{1}{2} + \dfrac{a}{3}$

18. $\dfrac{2}{3} + \dfrac{2a}{5}$

19. $\dfrac{x}{x+1} + \dfrac{3}{4}$

20. $\dfrac{x}{x-3} + \dfrac{1}{3}$

21. $\dfrac{3}{x-2} + \dfrac{2}{5x-10}$

22. $\dfrac{5}{2x-6} + \dfrac{3}{x-3}$

23. $\dfrac{4}{3x+12} - \dfrac{x}{x+4}$

24. $\dfrac{2}{5x-25} - \dfrac{x}{x-5}$

25. $\dfrac{6}{x(x-2)} + \dfrac{3}{x}$

26. $\dfrac{10}{x(x+5)} - \dfrac{2}{x}$

27. $\dfrac{4}{a} - \dfrac{12}{a^2 + 3a}$

28. $\dfrac{5}{a} + \dfrac{20}{a^2 - 4a}$

29. $\dfrac{2}{x+5} + \dfrac{1}{x^2 - 25}$

30. $\dfrac{4}{x^2 - 1} + \dfrac{3}{x+1}$

31. $\dfrac{1}{x-3} - \dfrac{6}{x^2 - 9}$

32. $\dfrac{1}{x-1} - \dfrac{2}{x^2 - 1}$

33. $\dfrac{5a}{a^2 - a - 6} - \dfrac{2}{a-3}$

34. $\dfrac{3a}{a^2 - 5a - 6} - \dfrac{4}{a+1}$

35. $\dfrac{2}{2x-6} - \dfrac{5}{x^2 - 9}$

36. $\dfrac{x}{3x-3} - \dfrac{1}{x^2 - 2x + 1}$

37. $\dfrac{5}{a} + \dfrac{2}{a+1}$

38. $\dfrac{a}{a-3} + \dfrac{7}{a}$

39. $\dfrac{8}{x^2 - 16} - \dfrac{7}{x^2 - x - 12}$

40. $\dfrac{6}{x^2 - 9} - \dfrac{5}{x^2 - x - 6}$

41. $\dfrac{4y}{y^2 + 6y + 5} - \dfrac{3y}{y^2 + 5y + 4}$

42. $\dfrac{3y}{y^2 + 7y + 10} - \dfrac{2y}{y^2 + 6y + 8}$

Review Problems   The problems below review material we covered in Section 5.7. Reviewing these problems will help you with some of the material in the next section.

Solve each quadratic equation.

43. $x^2 + 5x + 6 = 0$

44. $x^2 - 5x + 6 = 0$

45. $x^2 - x - 6 = 0$

46. $x^2 + x - 6 = 0$

47. $x^2 + x - 20 = 0$

48. $x^2 - x - 20 = 0$

49. $x^2 - 9x + 20 = 0$

50. $x^2 + 9x + 20 = 0$

6.5
Equations Involving Rational Expressions

   The first step in solving an equation that contains one or more rational expressions is to find the LCD for all denominators in the equation. Once the LCD has been found, we multiply both sides of the equation by it. The resulting equation should be equivalent to the original one (unless we inadvertently multiplied by zero) and free from any denominators except the number 1.

▼ **Example 1** Solve $\dfrac{x}{3} + \dfrac{5}{2} = \dfrac{1}{2}$ for x.

SOLUTION The LCD for 3 and 2 is 6. If we multiply both sides by 6, we have

$$6\left(\frac{x}{3} + \frac{5}{2}\right) = 6\left(\frac{1}{2}\right) \qquad \text{Multiply both sides by 6}$$

$$6\left(\frac{x}{3}\right) + 6\left(\frac{5}{2}\right) = 6\left(\frac{1}{2}\right) \qquad \text{Distributive property}$$

$$2x + 15 = 3$$

$$2x = -12$$

$$x = -6$$

We can check our solution by replacing x with -6 in the original equation:

$$-\frac{6}{3} + \frac{5}{2} = \frac{1}{2}$$

$$\frac{1}{2} = \frac{1}{2} \qquad\qquad\qquad ▲$$

Multiplying both sides of an equation containing fractions by the LCD clears the equation of all denominators, because the LCD has the property that all the denominators will divide it evenly.

▼ **Example 2** Find the solution set for $\dfrac{3}{x-1} = \dfrac{3}{5}$.

SOLUTION The LCD for $(x - 1)$ and 5 is $5(x - 1)$. Multiplying both sides by $5(x - 1)$ we have

$$5(x-1)\cdot\frac{3}{x-1} = 5(x-1)\cdot\frac{3}{5}$$

$$5\cdot 3 = (x-1)\cdot 3$$

$$15 = 3x - 3$$

$$18 = 3x$$

$$6 = x$$

If we substitute $x = 6$ into the original equation, we have

$$\frac{3}{6-1} = \frac{3}{5}$$

$$\frac{3}{5} = \frac{3}{5}$$

The solution set is $\{6\}$.   ▲

▼ **Example 3** Solve $1 - \dfrac{5}{x} = \dfrac{-6}{x^2}$.

SOLUTION The LCD is x^2. Multiplying both sides by x^2, we have

$$x^2\left(1 - \frac{5}{x}\right) = x^2\left(\frac{-6}{x^2}\right) \qquad \text{Multiply both sides by } x^2$$

$$x^2(1) - x^2\left(\frac{5}{x}\right) = x^2\left(\frac{-6}{x^2}\right) \qquad \begin{array}{l}\text{Apply distributive property} \\ \quad \text{to the left side}\end{array}$$

$$x^2 - 5x = -6 \qquad \text{Simplify each side}$$

We have a quadratic equation, which we write in standard form, factor, and solve as we did in Section 5.6.

$$x^2 - 5x + 6 = 0 \qquad \text{Standard form}$$

$$(x - 2)(x - 3) = 0 \qquad \text{Factor}$$

$$x - 2 = 0 \quad \text{or} \quad x - 3 = 0 \qquad \text{Set factors to 0}$$

$$x = 2 \quad \text{or} \qquad x = 3$$

The two possible solutions are 2 and 3. Checking each in the original equation, we find they both give true statements. They are both solutions to the original equation.

$$\text{Check } x = 2. \qquad \text{Check } x = 3.$$

$$1 - \frac{5}{2} = \frac{-6}{4} \qquad 1 - \frac{5}{3} = \frac{-6}{9}$$

$$\frac{2}{2} - \frac{5}{2} = -\frac{3}{2} \qquad \frac{3}{3} - \frac{5}{3} = -\frac{2}{3}$$

$$-\frac{3}{2} = -\frac{3}{2} \qquad -\frac{2}{3} = -\frac{2}{3} \qquad \qquad ▲$$

▼ **Example 4**   Solve $\dfrac{x}{x^2 - 9} - \dfrac{3}{x - 3} = \dfrac{1}{x + 3}$.

SOLUTION   The factors of $x^2 - 9$ are $(x + 3)(x - 3)$. The LCD, then, is $(x + 3)(x - 3)$:

$$(x + 3)(x - 3) \cdot \dfrac{x}{(x + 3)(x - 3)} + (x + 3)(x - 3) \cdot \dfrac{-3}{x - 3}$$

$$= (x + 3)(x - 3) \cdot \dfrac{1}{x + 3}$$

$$x + (x + 3)(-3) = (x - 3)1$$

$$x + (-3x) + (-9) = x - 3$$

$$-2x - 9 = x - 3$$

$$x = -2$$

The solution is $x = -2$. It checks when replaced for x in the original equation.   ▲

▼ **Example 5**   Solve $\dfrac{x}{x - 3} + \dfrac{3}{2} = \dfrac{3}{x - 3}$.

SOLUTION   We begin by multiplying each term on both sides of the equation by $2(x - 3)$.

$$2(x - 3) \cdot \dfrac{x}{x - 3} + 2(x - 3) \cdot \dfrac{3}{2} = 2(x - 3) \cdot \dfrac{3}{x - 3}$$

$$2x + (x - 3) \cdot 3 = 2 \cdot 3$$

$$2x + 3x - 9 = 6$$

$$5x - 9 = 6$$

$$5x = 15$$

$$x = 3$$

Our only possible solution is $x = 3$. If we substitute $x = 3$ into our original equation, we get

$$\dfrac{3}{3 - 3} + \dfrac{3}{2} = \dfrac{3}{3 - 3}$$

$$\dfrac{3}{0} + \dfrac{3}{2} = \dfrac{3}{0}$$

Two of the terms are undefined, so the equation is meaningless. What has happened is that we multiplied both sides of the original equation by zero. The equation produced by doing this is not equivalent to our original equation. We must always check our solution when we multiply both sides of an equation by an expression containing the variable in order to make sure we have not multiplied both sides by zero.

Our original equation has no solutions. That is, there is no real number x such that

$$\frac{x}{x-3} + \frac{3}{2} = \frac{3}{x-3}$$

The solution set is $\emptyset$.   ▲

▼ **Example 6**   Solve $\dfrac{a+4}{a^2+5a} = \dfrac{-2}{a^2-25}$ for a.

SOLUTION   Factoring each denominator, we have

$$a^2 + 5a = a(a+5)$$

$$a^2 - 25 = (a+5)(a-5)$$

The LCD is $a(a+5)(a-5)$. Multiplying both sides of the equation by the LCD gives us

$$a(a+5)(a-5) \cdot \frac{a+4}{a(a+5)} = \frac{-2}{(a+5)(a-5)} \cdot a(a+5)(a-5)$$

$$(a-5)(a+4) = -2a$$

$$a^2 - a - 20 = -2a$$

The result is a quadratic equation, which we write in standard form, factor, and solve:

$$a^2 + a - 20 = 0 \qquad \text{Add } 2a \text{ to both sides}$$

$$(a+5)(a-4) = 0 \qquad \text{Factor}$$

$$a + 5 = 0 \quad \text{or} \quad a - 4 = 0 \qquad \text{Set each factor to 0}$$

$$a = -5 \quad \text{or} \qquad a = 4$$

The two possible solutions are -5 and 4. There is no problem with the 4. It checks when substituted for a in the original equation. However, -5 is not a solution. Substituting -5 into the original equation gives

$$\frac{-5 + 4}{(-5)^2 + 5(-5)} = \frac{-2}{(-5)^2 - 25}$$

$$\frac{-1}{0} = \frac{-2}{0}$$

This indicates -5 is not a solution. The solution is 4. ▲

Solve the following equations. Be sure to check each answer in the original equation if you multiply both sides by an expression that contains the variable.

1. $\dfrac{x}{3} + \dfrac{1}{2} = -\dfrac{1}{2}$

2. $\dfrac{x}{2} + \dfrac{4}{3} = -\dfrac{2}{3}$

3. $\dfrac{4}{a} = \dfrac{1}{5}$

4. $\dfrac{2}{3} = \dfrac{6}{a}$

5. $\dfrac{3}{x} + 1 = \dfrac{2}{x}$

6. $\dfrac{4}{x} + 3 = \dfrac{1}{x}$

7. $\dfrac{3}{a} - \dfrac{2}{a} = \dfrac{1}{5}$

8. $\dfrac{7}{a} + \dfrac{1}{a} = 2$

9. $\dfrac{3}{x} + 2 = \dfrac{1}{2}$

10. $\dfrac{5}{x} + 3 = \dfrac{4}{3}$

11. $\dfrac{1}{y} - \dfrac{1}{2} = -\dfrac{1}{4}$

12. $\dfrac{3}{y} - \dfrac{4}{5} = -\dfrac{1}{5}$

13. $1 - \dfrac{8}{x} = \dfrac{-15}{x^2}$

14. $1 - \dfrac{3}{x} = \dfrac{-2}{x^2}$

15. $\dfrac{x}{2} - \dfrac{4}{x} = -\dfrac{7}{2}$

16. $\dfrac{x}{2} - \dfrac{5}{x} = -\dfrac{3}{2}$

17. $\dfrac{x - 3}{2} + \dfrac{2x}{3} = \dfrac{5}{6}$

18. $\dfrac{x - 2}{3} + \dfrac{5x}{2} = 5$

19. $\dfrac{x + 1}{3} + \dfrac{x - 3}{4} = \dfrac{1}{6}$

20. $\dfrac{x + 2}{3} + \dfrac{x - 1}{5} = -\dfrac{3}{5}$

21. $\dfrac{6}{x + 2} = \dfrac{3}{5}$

22. $\dfrac{4}{x + 3} = \dfrac{1}{2}$

23. $\dfrac{3}{y - 2} = \dfrac{2}{y - 3}$

24. $\dfrac{5}{y + 1} = \dfrac{4}{y + 2}$

25. $\dfrac{x}{x-2} + \dfrac{2}{3} = \dfrac{2}{x-2}$ **26.** $\dfrac{x}{x-5} + \dfrac{1}{5} = \dfrac{5}{x-5}$

27. $\dfrac{x}{x-2} + \dfrac{3}{2} = \dfrac{9}{2(x-2)}$ **28.** $\dfrac{x}{x+1} + \dfrac{4}{5} = \dfrac{-14}{5(x+1)}$

29. $\dfrac{5}{x+2} + \dfrac{1}{x+3} = \dfrac{-1}{x^2+5x+6}$ **30.** $\dfrac{3}{x-1} + \dfrac{2}{x+3} = \dfrac{-3}{x^2+2x-3}$

31. $\dfrac{8}{x^2-4} + \dfrac{3}{x+2} = \dfrac{1}{x-2}$ **32.** $\dfrac{10}{x^2-25} - \dfrac{1}{x-5} = \dfrac{3}{x+5}$

33. $\dfrac{a}{2} + \dfrac{3}{a-3} = \dfrac{a}{a-3}$ **34.** $\dfrac{a}{2} + \dfrac{4}{a-4} = \dfrac{a}{a-4}$

35. $\dfrac{6}{y^2-4} = \dfrac{4}{y^2+2y}$ **36.** $\dfrac{2}{y^2-9} = \dfrac{5}{y^2-3y}$

37. $\dfrac{2}{a^2-9} = \dfrac{3}{a^2+a-12}$ **38.** $\dfrac{2}{a^2-1} = \dfrac{6}{a^2-2a-3}$

39. $\dfrac{3x}{x-5} - \dfrac{2x}{x+1} = \dfrac{-42}{x^2-4x-5}$ **40.** $\dfrac{4x}{x-4} - \dfrac{3x}{x-2} = \dfrac{-3}{x^2-6x+8}$

41. $\dfrac{2x}{x+2} = \dfrac{x}{x+3} - \dfrac{3}{x^2+5x+6}$ **42.** $\dfrac{3x}{x-4} = \dfrac{2x}{x-3} + \dfrac{6}{x^2-7x+12}$

Review Problems   The problems below review material we covered in Section 3.2.

Graph each straight line.

43. $x + y = 3$ **44.** $x - y = 3$
45. $4x - 3y = 12$ **46.** $3x + 4y = 12$
47. $y = 2x - 3$ **48.** $y = 3x - 2$
49. $y = 2x$ **50.** $y = -2x$

6.6
Word Problems

In this section we will solve some word problems whose equations involve rational expressions. Like the other word problems we have encountered, the more you work with them the easier they become.

▼   **Example 1**   One number is twice another. The sum of their reciprocals is $\frac{9}{2}$. Find the two numbers.

SOLUTION Let $x =$ the smaller number. The larger then must be $2x$. Their reciprocals are $1/x$ and $1/2x$, respectively. An equation that describes the situation is

$$\frac{1}{x} + \frac{1}{2x} = \frac{9}{2}$$

We can multiply both sides by the LCD $2x$ and then solve the resulting equation:

$$2x\left(\frac{1}{x}\right) + 2x\left(\frac{1}{2x}\right) = 2x\left(\frac{9}{2}\right)$$

$$2 + 1 = 9x$$

$$3 = 9x$$

$$x = \frac{3}{9}$$

$$x = \frac{1}{3}$$

The smaller number is $\frac{1}{3}$. The other number is twice as large, or $\frac{2}{3}$. If we add their reciprocals, we have

$$\frac{3}{1} + \frac{3}{2} = \frac{6}{2} + \frac{3}{2}$$

$$= \frac{9}{2}$$

The solutions check with the original problem.   ▲

▼ **Example 2** A boat travels 30 miles up a river in the same amount of time it takes to travel 50 miles down the same river. If the current is 5 mph, what is the speed of the boat in still water?

SOLUTION Let $x =$ speed of the boat in still water. Then $x + 5 =$ speed of the boat downstream and $x - 5 =$ speed of the boat upstream.

With all problems involving distance, rate, and time, the basic formula is distance $=$ rate $\times$ time, or $d = r \cdot t$. The formula has two other, equivalent forms: $r = d/t$ and $t = d/r$.

Reading back over the problem, we see that the time upstream and the time downstream are equal:

$$\text{time (downstream)} = \text{time (upstream)}$$

and since

$$\text{time} = \frac{\text{distance}}{\text{rate}}$$

we have

$$\frac{\text{distance}}{\text{rate}}\text{(downstream)} = \frac{\text{distance}}{\text{rate}}\text{(upstream)}$$

$$\frac{50}{x + 5} = \frac{30}{x - 5}$$

The LCD is $(x + 5)(x - 5)$. We multiply both sides of the equation by the LCD to clear it of all denominators. Here is the solution:

$$(x + 5)(x - 5) \cdot \frac{50}{x + 5} = (x + 5)(x - 5) \cdot \frac{30}{x - 5}$$

$$(x - 5)50 = (x + 5)30$$

$$50x - 250 = 30x + 150$$

$$20x = 400$$

$$x = 20$$

The speed of the boat in still water is 20 mph.   ▲

▼ **Example 3** An inlet pipe can fill a water tank in 10 hours, while an outlet pipe can empty the same tank in 15 hours. By mistake, both pipes are left open. How long will it take to fill the water tank with both pipes open?

SOLUTION Let $x =$ amount of time to fill the tank with both pipes open.

One method of solving this type of problem is to think in terms of how much of the job is done by a pipe in 1 hour:

1. If the inlet pipe fills the tank in 10 hours, then in 1 hour the inlet pipe fills $\frac{1}{10}$ of the tank.

2. If the outlet pipe empties the tank in 15 hours, then in 1 hour the outlet pipe empties $\frac{1}{15}$ of the tank.

3. If it takes x hours to fill the tank with both pipes open, then in 1 hour the tank is $1/x$ full.

Here is how we set up the equation. *In one hour,*

$$\frac{1}{10} \quad - \quad \frac{1}{15} \quad = \quad \frac{1}{x}$$

Amount of water let	Amount of water let	Total amount of
in by inlet pipe	out by outlet pipe	water in tank

The LCD for our equation is $30x$. We multiply both sides by the LCD and solve:

$$30x\left(\frac{1}{10}\right) - 30x\left(\frac{1}{15}\right) = 30x\left(\frac{1}{x}\right)$$

$$3x - 2x = 30$$

$$x = 30$$

It takes 30 hours with both pipes open to fill the tank.   ▲

In solving any problem of this type, we have to assume that whatever is doing the work (whether it is a pipe, a person, or a machine) is working at a constant rate. That is, as much work gets done in the first hour as is done in the last or any other hour.

Problem Set 6.6

1. One number is three times as large as another. The sum of their reciprocals is $\frac{16}{3}$. Find the two numbers.
2. If $\frac{3}{5}$ of a certain number is subtracted from the number, the result is 6. Find the number.
3. If $\frac{3}{5}$ is added to twice the reciprocal of a number, the result is 1. Find the number.
4. The sum of a number and its reciprocal is $\frac{13}{6}$. Find the number.
5. The sum of a number with 10 times its reciprocal is 7. Find the number.
6. If a certain number is added to both the numerator and denominator of the fraction $\frac{7}{9}$, the result is $\frac{5}{7}$. Find the number.
7. The numerator of a certain fraction is two more than the denominator. If $\frac{1}{3}$ is added to the fraction, the result is 2. Find the fraction.
8. The sum of the reciprocals of two consecutive even integers is $\frac{5}{12}$. Find the integers.
9. The sum of the reciprocals of two consecutive integers is $\frac{7}{12}$. Find the two integers.
10. A boat travels 26 miles up a river in the same amount of time it takes to travel 38 miles down the same river. If the current is 3 mph, what is the speed of the boat in still water?
11. A boat can travel 9 miles up a river in the same amount of time it takes

to travel 11 miles down the same river. If the current is 2 mph, what is the speed of the boat in still water?

12. An airplane flying against the wind travels 140 miles in the same amount of time it would take the same plane to travel 160 miles with the wind. If the wind speed is a constant 20 mph, how fast would the plane travel in still air?

13. An airplane flying against the wind travels 500 miles in the same amount of time that it would take to travel 600 miles with the wind. If the speed of the wind is 50 mph, what is the speed of the plane in still air?

14. One plane can travel 20 mph faster than another. One of them goes 285 miles in the same time it takes the other to go 255 miles. What are their speeds?

15. One car travels 300 miles in the same amount of time it takes a second car traveling 5 mph slower than the first to go 275 miles. What are the speeds of the cars?

16. Train A is 15 mph slower than train B. Train B travels 180 miles in the same amount of time it takes train A to travel 135 miles. How fast is train A?

17. An inlet pipe can fill a pool in 12 hours, while an outlet pipe can empty it in 15 hours. If both pipes are left open, how long will it take to fill the pool?

18. A water tank can be filled in 20 hours by an inlet pipe and emptied in 25 hours by an outlet pipe. How long will it take to fill the tank if both pipes are left open?

19. A bathtub can be filled by the cold water faucet in 10 minutes and by the hot water faucet in 12 minutes. How long does it take to fill the tub if both faucets are open?

20. A water faucet can fill a sink in 6 minutes, while the drain can empty it in 4 minutes. If the sink is full, how long will it take to empty if both the faucet and the drain are open?

Review Problems   The problems below review material we covered in Sections 3.4 and 3.5.

Solve each system of equations.

21. $2x + y = 3$
   $3x - y = 7$

22. $3x - y = -6$
   $4x + y = -8$

23. $4x - 5y = 1$
   $x - 2y = -2$

24. $6x - 4y = 2$
   $2x + y = 10$

25. $5x + 2y = 7$
   $y = 3x - 2$

26. $-7x - 5y = -1$
   $y = x + 5$

27. $4x - 9y = 2$
   $y = x - 3$

28. $3x - 5y = 4$
   $y = x - 2$

A complex fraction is a fraction or rational expression that contains other fractions in its numerator or denominator. Each of the following is a complex fraction:

$$\frac{\frac{1}{2}}{\frac{2}{3}} \qquad \frac{x + \dfrac{1}{y}}{y + \dfrac{1}{x}} \qquad \frac{\dfrac{a+1}{a^2 - 9}}{\dfrac{2}{a+3}}$$

We will begin this section by simplifying the first complex fraction given above. Before we do, though, let's agree on some vocabulary. So that we won't have to use phrases such as the numerator of the denominator, let's call the numerator of a complex fraction the *top* and the denominator of a complex fraction the *bottom*.

▼ **Example 1** Simplify $\dfrac{\frac{1}{2}}{\frac{2}{3}}$.

SOLUTION There are two methods we can use to solve this problem.

   METHOD 1   We can treat this as a division problem. Instead of dividing by $\frac{2}{3}$, we can multiply by its reciprocal $\frac{3}{2}$.

$$\frac{\frac{1}{2}}{\frac{2}{3}} = \frac{1}{2} \cdot \frac{3}{2} = \frac{3}{4}$$

   METHOD 2   We can multiply the top and bottom of this complex fraction by the LCD for both fractions. In this case, the LCD is 6:

$$\frac{\frac{1}{2}}{\frac{2}{3}} = \frac{6 \cdot \dfrac{1}{2}}{6 \cdot \dfrac{2}{3}} = \frac{3}{4}$$

Using either method, we obtain the same result.   ▲

▼ **Example 2** Simplify

$$\frac{\dfrac{2x^3}{y^2}}{\dfrac{4x}{y^5}}$$

SOLUTION

METHOD 1   Instead of dividing by $4x/y^5$ we can multiply by its reciprocal, $y^5/4x$:

$$\frac{\dfrac{2x^3}{y^2}}{\dfrac{4x}{y^5}} = \frac{2x^3}{y^2} \cdot \frac{y^5}{4x} = \frac{x^2y^3}{2}$$

METHOD 2   The LCD for each rational expression is y^5. Multiplying the top and bottom of the complex fraction by y^5, we have

$$\frac{\dfrac{2x^3}{y^2}}{\dfrac{4x}{y^5}} = \frac{y^5 \cdot \dfrac{2x^3}{y^2}}{y^5 \cdot \dfrac{4x}{y^5}} = \frac{2x^3y^3}{4x} = \frac{x^2y^3}{2}$$

Again the result is the same, whether we use Method 1 or Method 2.   ▲

▼ **Example 3**   Simplify

$$\frac{x + \dfrac{1}{y}}{y + \dfrac{1}{x}}$$

SOLUTION   To apply Method 1 as we did in the first two examples, we would have to simplify the top and bottom separately to obtain a single rational expression for both before we could multiply by the reciprocal. It is much easier, in this case, to multiply the top and bottom by the LCD xy:

$$\frac{x + \dfrac{1}{y}}{y + \dfrac{1}{x}} = \frac{xy\left(x + \dfrac{1}{y}\right)}{xy\left(y + \dfrac{1}{x}\right)} \qquad \text{Multiply top and bottom by } xy$$

$$= \frac{xy \cdot x + xy \cdot \dfrac{1}{y}}{xy \cdot y + xy \cdot \dfrac{1}{x}} \qquad \text{Distributive property}$$

$$= \frac{x^2y + x}{xy^2 + y} \qquad \text{Simplify}$$

We can factor an x from $x^2y + x$ and a y from $xy^2 + y$ and then reduce to lowest terms:

$$= \frac{x(xy + 1)}{y(xy + 1)}$$

$$= \frac{x}{y}$$

▲

▼ **Example 4** Simplify

$$\frac{1 - \dfrac{4}{x^2}}{1 - \dfrac{1}{x} - \dfrac{6}{x^2}}$$

SOLUTION The simplest way to simplify this complex fraction is to multiply the top and bottom by the LCD, x^2:

$$\frac{1 - \dfrac{4}{x^2}}{1 - \dfrac{1}{x} - \dfrac{6}{x^2}} = \frac{x^2\left(1 - \dfrac{4}{x^2}\right)}{x^2\left(1 - \dfrac{1}{x} - \dfrac{6}{x^2}\right)} \qquad \text{Multiply top and bottom by } x^2$$

$$= \frac{x^2 \cdot 1 - x^2 \cdot \dfrac{4}{x^2}}{x^2 \cdot 1 - x^2 \cdot \dfrac{1}{x} - x^2 \cdot \dfrac{6}{x^2}} \qquad \text{Distributive property}$$

$$= \frac{x^2 - 4}{x^2 - x - 6} \qquad \text{Simplify}$$

$$= \frac{(x - 2)(x + 2)}{(x - 3)(x + 2)} \qquad \text{Factor}$$

$$= \frac{x - 2}{x - 3} \qquad \text{Reduce}$$

▲

Simplify each complex fraction.

1. $\dfrac{\frac{3}{4}}{\frac{1}{8}}$ **2.** $\dfrac{\frac{1}{3}}{\frac{5}{6}}$ **3.** $\dfrac{\frac{2}{3}}{4}$ **4.** $\dfrac{5}{\frac{1}{2}}$

5. $\dfrac{\frac{x^2}{y}}{\frac{x}{y^3}}$ **6.** $\dfrac{\frac{x^5}{y^3}}{\frac{x^2}{y^8}}$ **7.** $\dfrac{\frac{4x^3}{y^6}}{\frac{8x^2}{y^7}}$ **8.** $\dfrac{\frac{6x^4}{y}}{\frac{2x}{y^5}}$

9. $\dfrac{y + \dfrac{1}{x}}{x + \dfrac{1}{y}}$ **10.** $\dfrac{y - \dfrac{1}{x}}{x - \dfrac{1}{y}}$ **11.** $\dfrac{1 + \dfrac{1}{a}}{1 - \dfrac{1}{a}}$ **12.** $\dfrac{\dfrac{1}{a} - 1}{\dfrac{1}{a} + 1}$

13. $\dfrac{\dfrac{x + 1}{x^2 - 9}}{\dfrac{2}{x + 3}}$ **14.** $\dfrac{\dfrac{3}{x - 5}}{\dfrac{x + 1}{x^2 - 25}}$ **15.** $\dfrac{\dfrac{1}{a + 2}}{\dfrac{1}{a^2 - a - 6}}$ **16.** $\dfrac{\dfrac{1}{a^2 + 5a + 6}}{\dfrac{1}{a + 3}}$

17. $\dfrac{1 - \dfrac{9}{y^2}}{1 - \dfrac{1}{y} - \dfrac{6}{y^2}}$ **18.** $\dfrac{1 - \dfrac{4}{y^2}}{1 - \dfrac{2}{y} - \dfrac{8}{y^2}}$

19. $\dfrac{\dfrac{1}{y} + \dfrac{1}{x}}{\dfrac{1}{xy}}$ **20.** $\dfrac{\dfrac{xy}{}}{\dfrac{1}{y} - \dfrac{1}{x}}$

21. $\dfrac{1 - \dfrac{1}{a^2}}{1 - \dfrac{1}{a}}$ **22.** $\dfrac{1 + \dfrac{1}{a}}{1 - \dfrac{1}{a^2}}$

23. $\dfrac{\dfrac{1}{10x} - \dfrac{y}{10x^2}}{\dfrac{1}{10} - \dfrac{y}{10x}}$ **24.** $\dfrac{\dfrac{1}{2x} + \dfrac{y}{2x^2}}{\dfrac{1}{4} + \dfrac{y}{4x}}$

25. $\dfrac{\dfrac{1}{a + 1} + 2}{\dfrac{1}{a + 1} + 3}$ **26.** $\dfrac{\dfrac{2}{a + 1} + 3}{\dfrac{3}{a + 1} + 4}$

Review Problems The problems below review material we covered in Section 2.6.

Solve each inequality.

27. $2x + 3 < 5$ **28.** $3x - 2 > 7$

29. $-3x \leq 21$ **30.** $-5x \geq -10$

31. $-2x + 8 > -4$ **32.** $-4x - 1 < 11$

33. $4 - 2(x + 1) \geq -2$ **34.** $6 - 2(x + 3) \leq -8$

Chapter 6 Summary and Review

RATIONAL NUMBERS [6.1]

Any number that can be put in the form a/b, where a and b are integers ($b \neq 0$), is called a rational number.

Multiplying or dividing the numerator and denominator of a rational number by the same nonzero number never changes the value of the rational number.

RATIONAL EXPRESSIONS [6.1]

Any expression of the form P/Q, where P and Q are polynomials ($Q \neq 0$), is a rational expression.

Multiplying or dividing the numerator and denominator of a rational expression by the same nonzero quantity always produces a rational expression equivalent to the original one.

LONG DIVISION WITH POLYNOMIALS [6.2]

If division with polynomials cannot be accomplished by dividing out factors common to the numerator and denominator, then we use a process similar to long division with whole numbers. The steps in the process are: estimate, multiply, subtract, and bring down the next term. The divisors in all the long division problems in this chapter were binomials. For division by a monomial, see Section 4.7.

MULTIPLICATION [6.3]

To multiply two rational numbers or two rational expressions, multiply numerators, multiply denominators, and divide out any factors common to the numerator and denominator:

For rational numbers: $\dfrac{a}{b}$ and $\dfrac{c}{d}$, $\dfrac{a}{b} \cdot \dfrac{c}{d} = \dfrac{ac}{bd}$

For rational expressions: $\dfrac{P}{Q}$ and $\dfrac{R}{S}$, $\dfrac{P}{Q} \cdot \dfrac{R}{S} = \dfrac{PR}{QS}$

Examples

1.   We can reduce $\frac{6}{8}$ to lowest terms by dividing the numerator and denominator by their greatest common factor 2:

$$\frac{6}{8} = \frac{\cancel{2} \cdot 3}{\cancel{2} \cdot 4} = \frac{3}{4}$$

2.   We reduce rational expressions to lowest terms by factoring the numerator and denominator and then dividing out any factors they have in common.

$$\frac{x - 3}{x^2 - 9} = \frac{\cancel{x - 3}}{(\cancel{x - 3})(x + 3)} = \frac{1}{x + 3}$$

3. $x - 3 \overline{\smash{)}\, x^2 - 5x + 8} \quad \dfrac{x - 2}{}$

$$\begin{array}{r} x - 2 \\ x-3\overline{\smash{)}x^2 - 5x + 8} \\ \underline{-+} \\ \cancel{x^2} \cancel{+} 3x \quad \downarrow \\ \hline -2x + 8 \\ \underline{+-} \\ \cancel{+}\, 2x \cancel{+} 6 \\ \hline 2 \end{array}$$

4. $\dfrac{x - 1}{x^2 + 2x - 3} \cdot \dfrac{x^2 - 9}{x - 2}$

$$= \frac{x - 1}{(x + 3)(x - 1)} \cdot \frac{(x - 3)(x + 3)}{x - 2}$$

$$= \frac{x - 3}{x - 2}$$

5. $\dfrac{2x}{x^2 - 25} \div \dfrac{4}{x - 5}$

$= \dfrac{2x}{(x - 5)(x + 5)} \cdot \dfrac{x - 5}{4}$

$= \dfrac{x}{2(x + 5)}$

6. $\dfrac{3}{x - 1} + \dfrac{x}{2}$

$= \dfrac{3}{x - 1} \cdot \dfrac{2}{2} + \dfrac{x}{2} \cdot \dfrac{x - 1}{x - 1}$

$= \dfrac{6}{2(x - 1)} + \dfrac{x^2 - x}{2(x - 1)}$

$= \dfrac{x^2 - x + 6}{2(x - 1)}$

7. $\dfrac{x}{x^2 - 4} - \dfrac{2}{x^2 - 4}$

$= \dfrac{x - 2}{x^2 - 4}$

$= \dfrac{x - 2}{(x - 2)(x + 2)}$

$= \dfrac{1}{x + 2}$

8. Solve $\dfrac{1}{2} + \dfrac{3}{x} = 5$.

$2x\left(\dfrac{1}{2}\right) + 2x\left(\dfrac{3}{x}\right) = 2x(5)$

$x + 6 = 10x$

$x = \dfrac{2}{3}$

DIVISION [6.3]

To divide by a rational number or rational expression, simply multiply by its reciprocal:

For rational numbers: $\dfrac{a}{b}$ and $\dfrac{c}{d}$, $\dfrac{a}{b} \div \dfrac{c}{d} = \dfrac{a}{b} \cdot \dfrac{d}{c}$

For rational expressions: $\dfrac{P}{Q}$ and $\dfrac{R}{S}$, $\dfrac{P}{Q} \div \dfrac{R}{S} = \dfrac{P}{Q} \cdot \dfrac{S}{R}$

ADDITION [6.4]

To add two rational numbers or rational expressions, find a common denominator, change each expression to an equivalent expression having the common denominator, then add numerators and reduce if possible:

For rational numbers: $\dfrac{a}{c}$ and $\dfrac{b}{c}$, $\dfrac{a}{c} + \dfrac{b}{c} = \dfrac{a + b}{c}$

For rational expressions: $\dfrac{P}{S}$ and $\dfrac{Q}{S}$, $\dfrac{P}{S} + \dfrac{Q}{S} = \dfrac{P + Q}{S}$

SUBTRACTION [6.4]

To subtract a rational number or rational expression, simply add its opposite:

For rational numbers:

$$\dfrac{a}{c} \text{ and } \dfrac{b}{c}, \quad \dfrac{a}{c} - \dfrac{b}{c} = \dfrac{a}{c} + \left(\dfrac{-b}{c}\right)$$

For rational expressions:

$$\dfrac{P}{S} \text{ and } \dfrac{Q}{S}, \quad \dfrac{P}{S} - \dfrac{Q}{S} = \dfrac{P}{S} + \left(\dfrac{-Q}{S}\right)$$

EQUATIONS [6.5]

To solve equations involving rational expressions, first find the least common denominator (LCD) for all denominators. Then multiply both sides by the LCD and solve as usual. Be sure to check all solutions in the original equation to be sure there are no undefined terms.

COMPLEX FRACTIONS [6.7]

A rational expression that contains a fraction in its numerator or denominator is called a complex fraction. The most common method of simplifying a complex fraction is to multiply the top and bottom by the LCD for all denominators.

9. $$\dfrac{1 - \dfrac{4}{x}}{x - \dfrac{16}{x}} = \dfrac{x\left(1 - \dfrac{4}{x}\right)}{x\left(x - \dfrac{16}{x}\right)}$$

$$= \dfrac{x - 4}{x^2 - 16}$$

$$= \dfrac{x - 4}{(x - 4)(x + 4)}$$

$$= \dfrac{1}{x + 4}$$

COMMON MISTAKES

1. Trying to reduce by dividing top and bottom of a rational expression by a quantity that is not a factor of both the top and bottom:

$$\dfrac{\overset{2}{\cancel{x^2 - \cancel{4x} + \cancel{4}}}}{\underset{3 \quad 2}{\cancel{x^2 - \cancel{6x} + \cancel{8}}}} \qquad \text{Mistake}$$

This makes no sense at all. The numerator and denominator must be factored completely before any factors common to the numerator and denominator can be recognized:

$$\dfrac{x^2 - 4x + 4}{x^2 - 6x + 8} = \dfrac{(x - 2)(x - 2)}{(x - 2)(x - 4)}$$

$$= \dfrac{x - 2}{x - 4}$$

2. Forgetting to check solutions to equations involving rational expressions. If you multiply both sides of an equation by an expression that contains the variable, you must check your solutions to be sure you have not multiplied by zero.

Chapter 6 Test

Work the following problems involving fractions. Be sure to reduce all answers to lowest terms.

1. $\frac{2}{3} \cdot \frac{5}{6}$ 2. $\frac{3}{4} \cdot \frac{8}{15}$

3. $\frac{5}{7} \div \frac{6}{7}$ 4. $\frac{1}{2} \div \frac{3}{4}$

5. $\frac{5}{6} - \frac{1}{6}$ 6. $\frac{3}{8} + \frac{1}{2}$

Work the following problems involving rational expressions. Be sure to reduce all answers to lowest terms.

7. $\dfrac{3x - 12}{4} \cdot \dfrac{8}{2x - 8}$

8. $\dfrac{x^2 - 49}{x + 1} \div \dfrac{x + 7}{x^2 - 1}$

9. $\dfrac{x^2 - 3x - 10}{x^2 - 8x + 15} \div \dfrac{3x^2 + 2x - 8}{x^2 + x - 12}$

10. $\dfrac{3}{x - 2} - \dfrac{6}{x - 2}$

11. $\dfrac{x}{x^2 - 9} + \dfrac{4}{4x - 12}$

12. $\dfrac{x}{x^2 - 1} - \dfrac{2x}{x^2 - 3x + 2}$

Use long division to divide.

13. $\dfrac{x^2 + 4x - 5}{x + 1}$

14. $\dfrac{x^3 - 2x + 3}{x - 3}$

Solve the following equations.

15. $\dfrac{7}{5} = \dfrac{x + 2}{3}$ 16. $\dfrac{10}{x + 4} = \dfrac{6}{x} - \dfrac{4}{x}$ 17. $\dfrac{3}{x - 2} - \dfrac{4}{x + 1} = \dfrac{5}{x^2 - x - 2}$

Solve the following problems.

18. The sum of a number and its reciprocal is $\frac{25}{12}$. Find the number.

19. An inlet pipe can fill a pool in 15 hours, while an outlet pipe can empty it in 12 hours. If the pool is full and both pipes are open, how long will it take to empty?

20. Simplify $\dfrac{1 - \dfrac{16}{x^2}}{1 - \dfrac{2}{x} - \dfrac{8}{x^2}}$.

Roots and Radicals

Chapter 7 is concerned with operations on and simplification of radicals. Most of the radical expressions we will work with will involve square roots. Occasionally we will encounter higher roots. Finding the square root of a number is the reverse of raising a number to the second power. For instance, the square root of 49 is the number we square to get 49, which is 7.

Since much of what we will do in this chapter is the reverse of what we did in Chapter 4, the more familiar we are with the properties of exponents, the better equipped we will be to handle this chapter.

The main ideas necessary for success in Chapter 7 are (1) properties of exponents and (2) operations on polynomials.

In Chapter 4 we developed notation (exponents) that would take us from a number to its square. If we wanted the square of 5, we wrote $5^2 = 25$. In this section we will use another type of notation that will take us in the reverse direction—from the square of a number back to the number itself.

In general we are interested in going from a number, say, 49, back to the number we squared to get 49. Since the square of 7 is 49, we say 7 is a square root of 49. The notation we use looks like this:

$$\sqrt{49}$$

7.1
Definitions and
Common Roots

Notation: In the expression $\sqrt{49}$, 49 is called the *radicand*, $\sqrt{}$ is the *radical sign*, and the complete expression $\sqrt{49}$ is called the *radical*.

DEFINITION If x represents any positive real number, then the expression $\sqrt{x}$ is the *positive square root* of x. It is the *positive* number we square to get x.

The expression $-\sqrt{x}$ is the *negative square root* of x. It is the *negative* number we square to get x.

Square Roots of Positive Numbers

Every positive number has two square roots, one positive and the other negative. Some books refer to the positive square root of a number as the principal root.

▼ **Example 1** The positive square root of 25 is 5 and can be written $\sqrt{25} = 5$. The negative square root of 25 is -5 and can be written $-\sqrt{25} = -5$.   ▲

If we want to consider the negative square root of a number, we must put a negative sign in front of the radical. It is a common mistake to think of $\sqrt{25}$ as meaning either 5 or -5. The expression $\sqrt{25}$ means the *positive* square root of 25, which is 5. If we want the negative square root, we must use a negative sign to begin with.

▼ **Example 2** The positive square root of 17 is written $\sqrt{17}$. The negative square root of 17 is written $-\sqrt{17}$.   ▲

We have no other exact representation for the two roots in Example 2. Since 17 itself is not a perfect square (the square of an integer), its two square roots, $\sqrt{17}$ and $-\sqrt{17}$, are irrational numbers. They have a place on the real number line but cannot be written as the ratio of two integers. The square roots of any number that is not itself a perfect square are irrational numbers.

▼ **Example 3**

Number	Positive Square Root	Negative Square Root	Roots Are
9	3	-3	Rational numbers
36	6	-6	Rational numbers
7	$\sqrt{7}$	$-\sqrt{7}$	Irrational numbers
22	$\sqrt{22}$	$-\sqrt{22}$	Irrational numbers
100	10	-10	Rational numbers

▲

The number 0 is the only real number with one square root. It is also its own square root.

$$\sqrt{0} = 0$$

Square Root of Zero

Negative numbers have square roots, but their square roots are not real numbers. They do not have a place on the real number line. We will consider square roots of negative numbers later in the book.

Square Roots of Negative Numbers

▼ **Example 4** The expression $\sqrt{-4}$ does not represent a real number, since there is no real number we can square and end up with −4. The same is true of square roots of any negative number. ▲

There are many other roots of numbers besides square roots, although square roots seem to be the most commonly used. The cube root of a number is the number we cube (raise to the third power) to get the original number. The cube root of 8 is 2, since $2^3 = 8$. The cube root of 27 is 3, since $3^3 = 27$. The notation for cube roots looks like this:

Other Roots

$$\sqrt[3]{8} = 2 \qquad \text{The 3 is called the index}$$
$$\sqrt[3]{27} = 3$$

We can go as high as we want with roots. The fourth root of 16 is 2 because $2^4 = 16$. We can write this in symbols as $\sqrt[4]{16} = 2$.

Here is a list of the most common roots. They are the roots that will come up most often in the remainder of the book and should be memorized.

Square Roots		Cube Roots	Fourth Roots
$\sqrt{1} = 1$	$\sqrt{49} = 7$	$\sqrt[3]{1} = 1$	$\sqrt[4]{1} = 1$
$\sqrt{4} = 2$	$\sqrt{64} = 8$	$\sqrt[3]{8} = 2$	$\sqrt[4]{16} = 2$
$\sqrt{9} = 3$	$\sqrt{81} = 9$	$\sqrt[3]{27} = 3$	$\sqrt[4]{81} = 3$
$\sqrt{16} = 4$	$\sqrt{100} = 10$	$\sqrt[3]{64} = 4$	$\sqrt[4]{625} = 5$
$\sqrt{25} = 5$	$\sqrt{121} \doteq 11$	$\sqrt[3]{125} = 5$	
$\sqrt{36} = 6$	$\sqrt{144} = 12$		

With even roots—square roots, fourth roots, sixth roots, and so on—we cannot have negative numbers under the radical sign. With odd roots, negative numbers under the radical sign do not cause problems.

In this chapter, unless we say otherwise, we will assume that all variables that appear under a radical sign represent positive numbers. That way we can simplify expressions involving radicals that contain variables. Here are some examples, some of which involve variables and some of which do not.

▼ **Example 5** Simplify $\sqrt{49x^2}$.

SOLUTION We are looking for the expression we square to get $49x^2$. Since the square of 7 is 49 and the square of x is x^2, we can square $7x$ and get $49x^2$:

$$\sqrt{49x^2} = 7x \quad \text{because} \quad (7x)^2 = 49x^2 \qquad ▲$$

▼ **Example 6** Simplify $\sqrt{\frac{25}{36}}$.

SOLUTION We are looking for the number we square to get $\frac{25}{36}$. The number is $\frac{5}{6}$:

$$\sqrt{\tfrac{25}{36}} = \tfrac{5}{6} \quad \text{because} \quad (\tfrac{5}{6})^2 = \tfrac{25}{36} \qquad ▲$$

▼ **Example 7** Simplify $\sqrt[3]{-8}$.

SOLUTION We are looking for the number we cube (raise to the third power) to get -8. The number is -2:

$$\sqrt[3]{-8} = -2 \quad \text{because} \quad (-2)^3 = -8 \qquad ▲$$

▼ **Example 8** Simplify $\sqrt[3]{125a^3}$.

SOLUTION We want the number we cube to get $125a^3$. That number is $5a$:

$$\sqrt[3]{125a^3} = 5a \quad \text{because} \quad (5a)^3 = 125a^3 \qquad ▲$$

Problem Set 7.1

Find the following roots. If the root does not exist as a real number, write "not a real number."

1. $\sqrt{9}$	**2.** $\sqrt{16}$	**3.** $-\sqrt{9}$
4. $-\sqrt{16}$	**5.** $\sqrt{-25}$	**6.** $\sqrt{-36}$
7. $-\sqrt{144}$	**8.** $\sqrt{256}$	**9.** $\sqrt{625}$
10. $-\sqrt{625}$	**11.** $\sqrt{-49}$	**12.** $\sqrt{-169}$
13. $-\sqrt{64}$	**14.** $-\sqrt{25}$	**15.** $-\sqrt{100}$
16. $\sqrt{121}$	**17.** $\sqrt{1225}$	**18.** $-\sqrt{1681}$
19. $\sqrt[4]{1}$	**20.** $-\sqrt[4]{81}$	**21.** $\sqrt[3]{-8}$
22. $\sqrt[3]{125}$	**23.** $-\sqrt[3]{125}$	**24.** $-\sqrt[3]{-8}$
25. $\sqrt[3]{-1}$	**26.** $-\sqrt[3]{-1}$	**27.** $\sqrt[3]{-27}$

28. $-\sqrt[3]{27}$ **29.** $-\sqrt[4]{16}$ **30.** $\sqrt[4]{-16}$

31. $\sqrt{\frac{1}{9}}$ **32.** $\sqrt{\frac{1}{4}}$

33. $\sqrt{\frac{4}{9}}$ **34.** $\sqrt{\frac{36}{49}}$

35. $\sqrt{\frac{16}{25}}$ **36.** $\sqrt{\frac{81}{100}}$

37. $\sqrt{\frac{121}{49}}$ **38.** $\sqrt{\frac{144}{25}}$

Assume all variables are positive and find the following roots.

39. $\sqrt{x^2}$ **40.** $\sqrt{a^2}$

41. $\sqrt{9x^2}$ **42.** $\sqrt{25x^2}$

43. $\sqrt{x^2y^2}$ **44.** $\sqrt{a^2b^2}$

45. $\sqrt{(a+b)^2}$ **46.** $\sqrt{(x+y)^2}$

47. $\sqrt{49x^2y^2}$ **48.** $\sqrt{81x^2y^2}$

49. $\sqrt[3]{x^3}$ **50.** $\sqrt[3]{a^3}$

51. $\sqrt[3]{8x^3}$ **52.** $\sqrt[3]{27x^3}$

Simplify each expression.

53. $\sqrt{9} + \sqrt{16}$ **54.** $\sqrt{64} + \sqrt{36}$

55. $\sqrt{9 + 16}$ **56.** $\sqrt{64 + 36}$

57. $\sqrt{144} + \sqrt{25}$ **58.** $\sqrt{25} - \sqrt{16}$

59. $\sqrt{144 + 25}$ **60.** $\sqrt{25 - 16}$

For each problem below, find a value of x that will make the equation a true statement.

61. $\sqrt{x} = 3$ **62.** $\sqrt{x} = 4$

63. $\sqrt{x} = 5$ **64.** $\sqrt{x} = 7$

65. $\sqrt{x} = 15$ **66.** $\sqrt{x} = 20$

67. $\sqrt[3]{x} = 3$ **68.** $\sqrt[4]{x} = 3$

69. $-\sqrt{x} = -2$ **70.** $-\sqrt{x} = -6$

71. $-\sqrt[3]{x} = -2$ **72.** $\sqrt[3]{x} = -2$

73. We know that the trinomial $x^2 + 6x + 9$ is the square of the binomial $x + 3$. That is, $x^2 + 6x + 9 = (x + 3)^2$. Use this fact to find $\sqrt{x^2 + 6x + 9}$.

74. Use the fact that $x^2 + 10x + 25 = (x + 5)^2$ to find $\sqrt{x^2 + 10x + 25}$.

75. Replace x with 2 in the expression $\sqrt{x^2 + 6x + 9}$ and in the expression $x + 3$. They should both simplify to the same number.

76. Replace x with 3 in the expression $\sqrt{x^2 + 6x + 9}$ and in the expression $x + 3$. Simplify both results.

77. Replace x with 4 in the expression $\sqrt{x^2 + 9}$ and in the expression $x + 3$. Simplify each result.

78. Replace x with 8 in the expression $\sqrt{x^2 + 36}$ and in the expression $x + 6$. Simplify both results.

Review Problems   The problems below review material we covered in Section 5.1.

Factor each number below into the product of prime factors.

79. 20	**80.** 60
81. 75	**82.** 12
83. 36	**84.** 160
85. 27	**86.** 125

**7.2
Properties
of Radicals**

In this section we will consider the first part of what is called simplified form for radical expressions. A radical expression is any expression containing a radical, whether it is a square root, cube root, or higher root. Simplified form for a radical expression is the form that is easiest to work with. The first step in putting a radical expression in simplified form is to take as much out from under the radical sign as possible. To do this, we must first develop two properties of radicals in general.

Consider the following two problems:

$$\sqrt{9 \cdot 16} = \sqrt{144} = 12$$
$$\sqrt{9} \cdot \sqrt{16} = 3 \cdot 4 = 12$$

Since the answers to both are equal, the original problems must also be equal. That is, $\sqrt{9 \cdot 16} = \sqrt{9} \cdot \sqrt{16}$. We can generalize this property as follows.

Property 1 for Radicals

If x and y represent nonnegative real numbers, then it is always true that

$$\sqrt{xy} = \sqrt{x}\,\sqrt{y}$$

In words: The square root of a product is the product of the square roots.

We can use this property to simplify radical expressions.

▼ Example 1

SOLUTION To simplify $\sqrt{20}$ we want to take as much out from under the radical sign as possible. We begin by looking for the largest perfect square that is a factor of 20. The largest perfect square that divides 20 is 4, so we write 20 as $4 \cdot 5$:

$$\sqrt{20} = \sqrt{4 \cdot 5}$$

Next we apply the first property of radicals and write

$$\sqrt{4 \cdot 5} = \sqrt{4}\,\sqrt{5}$$

And since $\sqrt{4} = 2$, we have

$$\sqrt{4}\,\sqrt{5} = 2\sqrt{5}$$

The expression $2\sqrt{5}$ is the simplified form of $\sqrt{20}$, since we have taken as much out from under the radical sign as possible.

Showing all the steps at once, we have

$$
\begin{aligned}
\sqrt{20} &= \sqrt{4 \cdot 5} && \text{Factor 20 into } 4 \cdot 5 \\
&= \sqrt{4}\,\sqrt{5} && \text{Property 1 for radicals} \\
&= 2\sqrt{5} && \sqrt{4} = 2
\end{aligned}
$$

▲

▼ Example 2   Simplify $\sqrt{75}$.

SOLUTION   Since 25 is the largest perfect square that divides 75, we have

$$
\begin{aligned}
\sqrt{75} &= \sqrt{25 \cdot 3} && \text{Factor 75 into } 25 \cdot 3 \\
&= \sqrt{25}\,\sqrt{3} && \text{Property 1 for radicals} \\
&= 5\sqrt{3} && \sqrt{25} = 5
\end{aligned}
$$

The expression $5\sqrt{3}$ is simplified form for $\sqrt{75}$, since we have taken as much out from under the radical sign as possible.   ▲

The next two examples involve square roots of expressions that contain variables. Remember, we are assuming that all variables that appear under a radical sign represent positive numbers.

▼ Example 3   Simplify $\sqrt{25x^3}$.

SOLUTION   The largest perfect square that is a factor of $25x^3$ is $25x^2$. We write $25x^3$ as $25x^2 \cdot x$ and apply Property 1:

$$
\begin{aligned}
\sqrt{25x^3} &= \sqrt{25x^2 \cdot x} && \text{Factor } 25x^3 \text{ into } 25x^2 \cdot x \\
&= \sqrt{25x^2}\,\sqrt{x} && \text{Property 1 for radicals} \\
&= 5x\sqrt{x} && \sqrt{25x^2} = 5x
\end{aligned}
$$

▲

▼ **Example 4** Simplify $\sqrt{18y^4}$.

SOLUTION   The largest perfect square that is a factor of $18y^4$ is $9y^4$. We write $18y^4$ as $9y^4 \cdot 2$ and apply Property 1:

$$
\begin{aligned}
\sqrt{18y^4} &= \sqrt{9y^4 \cdot 2} && \text{Factor } 18y^4 \text{ into } 9y^4 \cdot 2 \\
&= \sqrt{9y^4}\,\sqrt{2} && \text{Property 1 for radicals} \\
&= 3y^2\,\sqrt{2} && \sqrt{9y^4} = 3y^2
\end{aligned}
$$
▲

▼ **Example 5** Simplify $3\sqrt{32}$.

SOLUTION   We want to get as much out from under $\sqrt{32}$ as possible. Since 16 is the largest perfect square that divides 32, we have

$$
\begin{aligned}
3\sqrt{32} &= 3\sqrt{16 \cdot 2} && \text{Factor 32 into } 16 \cdot 2 \\
&= 3\sqrt{16}\,\sqrt{2} && \text{Property 1 for radicals} \\
&= 3 \cdot 4\sqrt{2} && \sqrt{16} = 4 \\
&= 12\sqrt{2} && 3 \cdot 4 = 12
\end{aligned}
$$
▲

The second property of radicals has to do with division. The property becomes apparent when we consider the following two problems:

$$
\sqrt{\frac{64}{16}} = \sqrt{4} = 2
$$

$$
\frac{\sqrt{64}}{\sqrt{16}} = \frac{8}{4} = 2
$$

Since the answers in each case are equal, the original problems must be also:

$$
\sqrt{\frac{64}{16}} = \frac{\sqrt{64}}{\sqrt{16}}
$$

Here is the property in general.

Property 2 for Radicals

If x and y both represent nonnegative real numbers and $y \neq 0$, then it is always true that

$$
\sqrt{\frac{x}{y}} = \frac{\sqrt{x}}{\sqrt{y}}
$$

In words: The square root of a quotient is the quotient of the square roots.

Both properties of radicals have been stated for square roots only. Each of the properties holds for higher roots also.

We can use Property 2 for radicals in much the same way as we used Property 1 to simplify radical expressions.

▼ **Example 6** Simplify $\sqrt{\dfrac{7}{4}}$.

SOLUTION The goal is the same here as it was in the first three examples. We want to take as much out from under the radical as possible.

$$\sqrt{\dfrac{7}{4}} = \dfrac{\sqrt{7}}{\sqrt{4}} \qquad \text{Property 2 for radicals}$$

$$= \dfrac{\sqrt{7}}{2} \qquad \sqrt{4} = 2 \qquad\qquad\blacktriangle$$

▼ **Example 7** Simplify $\sqrt{\dfrac{50x^3y^2}{49}}$.

SOLUTION We begin by taking the square roots of $50x^3y^2$ and 49 separately and then writing $\sqrt{49}$ as 7:

$$\sqrt{\dfrac{50x^3y^2}{49}} = \dfrac{\sqrt{50x^3y^2}}{\sqrt{49}} \qquad \text{Property 2}$$

$$= \dfrac{\sqrt{50x^3y^2}}{7} \qquad \sqrt{49} = 7$$

To simplify the numerator of this last expression we determine that the largest perfect square that is a factor of $50x^3y^2$ is $25x^2y^2$. Continuing, we have

$$= \dfrac{\sqrt{25x^2y^2 \cdot 2x}}{7} \qquad \text{Factor } 50x^3y^2 \text{ into } 25x^2y^2 \cdot 2x$$

$$= \dfrac{\sqrt{25x^2y^2}\,\sqrt{2x}}{7} \qquad \text{Property 1}$$

$$= \dfrac{5xy\,\sqrt{2x}}{7} \qquad \sqrt{25x^2y^2} = 5xy \qquad\blacktriangle$$

The two properties we have developed in this section are the two main properties of radicals. The properties hold for multiplication and division. There are no similar properties for addition and subtraction. That is, in general,

$$\sqrt{x + y} \neq \sqrt{x} + \sqrt{y}$$

The square root of a sum is not, in general, the sum of the square roots. It just doesn't work. If we try it with numbers 16 and 9, we can see what is wrong:

$$\sqrt{16 + 9} \overset{?}{=} \sqrt{16} + \sqrt{9}$$
$$\sqrt{25} \overset{?}{=} 4 + 3$$
$$5 \neq 7$$

Although it is obvious that the property doesn't hold for addition, it is a very common mistake for beginning algebra students to try using it.

Problem Set 7.2

Use Property 1 for radicals to simplify the following radical expressions as much as possible. Assume that all variables represent positive numbers.

1. $\sqrt{8}$ 2. $\sqrt{18}$

3. $\sqrt{12}$ 4. $\sqrt{27}$

5. $\sqrt{50x^2}$ 6. $\sqrt{32x^2}$

7. $\sqrt{45a^2b^2}$ 8. $\sqrt{128a^2b^2}$

9. $\sqrt{32x^4}$ 10. $\sqrt{48x^4}$

11. $-\sqrt{80}$ 12. $-\sqrt{125}$

13. $-\sqrt{28}$ 14. $-\sqrt{54}$

15. $\sqrt{242}$ 16. $\sqrt{243}$

17. $5\sqrt{45}$ 18. $3\sqrt{300}$

19. $-7\sqrt{450}$ 20. $-8\sqrt{288}$

21. $3\sqrt{50x}$ 22. $4\sqrt{18x}$

23. $7\sqrt{12x^2y}$ 24. $6\sqrt{20x^2y}$

Use Property 2 for radicals to simplify each of the following.

25. $\sqrt{\dfrac{6}{25}}$ 26. $\sqrt{\dfrac{5}{36}}$

27. $\sqrt{\dfrac{3}{49}}$ 28. $\sqrt{\dfrac{7}{64}}$

29. $\sqrt{\dfrac{16}{25}}$ 30. $\sqrt{\dfrac{81}{64}}$

31. $\sqrt{\dfrac{4}{9}}$

32. $\sqrt{\dfrac{49}{16}}$

33. $\sqrt{\dfrac{100x^2}{25}}$

34. $\sqrt{\dfrac{100x^2}{4}}$

35. $\sqrt{\dfrac{81a^2b^2}{9}}$

36. $\sqrt{\dfrac{64a^2b^2}{16}}$

Use combinations of Properties 1 and 2 of radicals to simplify the following problems as much as possible.

37. $\sqrt{\dfrac{50}{9}}$

38. $\sqrt{\dfrac{32}{49}}$

39. $\sqrt{\dfrac{75}{25}}$

40. $\sqrt{\dfrac{300}{4}}$

41. $\sqrt{\dfrac{128}{49}}$

42. $\sqrt{\dfrac{32}{64}}$

43. $\sqrt{\dfrac{288x}{25}}$

44. $\sqrt{\dfrac{28y}{81}}$

45. $\sqrt{\dfrac{54a^2}{25}}$

46. $\sqrt{\dfrac{243a^2}{49}}$

47. $\dfrac{3\sqrt{50}}{2}$

48. $\dfrac{5\sqrt{48}}{3}$

49. $\dfrac{7\sqrt{28y^2}}{3}$

50. $\dfrac{9\sqrt{243x^2}}{2}$

51. $\dfrac{5\sqrt{72a^2b^2}}{\sqrt{36}}$

52. $\dfrac{2\sqrt{27a^2b^2}}{\sqrt{9}}$

53. $\dfrac{-6\sqrt{8x^2y}}{\sqrt{4}}$

54. $\dfrac{-5\sqrt{32xy^2}}{\sqrt{25}}$

55. $\dfrac{-8\sqrt{12x^2y^3}}{\sqrt{100}}$

56. $\dfrac{-6\sqrt{18x^3y^2}}{\sqrt{81}}$

Review Problems   The problems below review material we covered in Section 4.6. Reviewing these problems will help you understand parts of the next section.

Multiply.

57. $(x - 5)(x + 5)$

58. $(x + 3)(x - 3)$

59. $(3x + 4)(3x - 4)$ **60.** $(5x - 2)(5x + 2)$
61. $(a - b)(a + b)$ **62.** $(x + y)(x - y)$
63. $(7y + 1)(7y - 1)$ **64.** $(6y + 1)(6y - 1)$

**7.3
Simplified Form
for Radicals**

A radical expression is in simplified form if it has three special characteristics. Radical expressions that are in simplified form are generally easier to work with.

DEFINITION   A radical expression is in *simplified form* if:

   1.   There are no perfect squares that are factors of the quantity under the square root sign, no perfect cubes of the quantity under the cube root sign, and so on. We want as little as possible under the radical sign.

   2.   There are no fractions under the radical sign.

   3.   There are no radicals in the denominator.

A radical expression that has these three characteristics is said to be in simplified form. As we will see, simplified form is not always the least complicated expression. In many cases, the simplified expression looks more complicated than the original expression. The important thing about simplified form for radicals is that simplified expressions are easier to work with.

The tools we will use to put radical expressions into simplified form are the properties of radicals. We list the properties again for clarity:

If a and b represent any two positive real numbers, then it is always true that

   1. $\sqrt{a}\sqrt{b} = \sqrt{a \cdot b}$

   2. $\dfrac{\sqrt{a}}{\sqrt{b}} = \sqrt{\dfrac{a}{b}}$

   3. $\sqrt{a}\sqrt{a} = (\sqrt{a})^2 = a$ This property comes directly from the definition of radicals

The following examples illustrate how we put a radical expression into simplified form using the three properties of radicals. Although the properties are stated for square roots only, they hold for all roots. [Property 3 written for cube roots would be $\sqrt[3]{a}\,\sqrt[3]{a}\,\sqrt[3]{a} = (\sqrt[3]{a})^3 = a$.]

▼ **Example 1**   Put $\sqrt{\tfrac{1}{2}}$ into simplified form.

SOLUTION   The expression $\sqrt{\frac{1}{2}}$ is not in simplified form because there is a fraction under the radical sign. We can change this by applying Property 2 for radicals:

$$\sqrt{\frac{1}{2}} = \frac{\sqrt{1}}{\sqrt{2}} \qquad \text{Property 2 for radicals}$$

$$= \frac{1}{\sqrt{2}} \qquad \sqrt{1} = 1$$

The expression $\frac{1}{\sqrt{2}}$ is not in simplified form because there is a radical sign in the denominator. If we multiply the numerator and denominator of $\frac{1}{\sqrt{2}}$ by $\sqrt{2}$, the denominator becomes $\sqrt{2} \cdot \sqrt{2} = 2$:

$$\frac{1}{\sqrt{2}} = \frac{1}{\sqrt{2}} \cdot \frac{\sqrt{2}}{\sqrt{2}} \qquad \begin{array}{l}\text{Multiply numerator and} \\ \text{denominator by } \sqrt{2}\end{array}$$

$$= \frac{\sqrt{2}}{2} \qquad\qquad\qquad \blacktriangle$$

If we check the expression $\frac{\sqrt{2}}{2}$ against our definition of simplified form for radicals, we find that all three rules hold. There are no perfect squares that are factors of 2. There are no fractions under the radical sign. No radicals appear in the denominator. The expression $\frac{\sqrt{2}}{2}$, therefore, must be in simplified form.

▼   **Example 2** Put the expression $\frac{3\sqrt{20}}{\sqrt{5}}$ into simplified form.

SOLUTION   Although there are many ways to begin this problem, we notice that 20 is divisible by 5. Using Property 2 for radicals as the first step, the expression is quickly put into simplified form.

$$\frac{3\sqrt{20}}{\sqrt{5}} = 3\sqrt{\frac{20}{5}} \qquad \text{Property 2 for radicals}$$

$$= 3\sqrt{4} \qquad \frac{20}{5} = 4$$

$$= 3 \cdot 2 \qquad \sqrt{4} = 2$$

$$= 6 \qquad\qquad\qquad \blacktriangle$$

▼ **Example 3**   Simplify $\sqrt{\dfrac{4x^3y^2}{3}}$.

SOLUTION   We begin by separating the numerator and denominator and then taking the perfect squares out of the numerator:

$$\sqrt{\frac{4x^3y^2}{3}} = \frac{\sqrt{4x^3y^2}}{\sqrt{3}} \qquad \text{Property 2 for radicals}$$

$$= \frac{\sqrt{4x^2y^2}\sqrt{x}}{\sqrt{3}} \qquad \text{Property 1 for radicals}$$

$$= \frac{2xy\sqrt{x}}{\sqrt{3}} \qquad \sqrt{4x^2y^2} = 2xy$$

The only thing keeping our expression from being in simplified form is the $\sqrt{3}$ in the denominator. We can take care of this by multiplying the numerator and denominator by $\sqrt{3}$:

$$\frac{2xy\sqrt{x}}{\sqrt{3}} = \frac{2xy\sqrt{x}}{\sqrt{3}} \cdot \frac{\sqrt{3}}{\sqrt{3}} \qquad \begin{array}{l}\text{Multiply numerator and}\\ \text{denominator by } \sqrt{3}\end{array}$$

$$= \frac{2xy\sqrt{3x}}{3} \qquad \sqrt{3}\cdot\sqrt{3} = \sqrt{9} = 3 \qquad ▲$$

Although the final expression may look more complicated than the original expression, it is in simplified form. The last step is called *rationalizing the denominator*. We have taken the radical out of the denominator and replaced it with a rational number.

▼ **Example 4**   Simplify $\dfrac{3}{\sqrt{5}-2}$.

SOLUTION   The only problem with this expression is the radical in the denominator. If we multiplied the numerator and denominator by $\sqrt{5}$, the second term in the denominator would become $2\sqrt{5}$ and we would still have a radical in the denominator. We have to try something new.

Recall from Chapter 4 how we multiplied binomials to get the difference of two squares:

$$(x + y)(x - y) = x^2 - y^2$$

If we multiply the numerator and denominator of our expression by $\sqrt{5} + 2$, the denominator becomes

$$(\sqrt{5} - 2)(\sqrt{5} + 2) = (\sqrt{5})^2 - 2^2 = 5 - 4 = 1$$

$$\frac{3}{\sqrt{5}-2} = \left(\frac{3}{\sqrt{5}-2}\right)\frac{(\sqrt{5}+2)}{(\sqrt{5}+2)} \qquad \text{Multiply numerator and}$$
$$\text{denominator by } (\sqrt{5}+2)$$

$$= \frac{3(\sqrt{5}+2)}{(\sqrt{5})^2 - 2^2}$$

$$= \frac{3\sqrt{5}+6}{5-4}$$

$$= \frac{3\sqrt{5}+6}{1}$$

$$= 3\sqrt{5}+6 \qquad\qquad\qquad \blacktriangle$$

We have rationalized the denominator and our expression is now in simplified form. The expressions $(\sqrt{5}-2)$ and $(\sqrt{5}+2)$ are called *conjugates* of each other.

▼ **Example 5**   Rationalize the denominator in the expression $\dfrac{2}{5-\sqrt{3}}$.

SOLUTION   We use the same procedure as in Example 4. Multiply the numerator and denominator by the conjugate of the denominator, which is $5 + \sqrt{3}$:

$$\left(\frac{2}{5-\sqrt{3}}\right)\left(\frac{5+\sqrt{3}}{5+\sqrt{3}}\right) = \frac{10+2\sqrt{3}}{5^2 - (\sqrt{3})^2}$$

$$= \frac{10+2\sqrt{3}}{25-3}$$

$$= \frac{10+2\sqrt{3}}{22}$$

The numerator and denominator of this last expression have a factor of 2 in common. We can reduce to lowest terms by dividing out the common factor 2. Continuing, we have

$$= \frac{2(5+\sqrt{3})}{2\cdot 11}$$

$$= \frac{5+\sqrt{3}}{11}$$

The final expression is in simplified form.   ▲

As a final example, we will work on a cube root. Most radical expressions we will encounter in this book will involve square roots. It is not difficult to extend our work with square roots to include cube roots or even higher roots.

▼ **Example 6** Simplify $\sqrt[3]{\dfrac{1}{4}}$.

SOLUTION We begin by separating the numerator and denominator:

$$\sqrt[3]{\frac{1}{4}} = \frac{\sqrt[3]{1}}{\sqrt[3]{4}} \qquad \text{Property 2 for radicals}$$

$$= \frac{1}{\sqrt[3]{4}} \qquad \sqrt[3]{1} = 1$$

To rationalize the denominator, we need to have a perfect cube under the cube root sign. If we multiply numerator and denominator by $\sqrt[3]{2}$, we will have $\sqrt[3]{4} \cdot \sqrt[3]{2} = \sqrt[3]{8}$ in the denominator:

$$\frac{1}{\sqrt[3]{4}} = \frac{1}{\sqrt[3]{4}} \cdot \frac{\sqrt[3]{2}}{\sqrt[3]{2}} \qquad \begin{array}{l}\text{Multiply numerator and} \\ \text{denominator by } \sqrt[3]{2}\end{array}$$

$$= \frac{\sqrt[3]{2}}{\sqrt[3]{8}} \qquad \sqrt[3]{4} \cdot \sqrt[3]{2} = \sqrt[3]{8}$$

$$= \frac{\sqrt[3]{2}}{2} \qquad \sqrt[3]{8} = 2$$

The final expression has no radical sign in the denominator and therefore is in simplified form. ▲

Problem Set 7.3

Put each of the following radical expressions into simplified form. Assume all variables represent positive numbers.

1. $\sqrt{\dfrac{1}{2}}$ 2. $\sqrt{\dfrac{1}{5}}$

3. $\sqrt{\dfrac{1}{3}}$ 4. $\sqrt{\dfrac{1}{6}}$

5. $\sqrt{\dfrac{2}{5}}$ 6. $\sqrt{\dfrac{3}{7}}$

7. $\sqrt{\dfrac{20}{3}}$ 8. $\sqrt{\dfrac{32}{5}}$

9. $\sqrt{\dfrac{45}{6}}$

10. $\sqrt{\dfrac{48}{7}}$

11. $\sqrt{\dfrac{1}{20}}$

12. $\sqrt{\dfrac{5}{32}}$

13. $\dfrac{10\sqrt{15}}{5\sqrt{3}}$

14. $\dfrac{4\sqrt{12}}{8\sqrt{3}}$

15. $\dfrac{6\sqrt{21}}{3\sqrt{7}}$

16. $\dfrac{8\sqrt{50}}{16\sqrt{2}}$

17. $\dfrac{-5\sqrt{32}}{\sqrt{3}}$

18. $\dfrac{-7\sqrt{27}}{\sqrt{2}}$

19. $\sqrt{\dfrac{4x^2y^2}{2}}$

20. $\sqrt{\dfrac{8x^3y}{3}}$

21. $\sqrt{\dfrac{16a^4}{5}}$

22. $\sqrt{\dfrac{50a^2b^3}{6}}$

23. $\sqrt{\dfrac{72a^5}{5}}$

24. $\sqrt{\dfrac{20ab^2}{6}}$

25. $\dfrac{2\sqrt{20x^2}}{3}$

26. $\dfrac{5\sqrt{98x^3y}}{2}$

27. $\dfrac{6\sqrt{54a^2b^2}}{5}$

28. $\dfrac{7\sqrt{75a^3b^2}}{6}$

29. $\dfrac{-3\sqrt{72x^3}}{\sqrt{2x}}$

30. $\dfrac{-2\sqrt{45x^4}}{\sqrt{5x}}$

31. $\sqrt[3]{\dfrac{1}{2}}$

32. $\sqrt[3]{\dfrac{1}{4}}$

33. $\sqrt[3]{\dfrac{1}{9}}$

34. $\sqrt[3]{\dfrac{1}{27}}$

35. $\sqrt[3]{\dfrac{3}{2}}$

36. $\sqrt[3]{\dfrac{7}{9}}$

37. $\dfrac{2}{\sqrt{3}-5}$

38. $\dfrac{3}{\sqrt{2}-4}$

39. $\dfrac{1}{\sqrt{5}-2}$

40. $\dfrac{-2}{\sqrt{3}-4}$

41. $\dfrac{-3}{\sqrt{6}+2}$

42. $\dfrac{-1}{\sqrt{5}+7}$

43. $\dfrac{5}{3 + \sqrt{2}}$

44. $\dfrac{7}{4 + \sqrt{7}}$

45. $\dfrac{8}{3 - \sqrt{2}}$

46. $\dfrac{6}{5 - \sqrt{6}}$

47. $\dfrac{6}{\sqrt{5} - \sqrt{2}}$

48. $\dfrac{5}{\sqrt{6} - \sqrt{3}}$

49. $\dfrac{-2}{\sqrt{2} + \sqrt{5}}$

50. $\dfrac{10}{\sqrt{7} + \sqrt{8}}$

51. The table of square roots inside the back cover gives the four-place decimal approximation of $\sqrt{2}$ as 1.414. Use this number to find approximations for the expressions $1/\sqrt{2}$ and $\sqrt{2}/2$. Round your answer to three places past the decimal point.

52. Look up the decimal approximation for $\sqrt{3}$ in the table of squares and square roots at the back of the book and use it to find decimal approximations for $1/\sqrt{3}$ and $\sqrt{3}/3$. Round your answers to three places past the decimal point.

53. Use the table of squares and square roots at the back of the book to find decimal approximations for $\sqrt{12}$ and $2\sqrt{3}$. Round your answers to three places past the decimal point.

54. Use the table at the back of the book to find decimal approximations for $\sqrt{50}$ and $5\sqrt{2}$. Round your answers to three places past the decimal point.

Review Problems   The problems below review material we covered in Section 2.1. Reviewing these problems will help you understand the next section.

Use the distributive property to combine the following.

55. $3x + 7x$

56. $3x - 7x$

57. $15x + 8x$

58. $15x - 8x$

59. $7a - 3a + 6a$

60. $25a + 3a - a$

61. $2y - 9y + 50y$

62. $4y - 8y + 30y$

**7.4
Addition and
Subtraction of
Radical
Expressions**

To add two or more radical expressions we apply the distributive property. Adding radical expressions is very similar to adding similar terms of polynomials.

▼   **Example 1**   Combine terms in the expression $3\sqrt{5} - 7\sqrt{5}$.

SOLUTION The two terms $3\sqrt{5}$ and $7\sqrt{5}$ each have $\sqrt{5}$ in common. Since $3\sqrt{5}$ means 3 times $\sqrt{5}$, or $3 \cdot \sqrt{5}$, we apply the distributive property:

$$3\sqrt{5} - 7\sqrt{5} = (3 - 7)\sqrt{5} \qquad \text{Distributive property}$$
$$= -4\sqrt{5} \qquad\qquad 3 - 7 = -4 \qquad \blacktriangle$$

Since we use the distributive property to add radical expressions, each expression must contain exactly the same radical.

▼ **Example 2** Combine terms in the expression $7\sqrt{2} - 3\sqrt{2} + 6\sqrt{2}$.

SOLUTION

$$7\sqrt{2} - 3\sqrt{2} + 6\sqrt{2} = (7 - 3 + 6)\sqrt{2} \qquad \text{Distributive property}$$
$$= 10\sqrt{2} \qquad\qquad \text{Addition} \qquad \blacktriangle$$

In Examples 1 and 2 each term was a radical expression in simplified form. If one or more terms is not in simplified form, we must put it into simplifed form and then combine terms, if possible. It is not always possible to combine terms containing radicals. Occasionally two or more of the terms will not have a radical in common. If there is a possibility of combining terms, it will always become apparent when each term is in simplified form.

RULE To combine two or more radical expressions, put each expression in simplified form and then apply the distributive property, if possible.

▼ **Example 3** Combine terms in the expression $3\sqrt{50} + 2\sqrt{32}$.

SOLUTION We begin by putting each term into simplified form:

$$3\sqrt{50} + 2\sqrt{32} = 3\sqrt{25}\sqrt{2} + 2\sqrt{16}\sqrt{2} \qquad \text{Property 1 for radicals}$$
$$= 3 \cdot 5\sqrt{2} + 2 \cdot 4\sqrt{2} \qquad \sqrt{25} = 5 \text{ and } \sqrt{16} = 4$$
$$= 15\sqrt{2} + 8\sqrt{2} \qquad\qquad \text{Multiplication}$$

Applying the distributive property to the last line, we have

$$15\sqrt{2} + 8\sqrt{2} = (15 + 8)\sqrt{2} \qquad \text{Distributive property}$$
$$= 23\sqrt{2} \qquad\qquad 15 + 8 = 23 \qquad \blacktriangle$$

▼ **Example 4** Combine terms in the expression $5\sqrt{75} + \sqrt{27} - \sqrt{3}$.

SOLUTION

$5\sqrt{75} + \sqrt{27} - \sqrt{3}$

$$
\begin{aligned}
&= 5\sqrt{25}\sqrt{3} + \sqrt{9}\sqrt{3} - \sqrt{3} &&\text{Property 1 for radicals}\\
&= 5\cdot 5\sqrt{3} + 3\sqrt{3} - \sqrt{3} &&\sqrt{25} = 5 \text{ and } \sqrt{9} = 3\\
&= 25\sqrt{3} + 3\sqrt{3} - \sqrt{3} &&5\cdot 5 = 25\\
&= (25 + 3 - 1)\sqrt{3} &&\text{Distributive property}\\
&= 27\sqrt{3} &&\text{Addition}
\end{aligned}
$$

▲

The most time-consuming part of combining most radical expressions is simplifying each term in the expression. Once this has been done, applying the distributive property is simple and fast.

▼ **Example 5** Combine terms in the expression

$\sqrt{20x^3} - 3x\sqrt{45x} + 10\sqrt{25x^2}$. (Assume x is a positive real number.)

SOLUTION

$\sqrt{20x^3} - 3x\sqrt{45x} + 10\sqrt{25x^2}$

$$
\begin{aligned}
&= \sqrt{4x^2}\sqrt{5x} - 3x\sqrt{9}\sqrt{5x} + 10\sqrt{25x^2}\\
&= 2x\sqrt{5x} - 3x\cdot 3\sqrt{5x} + 10\cdot 5x\\
&= 2x\sqrt{5x} - 9x\sqrt{5x} + 50x
\end{aligned}
$$

Each term is now in simplified form. The best we can do next is to combine the first two terms. The last term does not have the common radical $\sqrt{5x}$.

$$
\begin{aligned}
2x\sqrt{5x} - 9x\sqrt{5x} + 50x &= (2x - 9x)\sqrt{5x} + 50x\\
&= -7x\sqrt{5x} + 50x
\end{aligned}
$$

We have, in any case, succeeded in reducing the number of terms in our original problem.

▲

Problem Set 7.4

In each of the following problems, simplify each term, if necessary, and then use the distributive property to combine terms, if possible.

1. $3\sqrt{2} + 4\sqrt{2}$

2. $7\sqrt{3} + 2\sqrt{3}$

3. $9\sqrt{5} - 7\sqrt{5}$

4. $6\sqrt{7} - 10\sqrt{7}$

5. $\sqrt{3} + 6\sqrt{3}$

6. $\sqrt{2} + 10\sqrt{2}$

7. $6\sqrt{5} - \sqrt{5}$

8. $9\sqrt{11} - \sqrt{11}$

9. $14\sqrt{13} - \sqrt{13}$

10. $-2\sqrt{6} - 9\sqrt{6}$

11. $-3\sqrt{10} + 9\sqrt{10}$

12. $11\sqrt{11} + \sqrt{11}$

13. $5\sqrt{5} + \sqrt{5}$

14. $\sqrt{6} - 10\sqrt{6}$

15. $\sqrt{8} + 2\sqrt{2}$

16. $\sqrt{20} + 3\sqrt{5}$

17. $3\sqrt{3} - \sqrt{27}$

18. $4\sqrt{5} - \sqrt{80}$

19. $5\sqrt{12} - 10\sqrt{48}$

20. $3\sqrt{300} - 5\sqrt{27}$

21. $-\sqrt{75} - \sqrt{3}$

22. $5\sqrt{20} + 8\sqrt{80}$

23. $4\sqrt{75} - 8\sqrt{12}$

24. $5\sqrt{24} + 4\sqrt{150}$

25. $8\sqrt{8} + 5\sqrt{75}$

26. $9\sqrt{54} - 8\sqrt{24}$

27. $\sqrt{27} - 2\sqrt{12} + \sqrt{3}$

28. $\sqrt{20} + 3\sqrt{45} - \sqrt{5}$

29. $\sqrt{72} - \sqrt{8} + \sqrt{50}$

30. $\sqrt{24} - \sqrt{54} - \sqrt{150}$

31. $5\sqrt{7} + 2\sqrt{28} - 4\sqrt{63}$

32. $3\sqrt{3} - 5\sqrt{27} + 8\sqrt{75}$

33. $6\sqrt{48} - 2\sqrt{12} + 5\sqrt{27}$

34. $5\sqrt{50} + 8\sqrt{12} - \sqrt{32}$

35. $6\sqrt{48} - \sqrt{72} - 3\sqrt{300}$

36. $7\sqrt{44} - 8\sqrt{99} + \sqrt{176}$

All variables in the following problems represent positive real numbers. Simplify each term and combine, if possible.

37. $\sqrt{x^3} + x\sqrt{x}$

38. $2\sqrt{x} - 2\sqrt{4x}$

39. $5\sqrt{3a^2} - a\sqrt{3}$

40. $6a\sqrt{a} + 7\sqrt{a^3}$

41. $5\sqrt{8x^3} + x\sqrt{50x}$

42. $2\sqrt{27x^2} - x\sqrt{48}$

43. $3\sqrt{75x^3y} - 2x\sqrt{3xy}$

44. $9\sqrt{24x^3y^2} - 5x\sqrt{54x^2y}$

45. $\sqrt{20ab^2} - b\sqrt{45a}$

46. $4\sqrt{a^3b^2} - 5a\sqrt{ab^2}$

47. $9\sqrt{18x^3} - 2x\sqrt{48x}$

48. $8\sqrt{72x^2} - x\sqrt{8}$

49. $7\sqrt{50x^2y} + 8x\sqrt{8y} - 7\sqrt{32x^2y}$

50. $6\sqrt{44x^3y^3} - 8x\sqrt{99xy^3} - 6y\sqrt{176x^3y}$

51. Use the table of squares and square roots at the back of the book to find a decimal approximation of the expression $\sqrt{5} + \sqrt{3}$. Is the answer equal to the decimal approximation of $\sqrt{8}$ listed in the table?

52. Find a decimal approximation of the expression $\sqrt{5} - \sqrt{3}$ using the table at the back of the book. Is the answer equal to the decimal approximation of $\sqrt{2}$ listed in the table?

53. The statement below is false. Correct the right side to make the statement true.
$$4\sqrt{3} + 5\sqrt{3} = 9\sqrt{6}$$

54. The statement below is false. Correct the right side to make the statement true.
$$7\sqrt{5} - 3\sqrt{5} = 4\sqrt{25}$$

Review Problems   The problems below review material we covered in Sections 4.5 and 4.6. Reviewing these problems will help you in the next section.

Multiply.

55. $3(5x + 2y)$

56. $5(2x - 3y)$

57. $(x + 2)(x - 5)$

58. $(x - 3)(x - 4)$

59. $(3x + y)^2$

60. $(2x - 3y)^2$

61. $(3x - 4y)(3x + 4y)$

62. $(7x + 2y)(7x - 2y)$

**7.5
Multiplication
and Division
of Radicals**

Multiplication of radical expressions is accomplished in the same way as multiplication of polynomials. We apply the distributive property and then simplify, if possible.

▼ **Example 1** Multiply $\sqrt{5}(\sqrt{2} + \sqrt{5})$.

SOLUTION

$$\sqrt{5}(\sqrt{2} + \sqrt{5}) = \sqrt{5} \cdot \sqrt{2} + \sqrt{5} \cdot \sqrt{5} \qquad \text{Distributive property}$$
$$= \sqrt{10} + 5 \qquad \text{Multiplication} \qquad ▲$$

▼ **Example 2** Multiply $3\sqrt{2}(2\sqrt{5} + 5\sqrt{3})$.

SOLUTION

$$3\sqrt{2}(2\sqrt{5} + 5\sqrt{3})$$
$$= 3\sqrt{2} \cdot 2\sqrt{5} + 3\sqrt{2} \cdot 5\sqrt{3} \qquad \text{Distributive property}$$
$$= 3 \cdot 2 \cdot \sqrt{2}\sqrt{5} + 3 \cdot 5\sqrt{2}\sqrt{3} \qquad \text{Commutative property}$$
$$= 6\sqrt{10} + 15\sqrt{6} \qquad ▲$$

Each term in the last line is in simplified form so the problem is complete.

▼ **Example 3** Multiply $(\sqrt{5} + 2)(\sqrt{5} + 7)$.

SOLUTION We multiply using the FOIL method we used to multiply binomials:

$$(\sqrt{5} + 2)(\sqrt{5} + 7) = \sqrt{5} \cdot \sqrt{5} + 7\sqrt{5} + 2\sqrt{5} + 14$$
$$\qquad\qquad\qquad\quad \text{F} \qquad\quad \text{O} \qquad \text{I} \qquad \text{L}$$
$$= 5 + 9\sqrt{5} + 14$$
$$= 19 + 9\sqrt{5}$$

We must be careful not to try to simplify further by adding 19 and 9. We can add only radical expressions that have a common radical part; 19 and $9\sqrt{5}$ are not similar. ▲

▼ Example 4 Expand and simplify $(\sqrt{3} - 2)^2$.

SOLUTION Multiplying $\sqrt{3} - 2$ times itself, we have

$$(\sqrt{3} - 2)^2 = (\sqrt{3} - 2)(\sqrt{3} - 2)$$
$$= \sqrt{3}\sqrt{3} - 2\sqrt{3} - 2\sqrt{3} + 4$$
$$= 9 - 4\sqrt{3} + 4$$
$$= 13 - 4\sqrt{3} \qquad \blacktriangle$$

Division of radical expressions can be accomplished by rationalizing the denominator. In the process of removing all radicals from the denominator, we are actually doing division with radical expressions.

▼ Example 5 Rationalize the denominator in the expression $\dfrac{\sqrt{2} + \sqrt{3}}{\sqrt{2} - \sqrt{3}}$.

SOLUTION We remove the two radicals in the denominator by multiplying both the numerator and denominator by the conjugate of $\sqrt{2} - \sqrt{3}$, which is $\sqrt{2} + \sqrt{3}$:

$$\frac{\sqrt{2} + \sqrt{3}}{\sqrt{2} - \sqrt{3}} = \left(\frac{\sqrt{2} + \sqrt{3}}{\sqrt{2} - \sqrt{3}}\right)\frac{(\sqrt{2} + \sqrt{3})}{(\sqrt{2} + \sqrt{3})}$$

$$= \frac{\sqrt{2}\sqrt{2} + \sqrt{2}\sqrt{3} + \sqrt{3}\sqrt{2} + \sqrt{3}\sqrt{3}}{(\sqrt{2})^2 - (\sqrt{3})^2}$$

$$= \frac{2 + \sqrt{6} + \sqrt{6} + 3}{2 - 3}$$

$$= \frac{5 + 2\sqrt{6}}{-1}$$

$$= -(5 + 2\sqrt{6}) \quad \text{or} \quad -5 - 2\sqrt{6} \qquad \blacktriangle$$

We can think of this process as rationalizing the denominator, or we can say $\sqrt{2} + \sqrt{3}$ divided by $\sqrt{2} - \sqrt{3}$ is $-(5 + 2\sqrt{6})$.

▼ Example 6 Divide $\dfrac{2\sqrt{3} - \sqrt{5}}{\sqrt{5} - \sqrt{2}}$.

SOLUTION To remove the two radicals from the denominator, we multiply the numerator and denominator by the conjugate of $\sqrt{5} - \sqrt{2}$, which is $\sqrt{5} + \sqrt{2}$:

$$\frac{2\sqrt{3} - \sqrt{5}}{\sqrt{5} - \sqrt{2}} = \frac{(2\sqrt{3} - \sqrt{5})}{(\sqrt{5} - \sqrt{2})} \cdot \frac{(\sqrt{5} + \sqrt{2})}{(\sqrt{5} + \sqrt{2})}$$

$$= \frac{2\sqrt{3}\sqrt{5} + 2\sqrt{3}\sqrt{2} - \sqrt{5}\sqrt{5} - \sqrt{5}\sqrt{2}}{(\sqrt{5})^2 - (\sqrt{2})^2}$$

$$= \frac{2\sqrt{15} + 2\sqrt{6} - 5 - \sqrt{10}}{5 - 2}$$

$$= \frac{2\sqrt{15} + 2\sqrt{6} - 5 - \sqrt{10}}{3}$$

▲

Problem Set 7.5

Perform the following multiplications. All answers should be in simplified form for radical expressions.

1. $\sqrt{2}(\sqrt{3} - 1)$
2. $\sqrt{3}(\sqrt{5} + 2)$
3. $\sqrt{7}(\sqrt{8} - 5)$
4. $\sqrt{7}(\sqrt{32} + 3)$
5. $\sqrt{2}(\sqrt{3} + \sqrt{2})$
6. $\sqrt{5}(\sqrt{7} - \sqrt{5})$
7. $\sqrt{3}(2\sqrt{2} + \sqrt{3})$
8. $\sqrt{11}(3\sqrt{2} - \sqrt{11})$
9. $\sqrt{3}(2\sqrt{3} - \sqrt{5})$
10. $\sqrt{7}(\sqrt{14} - \sqrt{7})$
11. $2\sqrt{3}(\sqrt{2} + \sqrt{5})$
12. $3\sqrt{2}(\sqrt{3} + \sqrt{2})$
13. $5\sqrt{3}(2\sqrt{3} - 3\sqrt{2})$
14. $7\sqrt{6}(5\sqrt{6} - 6\sqrt{5})$
15. $(\sqrt{2} + \sqrt{3})(\sqrt{3} + \sqrt{2})$
16. $(\sqrt{5} + \sqrt{6})(\sqrt{6} + \sqrt{5})$
17. $(\sqrt{2} - 1)(\sqrt{2} + 3)$
18. $(\sqrt{2} + 5)(\sqrt{2} + 4)$
19. $(\sqrt{5} - 6)(\sqrt{5} + 6)$
20. $(\sqrt{7} + 2)(\sqrt{7} - 2)$
21. $(\sqrt{2} + 1)^2$
22. $(\sqrt{5} - 4)^2$
23. $(\sqrt{6} + \sqrt{3})^2$
24. $(2\sqrt{2} - \sqrt{3})^2$
25. $(5 - \sqrt{2})^2$
26. $(2 + \sqrt{8})^2$
27. $(3 + \sqrt{7})^2$
28. $(3 - \sqrt{8})^2$
29. $(\sqrt{3} + 1)(\sqrt{2} + 3)$
30. $(\sqrt{5} - 2)(\sqrt{3} + 4)$
31. $(\sqrt{7} + \sqrt{2})(\sqrt{2} + 1)$
32. $(\sqrt{3} + \sqrt{5})(\sqrt{6} - 4)$
33. $(\sqrt{11} - \sqrt{2})(\sqrt{2} + 6)$
34. $(3\sqrt{2} + 2)(\sqrt{3} + 1)$
35. $(5\sqrt{3} - 7)(\sqrt{2} + \sqrt{3})$
36. $(6\sqrt{2} + 4)(\sqrt{5} - \sqrt{6})$

Rationalize the denominator. All answers should be expressed in simplified form.

37. $\dfrac{\sqrt{3}}{\sqrt{5} - \sqrt{2}}$
38. $\dfrac{\sqrt{2}}{\sqrt{6} + \sqrt{3}}$
39. $\dfrac{\sqrt{5}}{\sqrt{5} + \sqrt{2}}$
40. $\dfrac{\sqrt{7}}{\sqrt{8} + \sqrt{7}}$

41. $\dfrac{\sqrt{2} + \sqrt{3}}{\sqrt{2} - \sqrt{3}}$

42. $\dfrac{\sqrt{5} - \sqrt{2}}{\sqrt{5} + \sqrt{2}}$

43. $\dfrac{\sqrt{7} - \sqrt{3}}{\sqrt{7} + \sqrt{3}}$

44. $\dfrac{\sqrt{11} - \sqrt{6}}{\sqrt{11} + \sqrt{6}}$

45. $\dfrac{2\sqrt{3} - 1}{\sqrt{3} + 2}$

46. $\dfrac{3\sqrt{5} + 2}{\sqrt{6} - 2}$

47. $\dfrac{2\sqrt{3} - \sqrt{2}}{\sqrt{5} + \sqrt{2}}$

48. $\dfrac{3\sqrt{5} + \sqrt{6}}{\sqrt{7} - \sqrt{8}}$

49. $\dfrac{5\sqrt{3} - 2\sqrt{2}}{\sqrt{3} + \sqrt{2}}$

50. $\dfrac{6\sqrt{5} - 7\sqrt{6}}{\sqrt{6} - \sqrt{5}}$

51. The statement below is incorrect. Correct the right side to make the statement true.

$$2(3\sqrt{5}) = 6\sqrt{15}$$

52. The statement below is incorrect. Correct the right side to make the statement true.

$$5(2\sqrt{6}) = 10\sqrt{30}$$

53. The statement below is incorrect. What is missing from the right side?

$$(\sqrt{3} + 7)^2 = 3 + 49$$

54. The statement below is incorrect. What is missing from the right side?

$$(\sqrt{5} + \sqrt{2})^2 = 5 + 2$$

Review Problems   The problems below review material we covered in Sections 2.3 and 5.6. Reviewing these problems will help you understand the next section.

Solve each equation.

55. $2a - 3 = 81$

56. $2a + 5 = 41$

57. $2x + 5 = 4x - 3$

58. $3x + 1 = x - 9$

59. $x^2 + 5x - 6 = 0$

60. $x^2 + 5x + 6 = 0$

61. $x^2 - 3x = 0$

62. $x^2 + 5x = 0$

In order to solve equations that contain one or more radical expressions we need an additional property. From our work with exponents we know that if two quantities are equal, then so are the squares of those quantities. That is, for real numbers a and b

**7.6
Equations Involving
Radicals**

$$\text{if} \quad a = b$$
$$\text{then} \quad a^2 = b^2$$

The only problem with squaring both sides of an equation is that occasionally we will change a false statement into a true statement. Let's take the false statement $3 = -3$ as an example. Squaring both sides, we have

$$3 = -3 \qquad \text{A false statement}$$
$$(3)^2 = (-3)^2 \qquad \text{Square both sides}$$
$$9 = 9 \qquad \text{A true statement}$$

We can avoid this problem by always checking our solutions if, at any time during the process of solving an equation, we have squared both sides of the equation. Here is how the property is stated.

Squaring Property of Equality

We can square both sides of an equation any time it is convenient to do so, as long as we check all solutions in the original equation.

▼ **Example 1** Solve for x: $\sqrt{x + 1} = 7$.

SOLUTION In order to solve this equation by our usual methods, we must first eliminate the radical sign. We can accomplish this by squaring both sides of the equation:

$$\sqrt{x + 1} = 7$$
$$(\sqrt{x + 1})^2 = 7^2 \qquad \text{Square both sides}$$
$$x + 1 = 49$$
$$x = 48$$

To check our solution, we substitute $x = 48$ into the original equation:

$$\sqrt{48 + 1} = 7$$
$$\sqrt{49} = 7$$
$$7 = 7 \qquad \text{A true statement}$$

The solution checks.   ▲

▼ **Example 2** Solve for x: $\sqrt{2x - 3} = -9$.

SOLUTION We square both sides and proceed as in Example 1:

$$\sqrt{2x - 3} = -9$$

$$(\sqrt{2x - 3})^2 = (-9)^2 \qquad \text{Square both sides}$$
$$2x - 3 = 81$$
$$2x = 84$$
$$x = 42$$

Checking our solution in the original equation, we have

$$\sqrt{2(42) - 3} = -9$$
$$\sqrt{84 - 3} = -9$$
$$\sqrt{81} = -9$$
$$9 = -9 \qquad \text{A false statement}$$

Our solution does not check because we end up with a false statement. ▲

Squaring both sides of the equation has produced what is called an extraneous solution. This happens occasionally when we use the squaring property of equality. We can always eliminate extraneous solutions by checking each solution with the original equation.

▼ **Example 3** Solve for a: $\sqrt{3a - 2} + 3 = 5$.

SOLUTION Before we can square both sides to eliminate the radical, we must isolate the radical on the left side of the equation. To do so, we add -3 to both sides:

$$\sqrt{3a - 2} + 3 = 5$$
$$\sqrt{3a - 2} = 2 \qquad \text{Add } -3 \text{ to both sides}$$
$$(\sqrt{3a - 2})^2 = 2^2 \qquad \text{Square both sides}$$
$$3a - 2 = 4$$
$$3a = 6$$
$$a = 2$$

Checking $a = 2$ in the original equation, we have

$$\sqrt{3 \cdot 2 - 2} + 3 = 5$$
$$\sqrt{4} + 3 = 5$$
$$2 + 3 = 5$$
$$5 = 5 \qquad \text{A true statement} \qquad ▲$$

▼ **Example 4** Solve for x: $\sqrt{x + 15} = x + 3$.

SOLUTION We begin by squaring both sides:

$$(\sqrt{x+15})^2 = (x+3)^2 \qquad \text{Square both sides}$$
$$x + 15 = x^2 + 6x + 9$$

We have a quadratic equation. We put it into standard form by adding $-x$ and -15 to both sides. Then we factor and solve as usual.

$$0 = x^2 + 5x - 6 \qquad\qquad \text{Standard form}$$
$$0 = (x+6)(x-1) \qquad\qquad \text{Factor}$$
$$x + 6 = 0 \quad \text{or} \quad x - 1 = 0 \qquad \text{Set factors to 0}$$
$$x = -6 \quad \text{or} \qquad x = 1$$

We check each solution in the original equation:

Check -6 Check 1

$\sqrt{-6+15} = -6 + 3$ $\sqrt{1+15} = 1 + 3$

$\sqrt{9} = -3$ $\sqrt{16} = 4$

$3 = -3$ $4 = 4$

A false statement   A true statement

Since $x = -6$ does not check in the original equation, it cannot be a solution. The only solution is $x = 1$. ▲

Problem Set 7.6

Solve each equation by applying the squaring property of equality. Be sure to check all solutions in the original equation.

1. $\sqrt{x+1} = 2$
2. $\sqrt{x-3} = 4$
3. $\sqrt{x+5} = 7$
4. $\sqrt{x+8} = 5$
5. $\sqrt{x-9} = -6$
6. $\sqrt{x+10} = -3$
7. $\sqrt{x-5} = -4$
8. $\sqrt{x+7} = -5$
9. $\sqrt{x-8} = 0$
10. $\sqrt{x-9} = 0$
11. $\sqrt{2x+1} = 3$
12. $\sqrt{2x-5} = 7$
13. $\sqrt{2x-3} = -5$
14. $\sqrt{3x-8} = -4$
15. $\sqrt{3x+6} = 2$
16. $\sqrt{5x-1} = 5$
17. $2\sqrt{x} = 10$
18. $3\sqrt{x} = 9$
19. $3\sqrt{a} = 6$
20. $2\sqrt{a} = 12$
21. $\sqrt{3x+4} - 3 = 2$
22. $\sqrt{2x-1} + 2 = 5$
23. $\sqrt{5y-4} - 2 = 4$
24. $\sqrt{3y+1} + 7 = 2$

25. $\sqrt{2x + 1} + 5 = 2$

26. $\sqrt{6x - 8} - 1 = 3$

27. $\sqrt{x + 3} = x - 3$

28. $\sqrt{x - 3} = x - 3$

29. $\sqrt{a + 2} = a + 2$

30. $\sqrt{a + 10} = a - 2$

31. $\sqrt{2x + 9} = x + 5$

32. $\sqrt{x + 6} = x + 4$

33. $\sqrt{y - 4} = y - 6$

34. $\sqrt{2y + 13} = y + 7$

35. $\sqrt{5x + 1} = x + 1$

36. $\sqrt{6x + 1} = x - 1$

Review Problems   The problems below review material we covered in Sections 4.1 and 4.2.

Simplify each expression.

37. $x^3 \cdot x^5$

38. $(x^5)^3$

39. $\dfrac{x^5}{x^3}$

40. $(3x^2y^5)^4$

41. 2^{-3}

42. 3^{-2}

Chapter 7 Summary and Review

Examples

ROOTS [7.1]

Every positive real number x has two square roots, one positive and one negative. The positive square root is written $\sqrt{x}$. The negative square root of x is written $-\sqrt{x}$. In both cases the square root of x is a number we square to get x.

The cube root of x is written $\sqrt[3]{x}$ and is the number we cube to get x.

1.   The two square roots of 9 are 3 and −3:

$$\sqrt{9} = 3 \quad \text{and} \quad -\sqrt{9} = -3$$

NOTATION [7.1]

In the expression $\sqrt[3]{8}$, 8 is called the *radicand*, 3 is the *index*, $\sqrt{}$ is called the *radical sign*, and the whole expression $\sqrt[3]{8}$ is called the *radical*.

2.

Index $\longrightarrow$ $\sqrt[3]{24}$ $\longleftarrow$ Radical sign
 $\longleftarrow$ Radicand

Radical

PROPERTIES OF RADICALS [7.2]

If a and b represent positive real numbers, then

1. $\sqrt{a}\,\sqrt{b} = \sqrt{ab}$ The product of the square roots is the square root of the product

3.

(a) $\sqrt{5} \cdot \sqrt{5} = (\sqrt{5})^2 = 5$

(b) $\sqrt{3} \cdot \sqrt{2} = \sqrt{3 \cdot 2} = \sqrt{6}$

(c) $\dfrac{\sqrt{12}}{\sqrt{3}} = \sqrt{\dfrac{12}{3}} = \sqrt{4} = 2$

2. $\dfrac{\sqrt{a}}{\sqrt{b}} = \sqrt{\dfrac{a}{b}}$ The quotient of the square roots is the square root of the quotient

3. $\sqrt{a} \cdot \sqrt{a} = (\sqrt{a})^2 = a$ This property shows that squaring and square roots are inverse operations

4. Simplify $\sqrt{20}$ and $\sqrt{\tfrac{2}{3}}$.

$$\sqrt{20} = \sqrt{4 \cdot 5} = \sqrt{4}\sqrt{5} = 2\sqrt{5}$$

$$\sqrt{\frac{2}{3}} = \frac{\sqrt{2}}{\sqrt{3}} = \frac{\sqrt{2}}{\sqrt{3}} \cdot \frac{\sqrt{3}}{\sqrt{3}} = \frac{\sqrt{6}}{3}$$

SIMPLIFIED FORM FOR RADICALS [7.3]

A radical expression is in simplified form if:

1. There are no perfect squares as factors of the quantity under the square root sign.
2. There are no fractions under the radical sign.
3. There are no radicals in the denominator.

5. (a) $5\sqrt{7} + 3\sqrt{7} = 8\sqrt{7}$

   (b) $2\sqrt{18} - 3\sqrt{50}$
   $= 2 \cdot 3\sqrt{2} - 3 \cdot 5\sqrt{2}$
   $= 6\sqrt{2} - 15\sqrt{2}$
   $= -9\sqrt{2}$

ADDITION AND SUBTRACTION OF RADICAL EXPRESSIONS [7.4]

We add and subtract radical expressions by using the distributive property to combine terms that have the same radical parts. If the radicals are not in simplified form, we begin by writing them in simplified form and then combining similar terms, if possible.

6. (a) $\sqrt{3}(\sqrt{5} - \sqrt{3}) = \sqrt{15} - 3$

   (b) $(\sqrt{7} + 3)(\sqrt{7} - 5)$
   $= 7 - 5\sqrt{7} + 3\sqrt{7} - 15$
   $= -8 - 2\sqrt{7}$

MULTIPLICATION OF RADICAL EXPRESSIONS [7.5]

We multiply radical expressions by applying the distributive property or the FOIL method.

7. $\dfrac{7}{\sqrt{5} - \sqrt{3}}$

$= \dfrac{7}{\sqrt{5} - \sqrt{3}} \cdot \dfrac{\sqrt{5} + \sqrt{3}}{\sqrt{5} + \sqrt{3}}$

$= \dfrac{7\sqrt{5} + 7\sqrt{3}}{2}$

DIVISION OF RADICAL EXPRESSIONS [7.5]

To divide by an expression like $\sqrt{5} - \sqrt{3}$, we multiply the numerator and denominator by its conjugate, $\sqrt{5} + \sqrt{3}$. This process is also called rationalizing the denominator.

8. Solve $\sqrt{x - 3} = 2$

$$(\sqrt{x - 3})^2 = 2^2$$
$$x - 3 = 4$$
$$x = 7$$

The solution checks in the original equation.

SQUARING PROPERTY OF EQUALITY [7.6]

We are free to square both sides of an equation whenever it is convenient, as long as we check all solutions in the original equation. We must check solutions because squaring both sides of an equation occasionally produces extraneous solutions.

COMMON MISTAKE

The most common mistake when working with radicals is to try to apply a property similar to Property 2 for radicals involving addition instead of multiplication. Here is an example:

$$\sqrt{16 + 9} = \sqrt{16} + \sqrt{9}$$

Although the above example looks like it may be true, it isn't. If we carry it out further, the mistake becomes obvious:

$$\sqrt{16 + 9} = \sqrt{16} + \sqrt{9}$$
$$\sqrt{25} = 4 + 3$$
$$5 = 7 \qquad \qquad \text{False}$$

Find the following roots.   Chapter 7 Test

1. $\sqrt{16}$ 2. $-\sqrt{36}$

3. The square roots of 49   4. $\sqrt[3]{27}$

5. $\sqrt[3]{-8}$ 6. $-\sqrt[4]{81}$

Put the following expressions into simplified form.

7. $\sqrt{75}$ 8. $\sqrt{32}$

9. $\sqrt{\frac{2}{3}}$ 10. $\dfrac{1}{\sqrt[3]{4}}$

11. $3\sqrt{50x^2}$ 12. $\sqrt{\dfrac{12x^2y^3}{5}}$

13. $\dfrac{2}{\sqrt{2} - 1}$ 14. $\dfrac{1}{\sqrt{3} - \sqrt{2}}$

Perform the indicated operations. All answers should be written in simplified form.

15. $5\sqrt{12} - 2\sqrt{27}$ 16. $\sqrt{3}(\sqrt{5} - 2)$

17. $(\sqrt{2} + 3)(\sqrt{2} - 1)$ 18. $(2\sqrt{5} - \sqrt{2})(3\sqrt{5} + 2\sqrt{2})$

19. $(\sqrt{5} - \sqrt{3})^2$ 20. $\dfrac{\sqrt{2} + \sqrt{3}}{\sqrt{2} - \sqrt{3}}$

21. $\dfrac{\sqrt{6} - \sqrt{7}}{\sqrt{6} - 2}$

Solve the following equations.

22. $\sqrt{x + 5} = 2$

23. $\sqrt{2x + 1} + 2 = 7$

24. $\sqrt{3x + 1} + 6 = 2$

25. $\sqrt{2x - 3} = x - 3$

More Quadratic Equations

Our goal in this chapter is to solve quadratic equations that cannot be factored. We are going to develop a method of solving quadratic equations that can be used on all quadratic equations, regardless of whether or not they are factorable. We will then apply this method to the general quadratic equation $ax^2 + bx + c = 0$ and arrive at what is known as the quadratic formula. We will end the chapter by considering the graphs of second-degree equations.

To be successful in this chapter you should have a working knowledge of (1) square roots, (2) binomial squares, and (3) graphing by the use of tables.

We will begin by solving some quadratic equations in a special form. Once this has been accomplished, we can proceed to a method of solution for quadratic equations called "completing the square."

Consider the equation $x^2 = 9$. Inspection shows that there are two solutions, $x = 3$ and $x = -3$. We arrive at these solutions by taking the square root of both sides of the original equation. Since we are taking the square root ourselves, we must be sure to include both the positive and the negative square roots. When all the work is shown, the problem looks like this:

8.1
More Quadratic Equations

$$x^2 = 9$$
$$\sqrt{x^2} = \pm\sqrt{9} \qquad \text{Take the square root of both sides}$$
$$x = \pm 3 \qquad \pm\sqrt{9} = \pm 3$$

The notation $x = \pm 3$ is shorthand for the expression $x = 3$ or $x = -3$.

We can take the square root of both sides of an equation any time we feel it is helpful. We must make sure, however, that we include both the positive and the negative square roots.

▼ **Example 1** Solve for x: $x^2 = 7$.

SOLUTION

$$x^2 = 7$$
$$\sqrt{x^2} = \pm\sqrt{7} \qquad \text{Take the square root of both sides}$$
$$x = \pm\sqrt{7}$$

The solution set is $\{\sqrt{7}, -\sqrt{7}\}$.   ▲

▼ **Example 2** Solve for a: $(a + 3)^2 = 16$.

SOLUTION   We proceed as we did in the last example. The square root of $(a + 3)^2$ is $(a + 3)$:

$$(a + 3)^2 = 16$$
$$a + 3 = \pm 4$$

At this point we add -3 to both sides to get

$$a = -3 \pm 4$$

which we can write as

$$a = -3 + 4 \quad \text{or} \quad a = -3 - 4$$
$$a = 1 \qquad \text{or} \quad a = -7$$

Our solution set is $\{1, -7\}$.   ▲

▼ **Example 3** Solve for x: $(3x - 2)^2 = 25$

SOLUTION

$$(3x - 2)^2 = 25$$
$$3x - 2 = \pm 5$$

Adding 2 to both sides, we have

$$3x = 2 \pm 5$$

Dividing both sides by 3 gives us

$$x = \frac{2 \pm 5}{3}$$

We separate this last equation into two separate statements:

$$x = \frac{2 + 5}{3} \quad \text{or} \quad x = \frac{2 - 5}{3}$$

$$x = \frac{7}{3} \quad \text{or} \quad x = \frac{-3}{3} = -1$$

The solution set is $\{\frac{7}{3}, -1\}$. ▲

▼ **Example 4** Solve for y: $(4y - 5)^2 = 6$.

SOLUTION

$$(4y - 5)^2 = 6$$
$$4y - 5 = \pm\sqrt{6}$$
$$4y = 5 \pm \sqrt{6} \qquad \text{Add 5 to both sides}$$
$$y = \frac{5 \pm \sqrt{6}}{4} \qquad \text{Divide both sides by 4}$$

Since $\sqrt{6}$ is irrational, we cannot simplify the expression further.
The solution set is $\left\{\dfrac{5 + \sqrt{6}}{4}, \dfrac{5 - \sqrt{6}}{4}\right\}$. ▲

▼ **Example 5** Solve for x: $(2x + 6)^2 = 8$.

SOLUTION

$$(2x + 6)^2 = 8$$
$$2x + 6 = \pm\sqrt{8}$$
$$2x + 6 = \pm 2\sqrt{2} \qquad \sqrt{8} = \sqrt{4 \cdot 2} = 2\sqrt{2}$$
$$2x = -6 \pm 2\sqrt{2} \qquad \text{Add } -6 \text{ to both sides}$$
$$x = \frac{-6 \pm 2\sqrt{2}}{2} \qquad \text{Divide both sides by 2}$$
$$x = -3 \pm \sqrt{2} \qquad \text{Reduce to lowest terms}$$

The solution set is $\{-3 + \sqrt{2}, -3 - \sqrt{2}\}$.
   We can check our two solutions in the original equation. Let's check our first solution, $-3 + \sqrt{2}$.

When
$$x = -3 + \sqrt{2}$$

the equation $(2x + 6)^2 = 8$

becomes $[2(-3 + \sqrt{2}) + 6]^2 = 8$

$(-6 + 2\sqrt{2} + 6)^2 = 8$

$(2\sqrt{2})^2 = 8$

$4 \cdot 2 = 8$

$8 = 8$ A true statement

The second solution, $-3 - \sqrt{2}$, checks also.   ▲

Solving quadratic equations in the form $(ax + b)^2 = c$ is the first step toward learning the *completing the square* method for solving general quadratic equations. The problems in this section are all similar in form, and therefore the steps in solving them are almost identical.

Problem Set 8.1   Solve each of the following equations using the methods learned in this section.

1. $x^2 = 9$ 2. $x^2 = 16$
3. $a^2 = 25$ 4. $a^2 = 36$
5. $x^2 = 7$ 6. $x^2 = 13$
7. $y^2 = 8$ 8. $y^2 = 75$
9. $x^2 = 32$ 10. $x^2 = 288$
11. $x^2 = 50$ 12. $x^2 = 27$
13. $a^2 = 48$ 14. $a^2 = 162$
15. $(x + 2)^2 = 4$ 16. $(x - 3)^2 = 16$
17. $(x + 1)^2 = 25$ 18. $(x + 3)^2 = 64$
19. $(a - 5)^2 = 75$ 20. $(a - 4)^2 = 32$
21. $(y + 1)^2 = 50$ 22. $(y - 5)^2 = 27$
23. $(x + 8)^2 = 24$ 24. $(x - 6)^2 = 12$
25. $(2x + 1)^2 = 25$ 26. $(3x - 2)^2 = 16$
27. $(4a - 5)^2 = 36$ 28. $(2a + 6)^2 = 64$
29. $(3y - 1)^2 = 25$ 30. $(5y - 4)^2 = 121$
31. $(6x + 2)^2 = 16$ 32. $(8x - 1)^2 = 9$
33. $(3x - 1)^2 = 27$ 34. $(2x + 5)^2 = 32$
35. $(3x + 2)^2 = 45$ 36. $(5x - 10)^2 = 75$
37. $(2y - 5)^2 = 8$ 38. $(4y - 6)^2 = 48$
39. $(11x + 1)^2 = 72$ 40. $(9x - 7)^2 = 28$

41. Check the solution $x = -1 + 5\sqrt{2}$ in the equation $(x + 1)^2 = 50$.
42. Check the solution $x = -8 + 2\sqrt{6}$ in the equation $(x + 8)^2 = 24$.
43. The square of the sum of a number and 3 is 16. Find the number. (There are two solutions.)

44. The square of the sum of twice a number and 3 is 25. Find the number. (There are two solutions.)

Review Problems   The problems below review material we covered in Sections 4.6 and 5.5. Reviewing these problems will help you with the material in the next section.

Multiply.

45. $(x + 3)^2$ **46.** $(x - 3)^2$
47. $(x - 5)^2$ **48.** $(x + 5)^2$
49. $(x + 7)^2$ **50.** $(x - 7)^2$

Factor.

51. $x^2 - 12x + 36$ **52.** $x^2 + 12x + 36$
53. $x^2 + 4x + 4$ **54.** $x^2 - 4x + 4$
55. $x^2 - 2x + 1$ **56.** $x^2 + 2x + 1$

In this section we will develop a method of solving quadratic equations that works whether or not the equation can be factored. Since we will be working with the individual terms of trinomials, we need some new definitions so that we can keep our vocabulary straight.

**8.2
Completing
the Square**

DEFINITION   In the trinomial $2x^2 + 3x + 4$, the first term, $2x^2$, is called the *quadratic term;* the middle term, $3x$, is called the *linear term;* and the last term, 4, is called the *constant term.*

Now consider the following list of perfect square trinomials and their corresponding binomial squares.

$$x^2 + 6x + 9 = (x + 3)^2$$
$$x^2 - 8x + 16 = (x - 4)^2$$
$$x^2 - 10x + 25 = (x - 5)^2$$
$$x^2 + 12x + 36 = (x + 6)^2$$

In each case the coefficient of x^2 is 1. A more important observation, however, stems from the relationship between the linear terms (middle terms) and the constant terms (last terms). Notice that the constant in each case is the square of half the coefficient of x in the linear term. That is:

1.   For the first trinomial, $x^2 + 6x + 9$, the last term, 9, is the square of half the coefficient of the middle term: $9 = (\frac{6}{2})^2$.

2. For the second trinomial, $x^2 - 8x + 16$, $16 = [\tfrac{1}{2}(-8)]^2$.
3. For the trinomial $x^2 - 10x + 25$, it also holds: $25 = [\tfrac{1}{2}(-10)]^2$.

Check and see that it also works for the last trinomial.

In summary, then, for every perfect square trinomial in which the coefficient of x^2 is 1, the last term is always the square of half the coefficient of the linear term. We can use this fact to build our own perfect square trinomials.

▼ **Examples**   Write the correct last term to each of the following expressions so each becomes a perfect square trinomial.

1. $x^2 - 2x$

> **SOLUTION**   The coefficient of the linear term is -2. If we take the square of half of -2, we get 1. Adding the 1 as the last term, we have the perfect square trinomial:
>
> $$x^2 - 2x + 1 = (x - 1)^2$$

2. $x^2 + 18x$

> **SOLUTION**   Half of 18 is 9, the square of which is 81. If we add 81 at the end, we have
>
> $$x^2 + 18x + 81 = (x + 9)^2$$

3. $x^2 + 3x$

> **SOLUTION**   Half of 3 is $\tfrac{3}{2}$, the square of which is $\tfrac{9}{4}$:
>
> $$x^2 + 3x + \tfrac{9}{4} = (x + \tfrac{3}{2})^2 \qquad\qquad ▲$$

We can use the above procedure, along with the method developed in Section 8.1, to solve some quadratic equations.

▼ **Example 4**   Solve $x^2 - 6x + 5 = 0$ by completing the square.

> **SOLUTION**   We begin by adding -5 to both sides of the equation. We want just $x^2 - 6x$ on the left side so that we can add on our own last term to get a perfect square trinomial:
>
> $$x^2 - 6x + 5 = 0$$
> $$x^2 - 6x \quad\;\; = -5 \qquad \text{Add } -5 \text{ to both sides}$$
>
> Now we can add 9 to both sides and the left side will be a perfect square:
>
> $$x^2 - 6x + 9 = -5 + 9$$
> $$(x - 3)^2 = 4$$

The last line is in the form of the equations we solved in Section 8.1. Taking the square root of both sides, we have

$$x - 3 = \pm 2$$
$$x = 3 \pm 2 \qquad \text{Add 3 to both sides}$$
$$x = 3 + 2 \quad \text{or} \quad x = 3 - 2$$
$$x = 5 \qquad \text{or} \quad x = 1$$

The solution set is $\{5, 1\}$. ▲

The above method of solution is called *completing the square*.

▼ **Example 5** Solve by completing the square: $2x^2 + 16x - 18 = 0$.

SOLUTION We begin by moving the constant term to the other side:

$$2x^2 + 16x - 18 = 0$$
$$2x^2 + 16x \quad = 18 \qquad \text{Add 18 to both sides}$$

In order to complete the square, we must be sure the coefficient of x^2 is 1. To accomplish this, we divide both sides by 2:

$$\frac{2x^2}{2} + \frac{16x}{2} = \frac{18}{2}$$

$$x^2 + 8x = 9$$

We now complete the square by adding the square of half the coefficient of the linear term to both sides:

$$x^2 + 8x + 16 = 9 + 16 \qquad \text{Add 16 to both sides}$$
$$(x + 4)^2 = 25$$
$$x + 4 = \pm 5 \qquad \text{Take the square root of both sides}$$
$$x = -4 \pm 5 \qquad \text{Add } -4 \text{ to both sides}$$
$$x = -4 + 5 \quad \text{or} \quad x = -4 - 5$$
$$x = 1 \qquad \text{or} \quad x = -9$$

The solution set arrived at by completing the square is $\{1, -9\}$. ▲

▼ **Example 6** Solve for y: $3y^2 - 9y + 3 = 0$.

SOLUTION

$$3y^2 - 9y + 3 = 0$$
$$3y^2 - 9y \quad = -3 \qquad \text{Add } -3 \text{ to both sides}$$
$$y^2 - 3y \quad = -1 \qquad \text{Divide by 3}$$

$$y^2 - 3y + \frac{9}{4} = -1 + \frac{9}{4} \qquad \text{Complete the square}$$

$$\left(y - \frac{3}{2}\right)^2 = \frac{5}{4}$$

$$y - \frac{3}{2} = \pm \frac{\sqrt{5}}{2} \qquad \text{Square root of both sides}$$

$$y = \frac{3}{2} \pm \frac{\sqrt{5}}{2} \qquad \text{Add } \tfrac{3}{2} \text{ to both sides}$$

$$y = \frac{3}{2} + \frac{\sqrt{5}}{2} \quad \text{or} \quad y = \frac{3}{2} - \frac{\sqrt{5}}{2}$$

$$y = \frac{3 + \sqrt{5}}{2} \quad \text{or} \quad y = \frac{3 - \sqrt{5}}{2}$$

The solution set is

$$\left\{ \frac{3 + \sqrt{5}}{2}, \frac{3 - \sqrt{5}}{2} \right\}$$

which can be written in a shorter form as

$$\left\{ \frac{3 \pm \sqrt{5}}{2} \right\} \qquad \blacktriangle$$

We will now summarize the above examples by listing the steps involved in solving quadratic equations by completing the square.

To Solve a Quadratic Equation by Completing the Square

Step 1: Put the equation in the form $ax^2 + bx = c$. This usually involves only moving the constant term to the opposite side.

Step 2: Make sure the coefficient of the squared term is 1. If it is not 1, simply divide both sides by whatever it is.

Step 3: Add the square of half the coefficient of the linear term to both sides of the equation.

Step 4: Write the left-hand side of the equation as a binomial square and solve, using the methods developed in Section 8.1.

Problem Set 8.2

Give the correct last term for each of the following expressions to ensure that the resulting trinomial is a perfect square trinomial.

1. $x^2 + 6x$
2. $x^2 - 10x$
3. $x^2 + 2x$
4. $x^2 + 14x$
5. $y^2 - 8y$
6. $y^2 + 12y$
7. $y^2 - 2y$
8. $y^2 - 6y$

9. $x^2 + 16x$

10. $x^2 - 4x$

11. $a^2 - 3a$

12. $a^2 + 5a$

13. $x^2 - 7x$

14. $x^2 - 9x$

15. $y^2 + y$

16. $y^2 - y$

17. $x^2 - \frac{3}{2}x$

18. $x^2 + \frac{2}{3}x$

Solve each of the following equations by completing the square. Follow the steps given at the end of this section.

19. $x^2 + 4x = 12$

20. $x^2 - 2x = 8$

21. $x^2 - 6x = 16$

22. $x^2 + 12x = -27$

23. $a^2 + 2a = 3$

24. $a^2 - 8a = -7$

25. $x^2 - 10x = 0$

26. $x^2 + 4x = 0$

27. $y^2 + 2y - 15 = 0$

28. $y^2 - 10y - 11 = 0$

29. $x^2 + 4x - 3 = 0$

30. $x^2 + 6x + 5 = 0$

31. $x^2 - 3x - 4 = 0$

32. $x^2 + 5x - 3 = 0$

33. $a^2 - 7a - 8 = 0$

34. $a^2 + 3a - 1 = 0$

35. $4x^2 + 8x - 4 = 0$

36. $3x^2 + 12x + 6 = 0$

37. $2x^2 + 2x - 4 = 0$

38. $4x^2 + 4x - 3 = 0$

39. $3y^2 - 6y - 9 = 0$

40. $5y^2 - 10y - 15 = 0$

41. $2x^2 - 2x - 1 = 0$

42. $3x^2 - 3x - 1 = 0$

43. $4a^2 - 4a + 1 = 0$

44. $2a^2 + 4a + 1 = 0$

45. $3y^2 - 9y - 2 = 0$

46. $5y^2 - 10y - 4 = 0$

47. The two solutions to Problem 29 are $-2 + \sqrt{7}$ and $-2 - \sqrt{7}$. The table of square roots at the back of the book gives the decimal approximation of $\sqrt{7}$ as 2.646. Use this approximation for $\sqrt{7}$ to find decimal approximations of $-2 + \sqrt{7}$ and $-2 - \sqrt{7}$.

48. The solutions to Problem 41 are $\dfrac{1 + \sqrt{3}}{2}$ and $\dfrac{1 - \sqrt{3}}{2}$. Use the decimal approximation of $\sqrt{3}$ given in the table at the back of the book to find decimal approximations of these two numbers.

49. One of the solutions to the equation in Problem 35 is $-1 + \sqrt{2}$. Check this solution in the original equation.

50. Check the solution $x = -1 - \sqrt{2}$ from Problem 35 in the original equation, $4x^2 + 8x - 4 = 0$.

51. Find the sum of the two solutions to the equation in Problem 29. (Add $-2 + \sqrt{7}$ and $-2 - \sqrt{7}$.)

52. Find the product of the two solutions to the equation in Problem 29.

Review Problems   The problems below review material we first covered in Section 2.1.

Find the value of each expression if $a = 2$, $b = 4$, and $c = -3$.

53. $2a$

54. b^2

55. $4ac$ **56.** $b^2 - 4ac$

57. $\sqrt{b^2 - 4ac}$ **58.** $-b + \sqrt{b^2 - 4ac}$

8.3 The Quadratic Formula

In this section we will derive the quadratic formula. It is one formula that you will use in almost all types of mathematics. We will first state the formula as a theorem and then prove it. The proof is based on the method of completing the square developed in the last section.

The Quadratic Theorem

For any quadratic equation in the form $ax^2 + bx + c = 0$, when a, b, and c are real numbers and $a \neq 0$, the two solutions are

$$x = \frac{-b + \sqrt{b^2 - 4ac}}{2a} \quad \text{and} \quad x = \frac{-b - \sqrt{b^2 - 4ac}}{2a}$$

PROOF   We will prove the theorem by completing the square on

$$ax^2 + bx + c = 0$$

Adding $-c$ to both sides, we have

$$ax^2 + bx = -c$$

To make the coefficient of x^2 one, we divide both sides by a:

$$\frac{ax^2}{a} + \frac{bx}{a} = -\frac{c}{a}$$

$$x^2 + \frac{b}{a}x = -\frac{c}{a}$$

Now, to complete the square, we add the square of half of $\dfrac{b}{a}$ to both sides:

$$x^2 + \frac{b}{a}x + \left(\frac{b}{2a}\right)^2 = -\frac{c}{a} + \left(\frac{b}{2a}\right)^2 \qquad \frac{1}{2} \text{ of } \frac{b}{a} \text{ is } \frac{b}{2a}$$

Let's simplify the right side separately:

$$-\frac{c}{a} + \left(\frac{b}{2a}\right)^2 = -\frac{c}{a} + \frac{b^2}{4a^2}$$

The common denominator is $4a^2$. We multiply the numerator and denominator of $-\dfrac{c}{a}$ by $4a$ to give it the common denominator. Then we combine numerators:

$$\frac{4a}{4a}\left(-\frac{c}{a}\right) + \frac{b^2}{4a^2} = -\frac{4ac}{4a^2} + \frac{b^2}{4a^2}$$

$$= \frac{-4ac + b^2}{4a^2}$$

$$= \frac{b^2 - 4ac}{4a^2}$$

Now back to the equation. We use our simplified expression for the right side:

$$x^2 + \frac{b}{a}x + \left(\frac{b}{2a}\right)^2 = \frac{b^2 - 4ac}{4a^2}$$

$$\left(x + \frac{b}{2a}\right)^2 = \frac{b^2 - 4ac}{4a^2}$$

Taking the square root of both sides, we have

$$x + \frac{b}{2a} = \pm\frac{\sqrt{b^2 - 4ac}}{2a}$$

$$x = \frac{-b}{2a} \pm \frac{\sqrt{b^2 - 4ac}}{2a} \qquad \text{Add } \frac{-b}{2a} \text{ to both sides}$$

$$x = \frac{-b \pm \sqrt{b^2 - 4ac}}{2a}$$

Our proof is now complete. What we have is this: If our equation is in the form $ax^2 + bx + c = 0$ (standard form), then the solution can always be found by using the quadratic formula:

$$x = \frac{-b \pm \sqrt{b^2 - 4ac}}{2a}$$

▼ **Example 1** Solve $x^2 - 5x - 6 = 0$ by using the quadratic formula.

SOLUTION To use the quadratic formula, we must make sure the equation is in standard form; identify a, b, and c; substitute them into the formula; and work out the arithmetic.
For the equation $x^2 - 5x - 6 = 0$, $a = 1$, $b = -5$, and $c = -6$:

$$x = \frac{-b \pm \sqrt{b^2 - 4ac}}{2a} = \frac{-(-5) \pm \sqrt{(-5)^2 - 4(1)(-6)}}{2(1)}$$

$$= \frac{5 \pm \sqrt{49}}{2}$$

$$= \frac{5 \pm 7}{2}$$

$$x = \frac{5 + 7}{2} \quad \text{or} \quad x = \frac{5 - 7}{2}$$

$$x = \frac{12}{2} \qquad\qquad x = -\frac{2}{2}$$

$$x = 6 \qquad\qquad x = -1$$

The solution set is $\{6, -1\}$. ▲

▼ **Example 2**   Solve for x: $2x^2 = -4x + 3$.

SOLUTION   In standard form the equation is $2x^2 + 4x - 3 = 0$, so $a = 2$, $b = 4$, and $c = -3$:

$$x = \frac{-b \pm \sqrt{b^2 - 4ac}}{2a}$$

$$= \frac{-4 \pm \sqrt{16 - 4(2)(-3)}}{2(2)}$$

$$= \frac{-4 \pm \sqrt{40}}{4}$$

$$= \frac{-4 \pm 2\sqrt{10}}{4}$$

We can reduce this last expression to lowest terms by factoring 2 from the numerator and denominator and then dividing it out:

$$x = \frac{\cancel{2}(-2 \pm \sqrt{10})}{\cancel{2} \cdot 2}$$

$$= \frac{-2 \pm \sqrt{10}}{2}$$

Our two solutions are $\dfrac{-2 + \sqrt{10}}{2}$ and $\dfrac{-2 - \sqrt{10}}{2}$. ▲

▼ **Example 3**   Solve for x: $(x - 2)(x + 3) = 5$.

SOLUTION   We must put the equation into standard form before we can use the formula:

$$(x - 2)(x + 3) = 5$$
$$x^2 + x - 6 = 5$$
$$x^2 + x - 11 = 0$$
$$a = 1, \quad b = 1, \quad c = -11$$

$$x = \frac{-1 \pm \sqrt{1^2 - 4(1)(-11)}}{2(1)}$$

$$= \frac{-1 \pm \sqrt{45}}{2}$$

$$= \frac{-1 \pm 3\sqrt{5}}{2}$$

The solution set is $\left\{ \dfrac{-1 + 3\sqrt{5}}{2}, \dfrac{-1 - 3\sqrt{5}}{2} \right\}$. ▲

Solve the following equations by using the quadratic formula.

1. $x^2 + 3x + 2 = 0$ 2. $x^2 - 5x + 6 = 0$
3. $x^2 + 5x + 6 = 0$ 4. $x^2 - 7x - 8 = 0$
5. $x^2 + 6x + 9 = 0$ 6. $x^2 - 10x + 25 = 0$
7. $x^2 + x - 12 = 0$ 8. $x^2 + 6x + 8 = 0$
9. $2x^2 + 5x + 3 = 0$ 10. $2x^2 + 3x - 20 = 0$
11. $6x^2 - x - 2 = 0$ 12. $6x^2 + 5x - 4 = 0$
13. $x^2 - 2x + 1 = 0$ 14. $x^2 + 2x - 3 = 0$
15. $x^2 - 5x - 7 = 0$ 16. $2x^2 - 6x - 8 = 0$
17. $3x^2 - 5x - 1 = 0$ 18. $2x^2 - 10x + 2 = 0$
19. $(x - 2)(x + 1) = 3$ 20. $(x - 8)(x + 7) = 5$
21. $(2x - 3)(x + 2) = 1$ 22. $(4x - 5)(x - 3) = 6$
23. $2x^2 - 3x - 5 = 0$ 24. $3x^2 - 4x - 5 = 0$
25. $2x^2 = -3x + 7$ 26. $5x^2 = -2x + 3$
27. $3x^2 = -4x + 2$ 28. $3x^2 = 4x + 2$
29. $2x^2 - 5 = 2x$ 30. $5x^2 + 1 = -10x$

31. Solve the equation $2x^3 + 3x^2 - 4x = 0$ by first factoring out the common factor, x, and then using the quadratic formula. There are three solutions to this equation.
32. Solve the equation $5y^3 - 10y^2 + 4y = 0$ by first factoring out the common factor, y, and then using the quadratic formula.
33. To apply the quadratic formula to the equation $3x^2 - 4x = 0$, you have to notice that $c = 0$. Solve the equation, using the quadratic formula.
34. Solve the equation $9x^2 - 16 = 0$ using the quadratic formula. (Notice $b = 0$.)

35. Solve the equation below by first multiplying both sides by the LCD and then applying the quadratic formula to the result.

$$\tfrac{1}{2}x^2 - \tfrac{1}{2}x - \tfrac{1}{6} = 0$$

36. Solve the equation below by first multiplying both sides by the LCD and then applying the quadratic formula to the result.

$$\tfrac{1}{2}y^2 - y - \tfrac{3}{2} = 0$$

Review Problems   The problems below review the material we covered in Sections 4.4, 4.5, and 7.5. Reviewing these problems will help you with the material in the next section.

Combine similar terms.

37. $(3 + 4x) + (2 - 6x)$ **38.** $(2 + 5x) - (3 + 7x) + (2 - x)$

Multiply.

39. $2x(3x + 4)$ **40.** $(3x + 2)(4x - 3)$

Rationalize the denominator.

41. $\dfrac{2}{3 + \sqrt{5}}$ **42.** $\dfrac{2 + \sqrt{3}}{2 - \sqrt{3}}$

8.4
Complex Numbers

In order to solve quadratic equations such as $x^2 = -4$, we need to introduce a new set of numbers. If we try to solve $x^2 = -4$ using real numbers, we always get no solution. There is no real number whose square is -4.

The new set of numbers is called *complex numbers* and is based on the following definition.

DEFINITION   The number i is a number such that $i = \sqrt{-1}$.

The first thing we notice about this definition is that i is not a real number. There are no real numbers that represent the square root of -1. The other observation we make about i is $i^2 = -1$. If $i = \sqrt{-1}$, then, squaring both sides, we must have $i^2 = -1$. The most common power of i is i^2. Whenever we see i^2, we can write it as -1. We are now ready for a definition of complex numbers.

DEFINITION   A complex number is any number that can be put in the form $a + bi$, where a and b are real numbers and $i = \sqrt{-1}$.

EXAMPLE   The following are complex numbers:

$$3 + 4i$$
$$\tfrac{1}{2} - 6i$$
$$8 + i\sqrt{2}$$
$$\tfrac{3}{4} - 2i\sqrt{5}$$

EXAMPLE   The number $4i$ is a complex number because $4i = 0 + 4i$.

EXAMPLE   The number 8 is a complex number because $8 = 8 + 0i$.

From the last example we can see that the real numbers are a subset of the complex numbers because any real number x can be written as $x + 0i$.

We add and subtract complex numbers according to the same procedure we used to add and subtract polynomials: we combine similar terms.

Addition and Subtraction of Complex Numbers

▼   **Example 1**   Combine $(3 + 4i) + (2 - 6i)$.

SOLUTION

$$
\begin{aligned}
(3 + 4i) + (2 - 6i) &= (3 + 2) + (4i - 6i) && \text{Commutative and} \\
&&& \text{associative properties} \\
&= 5 + (-2i) && \text{Combine similar terms} \\
&= 5 - 2i
\end{aligned}
$$
▲

▼   **Example 2**   Combine $(2 - 5i) - (3 + 7i) + (2 - i)$.

SOLUTION

$$
\begin{aligned}
(2 - 5i) - (3 + 7i) + (2 - i) &= 2 - 5i - 3 - 7i + 2 - i \\
&= (2 - 3 + 2) + (-5i - 7i - i) \\
&= 1 - 13i
\end{aligned}
$$
▲

Multiplication of complex numbers is very similar to multiplication of polynomials. We can simplify many answers by using the fact that $i^2 = -1$.

Multiplication of Complex Numbers

▼   **Example 3**   Multiply $4i(3 + 5i)$.

SOLUTION

$$
\begin{aligned}
4i(3 + 5i) &= 4i(3) + 4i(5i) && \text{Distributive property} \\
&= 12i + 20i^2 && \text{Multiplication} \\
&= 12i + 20(-1) && i^2 = -1 \\
&= -20 + 12i
\end{aligned}
$$
▲

▼ **Example 4** Multiply $(3 + 2i)(4 - 3i)$.

SOLUTION

$(3 + 2i)(4 - 3i)$
$$= 3 \cdot 4 + 3(-3i) + 2i(4) + 2i(-3i) \qquad \text{FOIL method}$$
$$= 12 - 9i + 8i - 6i^2$$
$$= 12 - 9i + 8i - 6(-1) \qquad\qquad\qquad i^2 = -1$$
$$= (12 + 6) + (-9i + 8i)$$
$$= 18 - i \qquad\qquad\qquad\qquad\qquad\qquad ▲$$

Division of Complex Numbers

We divide complex numbers by applying the same process we used to rationalize denominators.

▼ **Example 5** Divide $\dfrac{2i}{3 - 4i}$.

SOLUTION We multiply numerator and denominator by the conjugate of the denominator, which is $3 + 4i$:

$$\left(\frac{2i}{3 - 4i}\right)\left(\frac{3 + 4i}{3 + 4i}\right) = \frac{6i + 8i^2}{9 - 16i^2}$$

$$= \frac{6i + 8(-1)}{9 - 16(-1)}$$

$$= \frac{-8 + 6i}{25} \qquad\qquad ▲$$

▼ **Example 6** Divide $\dfrac{2 + i}{5 + 2i}$.

SOLUTION The conjugate of the denominator is $5 - 2i$:

$$\left(\frac{2 + i}{5 + 2i}\right)\left(\frac{5 - 2i}{5 - 2i}\right) = \frac{10 - 4i + 5i - 2i^2}{25 - 4i^2}$$

$$= \frac{10 - 4i + 5i - 2(-1)}{25 - 4(-1)} \qquad i^2 = -1$$

$$= \frac{12 + i}{29} \qquad\qquad ▲$$

From the last two examples we notice that multiplying complex conjugates always results in a real number:

$$(a + bi)(a - bi) = a^2 - b^2 i^2$$
$$= a^2 - b^2(-1)$$
$$= a^2 + b^2$$

Combine the following complex numbers.

1. $(3 - 2i) + 3i$ 2. $(5 - 4i) - 8i$
3. $(6 + 2i) - 10i$ 4. $(8 - 10i) + 7i$
5. $(11 + 9i) - 9i$ 6. $(12 + 2i) + 6i$
7. $(3 + 2i) + (6 - i)$ 8. $(4 + 8i) - (7 + i)$
9. $(5 + 7i) - (6 + 8i)$ 10. $(11 + 6i) - (3 + 6i)$
11. $(9 - i) + (2 - i)$ 12. $(8 + 3i) - (8 - 3i)$
13. $(6 + i) - 4i - (2 - i)$ 14. $(3 + 2i) - 5i - (5 + 4i)$
15. $(6 - 11i) + 3i + (2 + i)$ 16. $(3 + 4i) - (5 + 7i) - (6 - i)$
17. $(2 + 3i) - (6 - 2i) + (3 - i)$ 18. $(8 + 9i) + (5 - 6i) - (4 - 3i)$

Multiply the following complex numbers.

19. $3(2 - i)$ 20. $4(5 + 3i)$
21. $2i(8 - 7i)$ 22. $-3i(2 + 5i)$
23. $(2 + i)(4 - i)$ 24. $(6 + 3i)(4 + 3i)$
25. $(2 + i)(3 - 5i)$ 26. $(4 - i)(2 - i)$
27. $(3 + 5i)(3 - 5i)$ 28. $(8 + 6i)(8 - 6i)$
29. $(2 + i)(2 - i)$ 30. $(3 + i)(3 - i)$

Divide the following complex numbers.

31. $\dfrac{2}{3 - 2i}$ 32. $\dfrac{3}{5 + 6i}$

33. $\dfrac{-3i}{2 + 3i}$ 34. $\dfrac{4i}{3 + i}$

35. $\dfrac{6i}{3 - i}$ 36. $\dfrac{-7i}{5 - 4i}$

37. $\dfrac{2 + i}{2 - i}$ 38. $\dfrac{3 + 2i}{3 - 2i}$

39. $\dfrac{4 + 5i}{3 - 6i}$ 40. $\dfrac{-2 + i}{5 + 6i}$

41. Use the FOIL method to multiply $(x + 3i)(x - 3i)$.
42. Use the FOIL method to multiply $(x + 5i)(x - 5i)$.
43. The opposite of i is $-i$. The reciprocal of i is $1/i$. Multiply the numerator and denominator of $1/i$ by i and simplify the result to see that the opposite of i and the reciprocal of i are the same number.
44. If $i^2 = -1$, what are i^3 and i^4. (*Hint:* $i^3 = i^2 \cdot i$.)

Review Problems   The problems below review material we covered in Section 8.1. Reviewing these problems will help you in the next section.

Solve each equation by taking the square root of each side.

45. $(x - 3)^2 = 25$ **46.** $(x - 2)^2 = 9$
47. $(2x - 6)^2 = 16$ **48.** $(2x + 1)^2 = 49$
49. $(x + 3)^2 = 12$ **50.** $(x + 3)^2 = 8$

**8.5
Complex Solutions
To Quadratic
Equations**

So far, the only problem we could have had in using the quadratic formula involves square roots of negative numbers. The quadratic formula tells us solutions to equations of the form $ax^2 + bx + c = 0$ are always

$$x = \frac{-b \pm \sqrt{b^2 - 4ac}}{2a}$$

The part of the quadratic formula under the radical sign is called the discriminant:

$$\text{Discriminant} = b^2 - 4ac$$

When the discriminant is negative, we have to deal with the square root of a negative number. Complex numbers are what we use to handle square roots of negative numbers. Example 1 shows how we write square roots of negative numbers as complex numbers. Remember, i is defined to be $\sqrt{-1}$.

▼ **Example 1**   Write the following radicals as complex numbers.

(a) $\sqrt{-4} = \sqrt{4(-1)} = \sqrt{4}\sqrt{-1} = 2i$
(b) $\sqrt{-36} = \sqrt{36(-1)} = \sqrt{36}\sqrt{-1} = 6i$
(c) $\sqrt{-7} = \sqrt{7(-1)} = \sqrt{7}\sqrt{-1} = i\sqrt{7}$
(d) $\sqrt{-75} = \sqrt{75(-1)} = \sqrt{75}\sqrt{-1} = 5i\sqrt{3}$ ▲

Note: In parts (c) and (d) of Example 1, we wrote i before the radical because it is less confusing that way. If we put i after the radical, it is sometimes mistaken for being under the radical.

Let's see how complex numbers relate to quadratic equations by looking at some examples of quadratic equations whose solutions are complex numbers.

▼ **Example 2**   Solve for x: $(x + 2)^2 = -9$.

SOLUTION   We can solve this equation by expanding the left side, putting the results into standard form, and then applying the quadratic formula. It is faster, however, simply to take the square root of both sides:

$$(x + 2)^2 = -9$$

$$x + 2 = \pm\sqrt{-9} \qquad \text{Square root of both sides}$$

$$x + 2 = \pm 3i \qquad \sqrt{-9} = \sqrt{9}\sqrt{-1} = 3i$$

$$x = -2 \pm 3i \qquad \text{Add } -2 \text{ to both sides}$$

The solution set contains two complex solutions. Notice that the two solutions are conjugates.

The solution set is $\{-2 + 3i, -2 - 3i\}$.   ▲

▼ **Example 3**   Solve for x: $x^2 - 2x = -5$.

SOLUTION   The easiest way to solve this equation is by using the quadratic formula. We begin by adding 5 to both sides in order to put the equation into standard form:

$$x^2 - 2x = -5$$

$$x^2 - 2x + 5 = 0 \qquad \text{Add 5 to both sides}$$

Applying the quadratic formula with $a = 1$, $b = -2$, and $c = 5$, we have

$$x = \frac{-(-2) \pm \sqrt{4 - 4(1)(5)}}{2(1)}$$

$$= \frac{2 \pm \sqrt{-16}}{2}$$

$$= \frac{2 \pm 4i}{2}$$

Dividing the numerator and denominator by 2, we have the two solutions

$$x = 1 \pm 2i$$

The two solutions are $1 + 2i$ and $1 - 2i$.
Let's check our first solution, $1 + 2i$, in the original equation:

$$(1 + 2i)^2 - 2(1 + 2i) = -5$$

$$1 + 4i + 4i^2 - 2 - 4i = -5$$

$$1 + 4i - 4 - 2 - 4i = -5$$

$$-5 = -5 \qquad \text{A true statement}$$

The second solution checks also.   ▲

▼   **Example 4**   Solve for x: $2x^2 + x + 3 = 0$.

$$x = \frac{-1 \pm \sqrt{1^2 - 4(2)(3)}}{2(2)}$$

$$= \frac{-1 \pm \sqrt{-23}}{4}$$

$$= \frac{-1 \pm i\sqrt{23}}{4}$$

The solution set is $\left\{ \dfrac{-1 + i\sqrt{23}}{4}, \dfrac{-1 - i\sqrt{23}}{4} \right\}$.   ▲

This completes our work with solving quadratic equations. We can solve any quadratic equation we come across. Factoring is probably still the fastest method of solution, but again, factoring works only if the equation is factorable. Applying the quadratic formula always produces solutions, whether the equation is factorable or not.

Problem Set 8.5

Write the following radicals as complex numbers.

1. $\sqrt{-16}$	2. $\sqrt{-25}$	3. $\sqrt{-49}$
4. $\sqrt{-81}$	5. $\sqrt{-6}$	6. $\sqrt{-10}$
7. $\sqrt{-11}$	8. $\sqrt{-19}$	9. $\sqrt{-32}$
10. $\sqrt{-288}$	11. $\sqrt{-50}$	12. $\sqrt{-45}$
13. $\sqrt{-8}$	14. $\sqrt{-24}$	15. $\sqrt{-48}$
16. $\sqrt{-27}$		

Solve the following quadratic equations. Use whatever method seems to fit the situation or is convenient for you.

17. $x^2 = 5x - 6$	18. $2x^2 = 5x + 3$
19. $x^2 - 4x = -4$	20. $x^2 - 4x = 4$
21. $2x^2 + 5x = 12$	22. $2x^2 + 30 = 16x$
23. $(x - 2)^2 = -4$	24. $(x - 5)^2 = -25$
25. $(2x + 1)^2 = -9$	26. $(4x - 2)^2 = -8$
27. $(6x - 3)^2 = -27$	28. $(8x + 4)^2 = -32$
29. $x^2 + x + 1 = 0$	30. $x^2 - 3x + 4 = 0$
31. $x^2 - 5x + 6 = 0$	32. $x^2 + 2x + 2 = 0$
33. $3x^2 + 2x + 1 = 0$	34. $4x^2 + x + 5 = 0$
35. $2x^2 = -3x + 2$	36. $3x^2 = -2x + 1$
37. $(x + 2)(x - 3) = 5$	38. $(x - 1)(x + 1) = 6$
39. $(2x + 3)(x + 4) = 1$	40. $(3x - 5)(2x + 1) = 4$

41. Is $x = 2 + 2i$ a solution to the equation $x^2 - 4x + 8 = 0$?
42. Is $x = 5 + 3i$ a solution to the equation $x^2 - 10x + 34 = 0$?
43. If one solution to a quadratic equation is $3 + 7i$, what do you think the other solution is?
44. If one solution to a quadratic equation is $4 - 2i$, what do you think the other solution is?

Review Problems   The problems below review material on graphing straight lines that we covered in Section 3.2. The next section is also about graphing, so reviewing these problems should help you get back into it.

Graph each straight line.

45. $x - y = 2$ **46.** $x + y = -2$
47. $2x + 4y = 8$ **48.** $2x - 4y = 8$
49. $y = x + 3$ **50.** $y = x + 1$

In this section we will graph equations of the form $y = ax^2 + bx + c$ and equations that can be put into this form. The graphs of this type of equation all have similar shapes.

**8.6
Graphing
Parabolas**

We will begin by graphing the simplest quadratic equation, $y = x^2$. To get the idea of the shape of this graph, we need to find some ordered pairs that are solutions. We can do this by setting up the following table:

x	$y = x^2$	y

We can choose any convenient numbers for x and then use the equation $y = x^2$ to find the corresponding values for y. Let's use the values -3, -2, -1, 0, 1, 2, and 3 for x and find corresponding values for y. Here is how the table looks when we let x have these values:

x	$y = x^2$	y
-3	$y = (-3)^2 = 9$	9
-2	$y = (-2)^2 = 4$	4
-1	$y = (-1)^2 = 1$	1
0	$y = 0^2 = 0$	0
1	$y = 1^2 = 1$	1
2	$y = 2^2 = 4$	4
3	$y = 3^2 = 9$	9

The table gives us the solutions $(-3, 9)$, $(-2, 4)$, $(-1, 1)$, $(0, 0)$, $(1, 1)$, $(2, 4)$, and $(3, 9)$ for the equation $y = x^2$. We plot each of the points on a rectangular coordinate system and draw a smooth curve between them, as shown here:

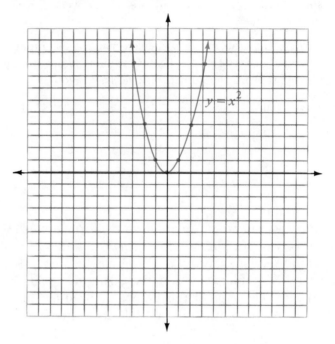

This graph is called a *parabola*. All equations of the form $y = ax^2 + bx + c$ $(a \neq 0)$ produce parabolas when graphed.

▼   **Example 1**   Graph the equation $y = x^2 - 3$.

SOLUTION   We begin by making a table using convenient values for x:

x	$y = x^2 - 3$	y
-2	$y = (-2)^2 - 3 = 4 - 3 = 1$	1
-1	$y = (-1)^2 - 3 = 1 - 3 = -2$	-2
0	$y = 0^2 - 3 = -3$	-3
1	$y = 1^2 - 3 = 1 - 3 = -2$	-2
2	$y = 2^2 - 3 = 4 - 3 = 1$	1

The table gives us the ordered pairs $(-2, 1)$, $(-1, -2)$, $(0, -3)$, $(1, -2)$, and $(2, 1)$ as solutions to $y = x^2 - 3$. The graph is as follows:

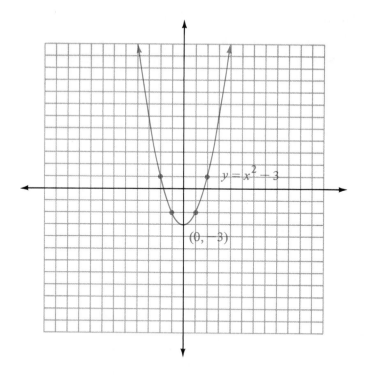

$y = x^2 - 3$

$(0, -3)$

▲

▼ **Example 2**   Graph $y = (x - 2)^2$.

SOLUTION

x	$y = (x - 2)^2$	y
-1	$y = (-1 - 2)^2 = (-3)^2 = 9$	9
0	$y = (0 - 2)^2 = (-2)^2 = 4$	4
1	$y = (1 - 2)^2 = (-1)^2 = 1$	1
2	$y = (2 - 2)^2 = 0^2 = 0$	0

We can continue the table if we feel more solutions will make the graph clearer.

3	$y = (3 - 2)^2 = 1^2 = 1$	1
4	$y = (4 - 2)^2 = 2^2 = 4$	4
5	$y = (5 - 2)^2 = 3^2 = 9$	9

Putting the results of the table onto a coordinate system, we have

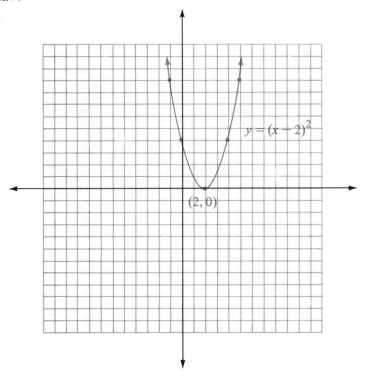

We are free to find as many ordered pairs as we need in order to get the shape of the graph correct. We are also free to use any convenient values for x when making up the table of ordered pairs.

▼ **Example 3**   Graph $y = (x + 1)^2 - 3$.

SOLUTION

x	$y = (x + 1)^2 - 3$	y
-5	$y = (-5 + 1)^2 - 3 = 16 - 3$	13
-3	$y = (-3 + 1)^2 - 3 = 4 - 3$	1
-1	$y = (-1 + 1)^2 - 3 = 0 - 3$	-3
1	$y = (1 + 1)^2 - 3 = 4 - 3$	1
3	$y = (3 + 1)^2 - 3 = 16 - 3$	13

Graphing the results of the table, we have

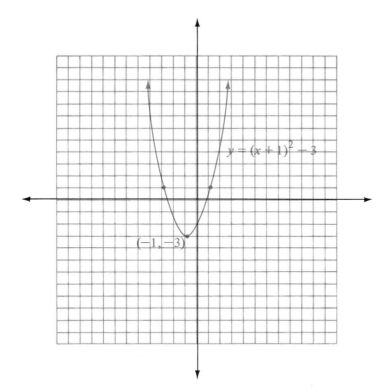

$y = (x + 1)^2 - 3$

$(-1, -3)$

To summarize this section, we graph second-degree equations by finding as many points as necessary to insure that the graph has the correct shape.

Graph each of the following equations by first making a table using the given values of x.

Problem Set 8.6

1. $y = x^2$ $x = -4, -2, 0, 2, 4$
2. $y = -x^2$ $x = -4, -2, 0, 2, 4$
3. $y = x^2 + 2$ $x = -3, -2, -1, 0, 1, 2, 3$
4. $y = x^2 - 1$ $x = -3, -1, 0, 1, 3$
5. $y = (x + 1)^2$ $x = -3, -2, -1, 0, 1, 2$
6. $y = (x - 3)^2$ $x = 0, 1, 2, 3, 4, 5, 6$

Graph each of the following equations by first making a table using convenient values for x.

7. $y = x^2 - 4$ 8. $y = x^2 + 2$
9. $y = x^2 + 5$ 10. $y = x^2 - 2$
11. $y = (x + 2)^2$ 12. $y = (x + 5)^2$

13. $y = (x - 5)^2$ 14. $y = (x + 3)^2$
15. $y = (x + 1)^2 - 2$ 16. $y = (x - 1)^2 + 2$
17. $y = (x + 2)^2 - 3$ 18. $y = (x - 2)^2 + 3$
19. $y = -2(x + 1)^2 + 2$ 20. $y = 2(x + 1)^2 - 2$

21. Graph the line $y = x + 2$ and the parabola $y = x^2$ on the same coordinate system. Name the points where the two graphs intersect.
22. Graph the line $y = x$ and the parabola $y = x^2 - 2$ on the same coordinate system. Name the points where the two graphs intersect.
23. Graph the parabola $y = 2x^2$ and the parabola $y = \frac{1}{2}x^2$ on the same coordinate system. (Remember, when you substitute a value for x into $y = 2x^2$, you square it first and then multiply by 2. If $x = 3$, then $y = 2 \cdot 3^2 = 2 \cdot 9 = 18$.)
24. Graph the parabola $y = 3x^2$ and the parabola $y = \frac{1}{3}x^2$ on the same coordinate system.

Review Problems   The problems below review material we covered in Sections 6.1 and 6.2.

Simplify each expression by factoring numerators and denominators and then dividing out any factors they have in common.

25. $\dfrac{x + 3}{x^2 - 9}$ 26. $\dfrac{x^2 - 25}{x + 5}$

27. $\dfrac{x^2 + 5x + 6}{x^2 + 7x + 12}$ 28. $\dfrac{x^2 + 5x + 6}{x^2 + 6x + 8}$

29. $\dfrac{x^2 - 36}{x^2 - 12x + 36}$ 30. $\dfrac{x^2 + 10x + 25}{x^2 - 25}$

Examples

1. $(x - 3)^2 = 25$
    $x - 3 = \pm 5$
    $x = 3 \pm 5$
    $x = -2$ or $x = 8$

Chapter 8 Summary and Review

SOLVING QUADRATRIC EQUATIONS OF THE FORM
$(ax + b)^2 = c$ [8.1]

We can solve equations of the form $(ax + b)^2 = c$ by taking the square root of both sides. We must remember to take both the positive and negative square roots of the right side when we do so.

COMPLETING THE SQUARE [8.2]

To complete the square on a quadratic equation as a method of solution we use the following steps:

Step 1: Move the constant term to one side and the variable terms to the other.

Step 2: Take the square of half the coefficient of the linear term and add it to both sides of the equation.

Step 3: Write the left side as a binomial square and then take the square root of both sides.

Step 4: Solve the resulting equation.

2. $x^2 - 6x + 2 = 0$

$x^2 - 6x = -2$

$x^2 - 6x + 9 = -2 + 9$

$(x - 3)^2 = 7$

$x - 3 = \pm\sqrt{7}$

$x = 3 \pm \sqrt{7}$

THE QUADRATIC FORMULA [8.3]

Any equation that is in the form $ax^2 + bx + c = 0$, where $a \neq 0$, has as its solutions

$$x = \frac{-b \pm \sqrt{b^2 - 4ac}}{2a}$$

The expression under the square root sign, $b^2 - 4ac$, is known as the discriminant. When the discriminant is negative, the solutions are complex numbers.

3. If $2x^2 + 3x - 4 = 0$

then $x = \dfrac{-3 \pm \sqrt{9 - 4(2)(-4)}}{2(2)}$

$= \dfrac{-3 \pm \sqrt{41}}{4}$

COMPLEX NUMBERS [8.4]

Any number that can be put in the form $a + bi$, where $i = \sqrt{-1}$, is called a complex number.

4. The numbers, 5, $3i$, $2 + 4i$, and $7 - i$ are all complex numbers.

ADDITION AND SUBTRACTION OF COMPLEX NUMBERS [8.4]

We add (or subtract) complex numbers by using the same procedure we used to add (or subtract) polynomials: we combine similar terms.

5. $(3 + 4i) + (6 - 7i)$

$= (3 + 6) + (4i - 7i)$

$= 9 - 3i$

MULTIPLICATION OF COMPLEX NUMBERS [8.4]

We multiply complex numbers in the same way we multiply binomials. The result, however, can be simplified further by substituting -1 for i^2 whenever it appears.

6. $(2 + 3i)(3 - i)$

$= 6 - 2i + 9i - 3i^2$

$= 6 + 7i + 3$

$= 9 + 7i$

DIVISION OF COMPLEX NUMBERS [8.4]

Division with complex numbers is accomplished with the method for rationalizing the denominator we developed while working with radical expressions. If the denominator has the form $a + bi$, we multiply both the numerator and denominator by its conjugate, $a - bi$.

7. $\dfrac{3}{2 + 5i}$

$= \dfrac{3}{2 + 5i} \cdot \dfrac{2 - 5i}{2 - 5i}$

$= \dfrac{6 - 15i}{29}$

8. Graph $y = x^2 - 2$.

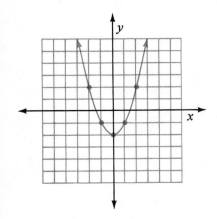

GRAPHING PARABOLAS [8.6]

The graph of an equation of the form $y = ax^2 + bx + c$, $a \neq 0$, is a parabola. To graph parabolas, we find as many points that satisfy the equation as we need to determine the position and shape of the graph.

COMMON MISTAKES

1. The most common mistake when working with complex numbers is to say $i = -1$. It does not; i is the *square root* of -1, not -1 itself.

2. The most common mistake when working with the quadratic formula is to try to identify the constants a, b, and c before putting the equation into standard form.

Chapter 8 Test

Solve the following quadratic equations.

1. $x^2 - 7x + 8 = 0$ **2.** $(x - 3)^2 = 12$
3. $(2x - 5)^2 = -75$ **4.** $x^2 - 6x - 6 = 0$
5. $2x^2 = 3x - 5$ **6.** $3x^2 = -2x + 1$
7. $(x + 2)(x - 1) = 6$ **8.** $9x^2 + 12x + 4 = 0$

Write as complex numbers.

9. $\sqrt{-9}$ **10.** $\sqrt{-121}$
11. $\sqrt{-72}$ **12.** $\sqrt{-18}$

Work the following problems involving complex numbers.

13. $(3i + 1) + (2 + 5i)$ **14.** $(6 - 2i) - (7 - 4i)$
15. $(2 + i)(2 - i)$ **16.** $(3 + 2i)(1 + i)$

17. $\dfrac{i}{3 - i}$ **18.** $\dfrac{2 + i}{2 - i}$

Graph the following equations.

19. $y = x^2 - 4$

20. $y = (x - 4)^2$

Additional Topics

In this chapter we will extend some of the topics we have already covered and preview some other topics that are common to intermediate algebra. The sections in this chapter do not all fit together in a smooth progression. Many of them can be studied individually and taken up in any order.

The review problems in this chapter cover the first eight chapters of the book. If you are getting ready to take a final exam in this course, doing the review problems at the end of each section should give you a good idea which of the first eight chapters you need to work on most.

There are three special quantities associated with straight lines and their graphs. One is the slope of the line. The other two are the *x*- and *y*-intercepts of the graph.

The idea behind developing what is called the slope of a line is to find a number we can associate with the steepness of the graph of a straight line.

**9.1
Slope and
y-Intercept**

Suppose we know the coordinates of two points on a straight line. Since we are trying to develop a general formula for the slope of a straight line, we will use general points—call the two points (x_1, y_1) and (x_2, y_2). They represent the coordinates of any two different points on our straight line. We define the *slope* of our line to be the ratio of the vertical change to the horizontal change as we move from point (x_1, y_1) to point (x_2, y_2) on the line.

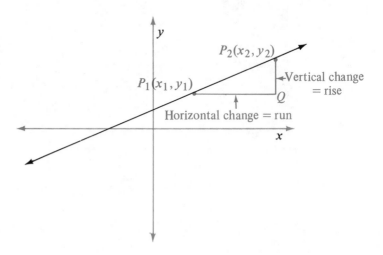

We call the vertical change the *rise* in the graph and the horizontal change the *run* in the graph. The slope, then, is

$$\text{Slope} = \frac{\text{vertical change}}{\text{horizontal change}} = \frac{\text{rise}}{\text{run}}$$

We would like to have a numerical value to associate with the rise in the graph and a numerical value to associate with the run in the graph. A quick study of the above figure shows that the coordinates of point Q must be (x_2, y_1), since Q is directly below point P_2 and right across from point P_1. We can draw our diagram again in the manner shown in the figure on the next page. It is apparent from this graph that the rise can be expressed as $(y_2 - y_1)$ and the run as $(x_2 - x_1)$. We usually denote the slope of a line by the letter m. Here is the complete definition of slope.

DEFINITION   If points (x_1, y_1) and (x_2, y_2) are any two different points, then the slope of the line on which they lie is:

$$\text{Slope} = m = \frac{\text{rise}}{\text{run}} = \frac{y_2 - y_1}{x_2 - x_1}$$

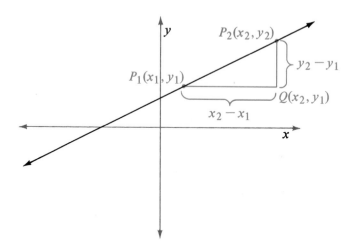

This definition of the slope of a line does just what we want it to do. If the line rises going from left to right, the slope will be positive. If the line falls from left to right, the slope will be negative. Also, the steeper the line, the larger numerical value the slope will have.

▼ **Example 1**　　Find the slope of the line between the points (3, 4) and (5, 7).

　　SOLUTION　　We can let

$$(x_1, y_1) = (3, 4)$$

and

$$(x_2, y_2) = (5, 7)$$

then

$$m = \frac{y_2 - y_1}{x_2 - x_1} = \frac{7 - 4}{5 - 3} = \frac{3}{2}$$

The slope is $\frac{3}{2}$. For every vertical change of 3 units, there will be a corresponding horizontal change of 2 units. (See the figure on the following page.)

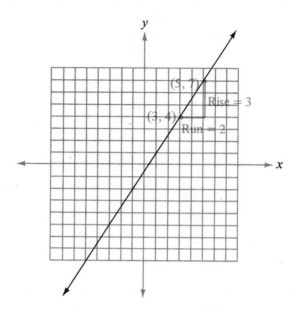

▼ **Example 2** Find the slope of the line through $(-2, 1)$ and $(5, -4)$.

    SOLUTION It makes no difference which ordered pair we call (x_1, y_1) and which we call (x_2, y_2).

$$\text{Slope} = m = \frac{y_2 - y_1}{x_2 - x_1} = \frac{-4 - 1}{5 - (-2)} = \frac{-5}{7}$$

    The slope is $-\frac{5}{7}$. Every vertical change of -5 units (down 5 units) is accompanied by a horizontal change of 7 units (to the right 7 units). ▲

DEFINITION The *x-intercept* of a straight line is the *x*-coordinate of the point where the graph crosses the *x*-axis. The *y-intercept* is defined similarly. It is the *y*-coordinate of the point where the graph crosses the *y*-axis.

    If the *x*-intercept is a, then the point $(a, 0)$ lies on the graph. (This is true because any point on the *x*-axis has a *y*-coordinate of 0.)

    If the *y*-intercept is b, then the point $(0, b)$ lies on the graph. (This is true because any point on the *y*-axis has an *x*-coordinate of 0.)

    Graphically the relationship looks like this:

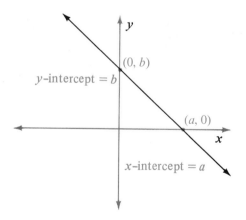

▼ **Example 3** Find the x- and y-intercepts for the graph of $3x - 2y = 6$.

SOLUTION To find where the graph crosses the x-axis, we let $y = 0$ (the y-coordinate of any point on the x-axis is 0). When $y = 0$, the equation $3x - 2y = 6$ becomes

$$3x - 2(0) = 6$$
$$3x = 6$$
$$x = 2$$

The graph crosses the x-axis at $(2, 0)$, which means the x-intercept is 2.

To find where the graph crosses the y-axis, we let $x = 0$. When $x = 0$, the equation $3x - 2y = 6$ becomes

$$3(0) - 2y = 6$$
$$-2y = 6$$
$$y = -3$$

The graph crosses the y-axis at $(0, -3)$, which means the y-intercept is -3. ▲

▼ **Example 4** Find the intercepts for $y = 2x - 3$.

SOLUTION When $y = 0$, the equation $y = 2x + 3$ becomes

$$0 = 2x - 3$$
$$2x = 3$$
$$x = \tfrac{3}{2}$$

The graph crosses the x-axis at $(\frac{3}{2}, 0)$. The x-intercept is $\frac{3}{2}$.

To find the y-intercept we let $x = 0$. When $x = 0$, the equation $y = 2x - 3$ becomes

$$y = 2(0) - 3$$
$$= -3$$

The graph crosses the y-axis at the point $(0, -3)$. The y-intercept is -3. ▲

Problem Set 9.1

Find the slope of the line through the following pairs of points.

1. $(3, 4)$, $(5, 8)$ **2.** $(1, 5)$, $(2, 9)$
3. $(2, 3)$, $(8, 9)$ **4.** $(3, 4)$, $(6, 10)$
5. $(6, 1)$, $(9, 7)$ **6.** $(3, -1)$, $(5, 5)$
7. $(3, 3)$, $(-5, -5)$ **8.** $(-2, 5)$, $(-3, 1)$
9. $(-5, 3)$, $(-6, 4)$ **10.** $(-1, 6)$, $(-5, 9)$
11. $(-3, -1)$, $(5, 6)$ **12.** $(-9, 8)$, $(2, -3)$
13. $(-3, -1)$, $(-5, -6)$ **14.** $(-2, -3)$, $(-8, -5)$
15. $(-3, -2)$, $(-1, -2)$ **16.** $(3, -4)$, $(8, -2)$
17. (a, b), (c, d) **18.** (r, s), (t, u)
19. $(a + b, 2)$, $(a - b, 1)$ **20.** $(a - b, a)$, $(a + b, b)$

21. Find x if the line through $(2, 4)$ and $(x, 6)$ has a slope of 2.
22. Find y if the line through $(3, y)$ and $(1, 5)$ has a slope of 4.
23. Find y if the line through $(7, -2)$ and $(8, y)$ has a slope of 5.
24. Find x if the line through $(x, 1)$, $(3, 7)$ has a slope of 6.

Find the x- and y-intercepts for the following equations.

25. $2x - y = 4$ **26.** $3x + y = 6$
27. $5x - 2y = 10$ **28.** $2x + 3y = 6$
29. $4x - 2y = 8$ **30.** $6x - 2y = 12$
31. $4x - 5y = 20$ **32.** $11x - y = 22$
33. $y = x - 4$ **34.** $y = -x + 5$
35. $y = 2x - 6$ **36.** $y = 3x + 9$
37. $y = 2x + 1$ **38.** $y = -3x - 1$
39. $y = 4x - 7$ **40.** $y = -3x + 4$
41. $y = -2x - 1$ **42.** $y = 4x + 8$

43. We can use the slope of a line to help graph the line. If a line has a slope of $\frac{2}{3}$, then, going from one point to another on the line, a vertical change of 2 will always be accompanied by a horizontal change of 3. Graph the line that goes through the point $(1, 1)$ and has a slope of $\frac{2}{3}$.

44. Graph the line that goes through the point $(-1, 1)$ and has a slope of $\frac{3}{5}$.

45. Graph the line that has the x-intercept 3 and the y-intercept -2. What is the slope of this line?

46. Graph the line that has the x-intercept 2 and the y-intercept -3. What is the slope of this line?

Review Problems   The problems below review material we covered in Chapter 1.

Simplify each expression.

47. $-3 + 4(-2)$

48. $8 - (-3)$

49. $-2(5) - 4(-3)$

50. $\dfrac{5 + 3(7 - 2)}{2(-3) - 4}$

51. $|-3|$

52. $-|-5|$

53. Use the associative property to simplify $-6(\frac{1}{3}x)$.

54. Which property of real numbers is illustrated by the statement $2 + 3 = 3 + 2$.

For the set $\{-3, -\frac{4}{5}, 0, \frac{5}{8}, 2, \sqrt{5}\}$, which numbers are

55. Integers

56. Rational numbers

In this section we will use some of our previous results to derive two important forms of the equation of a straight line. Both forms are used extensively in higher mathematics. Each has its own particular usefulness in specific situations.

Suppose we know a certain line l has a slope of m and contains the point (x_1, y_1). If the point (x, y) is any other point on line l, we can write the slope of l as

$$\frac{y - y_1}{x - x_1} = m$$

Multiplying both sides of this expression by $(x - x_1)$, we have

$$(x - x_1)\left(\frac{y - y_1}{x - x_1}\right) = m(x - x_1)$$

$$y - y_1 = m(x - x_1)$$

This last expression is known as the *point–slope* form of a straight line.

**9.2
The Point-Slope
and Slope-
Intercept Forms
of Straight Lines**

We can use it to find the equation of a line when we are given one point on the line and the slope of the line.

▼ **Example 1**   Find the equation of the line that has a slope of -3 and contains the point (4, 2).

SOLUTION   We have

$$(x_1, y_1) = (4, 2) \quad \text{and} \quad m = -3$$

The equation of the line is

$$y - y_1 = m(x - x_1)$$
$$y - 2 = -3(x - 4)$$

which we can simplify as follows:

$$y - 2 = -3x + 12$$
$$y = -3x + 14 \qquad\qquad ▲$$

▼ **Example 2**   Find the equation of the line with slope $\frac{2}{3}$ that contains the point (3, -1).

SOLUTION   We have the point

$$(x_1, y_1) = (3, -1)$$

and the slope

$$m = \tfrac{2}{3}$$

Using the point–slope form, we have

$$y - y_1 = m(x - x_1)$$
$$y - (-1) = \tfrac{2}{3}(x - 3)$$
$$y + 1 = \tfrac{2}{3}(x - 3)$$
$$3y + 3 = 2(x - 3) \qquad \text{Multiply both sides by 3}$$
$$3y + 3 = 2x - 6$$
$$3y = 2x - 9 \qquad\qquad ▲$$

To use the point–slope form of the equation of a line we need only enough information to find the slope and one point on the line. The rest involves simply substitution and simplification.

Slope-Intercept Form of a Straight Line

Suppose we are given the slope m and the y-intercept b for line l. Since the y-intercept is b, the point $(0, b)$ must be on l. We have the slope m and one point $(0, b)$. Substituting these quantities into the point–slope form, we have

$$y - b = m(x - 0)$$
$$y - b = mx$$
$$y = mx + b$$

This expression is known as the *slope–intercept* form of the equation of a straight line. When the equation has this form it is very easy to identify the slope and *y*-intercept. The slope is always the coefficient of *x* and the *y*-intercept is the constant term.

▼ **Example 3** Find the equation of the line with a slope of 5 and the *y*-intercept 3.

SOLUTION   We have

$$m = 5 \quad \text{and} \quad b = 3$$

Using the slope–intercept form, we have

$$y = mx + b$$
$$y = 5x + 3$$

It is that simple. If the slope is 5 and the *y*-intercept is 3, the equation is $y = 5x + 3$.   ▲

▼ **Example 4** Give the equation of the line with the slope $\frac{3}{5}$ and the *y*-intercept -2.

SOLUTION

$$m = \tfrac{3}{5} \quad \text{and} \quad b = -2$$

The equation must be

$$y = \tfrac{3}{5}x - 2$$ ▲

▼ **Example 5** Find the slope and the *y*-intercept for the equation $6x + 3y = 12$.

SOLUTION   We can identify the slope and the *y*-intercept easily if the equation is in the form $y = mx + b$. To put the equation in this form, we simply solve for *y*:

$$
\begin{aligned}
6x + 3y &= 12 \\
3y &= -6x + 12 \qquad \text{Add } -6x \text{ to both sides} \\
y &= -2x + 4 \qquad \text{Divide both sides by 3} \\
m &= -2 \quad \text{and} \quad b = 4
\end{aligned}
$$

The slope is -2 and the *y*-intercept is 4.   ▲

▼ **Example 6** Find the slope and the y-intercept for $5x - 2y = 3$.

SOLUTION Solving for y, we have

$$5x - 2y = 3$$
$$-2y = -5x + 3$$
$$y = \frac{-5x}{-2} + \frac{3}{-2}$$

$$y = \frac{5}{2}x - \frac{3}{2}$$

$$m = \frac{5}{2} \quad \text{and} \quad b = -\frac{3}{2}$$

The slope is $\frac{5}{2}$ and the y-intercept is $-\frac{3}{2}$. ▲

Problem Set 9.2

For each of the following, find the equation of the line that contains the given point and has the given slope.

1. $(3, 2)$; $m = 5$ **2.** $(1, 3)$; $m = 2$
3. $(5, 4)$; $m = 4$ **4.** $(2, 6)$; $m = 3$
5. $(4, 1)$; $m = -2$ **6.** $(5, 3)$; $m = -5$
7. $(2, -2)$; $m = 1$ **8.** $(3, -3)$; $m = 2$
9. $(-4, -7)$; $m = -2$ **10.** $(-3, -2)$; $m = -4$
11. $(3, 5)$; $m = \frac{1}{2}$ **12.** $(7, 6)$; $m = \frac{3}{5}$
13. $(-2, 5)$; $m = -\frac{4}{3}$ **14.** $(3, -4)$; $m = -\frac{5}{3}$
15. $(3, \frac{2}{3})$; $m = 5$ **16.** $(\frac{3}{4}, -2)$; $m = -3$

Find the equation of the line with the given slope and y-intercept.

17. $m = 2, b = 1$ **18.** $m = 3, b = 5$
19. $m = -4, b = 2$ **20.** $m = -3, b = 4$
21. $m = -1, b = -1$ **22.** $m = -2, b = -3$
23. $m = 1, b = \frac{2}{3}$ **24.** $m = 2, b = \frac{3}{4}$
25. $m = -\frac{1}{3}, b = -\frac{1}{2}$ **26.** $m = \frac{1}{2}, b = -\frac{2}{5}$
27. $m = -\frac{3}{4}, b = -\frac{1}{4}$ **28.** $m = -\frac{2}{5}, b = \frac{3}{5}$

For each of the following equations, determine the slope and y-intercept.

29. $y = 2x + 5$ **30.** $y = -3x + 1$
31. $y = 5x - 1$ **32.** $y = -2x - 7$
33. $3x + y = -2$ **34.** $2x - y = 3$
35. $5x - y = 4$ **36.** $5x - y = -2$
37. $4x + 2y = 8$ **38.** $6x - 2y = 24$
39. $2x + 3y = 6$ **40.** $4x - 2y = -8$
41. $2x + 3y = 5$ **42.** $3x - 2y = -4$

43. Find the equation of the line through the points (3, 5) and (1, −3) by first finding its slope and then using the slope and the point (3, 5) in the point–slope formula. Write your answer in slope–intercept form.

44. Find the equation of the line through the points (1, 4) and (3, −2). Write your answer in slope–intercept form.

45. Find the equation of the line through (−2, 0) and (0, 3). Write your answer in slope–intercept form.

46. Find the equation of the line through (−3, 0) and (0, 2). Write your answer in slope–intercept form.

Review Problems   The problems below review material we covered in Chapter 2.

Solve each equation.

47. $2x - 4 = 5x + 2$

48. $4(2x - 1) = 12$

49. $8 - 2(x + 7) = 2$

50. $2(x - 1) + 3(x + 4) = -5$

Solve each inequality and graph the solution set.

51. $2x - 3 < 5$

52. $3(x - 1) \leq 9$

53. $-4x \leq 20$

54. $9 - 6x > -3$

55. $-3 \leq 4x + 1 \leq 9$

56. Patrick is eight years older than his cousin Stacey. In five years the sum of their ages will be 28. How old are they now?

The idea of a function is very common in mathematics. Functions are studied in almost all branches of mathematics. They are the result of the process of working with quantities that vary with one another. That is, a change in one quantity brings about a change in another. The idea of a function is also a very simple concept mathematically.

9.3 Functions and Function Notation

DEFINITION   A *function* is any set of ordered pairs in which no two different ordered pairs have the same first coordinate. The *domain* of a function is the set of all first coordinates. The *range* of a function is the set of all second coordinates.

EXAMPLE   The set {(2, 3), (1, 5), (6, 3), (−3, 4)} is a function because it is a set of ordered pairs and no two different ordered pairs have the same first coordinates. The domain of this function is the set {2, 1, 6, −3}. The range is the set {3, 5, 4}.

In this example we listed all the ordered pairs of the function. Usually the ordered pairs are given in the form of a rule instead of a list.

EXAMPLE   The set $\{(x, y)|y = 2x + 1; x = 1, 2, 3, 4\}$ is an example of a function. The domain is the set $\{1, 2, 3, 4\}$. The range is the set $\{3, 5, 7, 9\}$, since these are the values of y that correspond to the x values 1, 2, 3, 4.

In this example we specified the function by giving the rule for obtaining ordered pairs, $y = 2x + 1$, and then listing the values the variable that x could assume, $x = 1, 2, 3, 4$. Most of the time, when a function is given by a rule, the domain is not specified. When this happens, we assume the domain includes all replacements for x except those that make the function undefined. For example, the domain of the function

$$y = \frac{3}{x}$$

is all real numbers except $x = 0$, because when x is 0 the expression $\frac{3}{0}$ is undefined. There is no value of y that corresponds to $x = 0$.

▼   **Example 1**   Specify the domain for $y = \dfrac{3}{x - 2}$.

SOLUTION   We can replace x with any real number except 2, because $x = 2$ makes the denominator zero. The domain is

$$\text{Domain} = \{x|x \text{ is real, } x \neq 2\} \qquad \blacktriangle$$

▼   **Example 2**   Specify the domain for $y = \dfrac{3}{x^2 - x - 6}$.

SOLUTION   We are interested in eliminating any values of x that will make the denominator zero:

$$x^2 - x - 6 = 0$$
$$(x - 3)(x + 2) = 0$$
$$x = 3 \quad \text{or} \quad x = -2$$

Since $x = 3$ and $x = -2$ both make the denominator zero, we cannot use them as replacements for x:

$$\text{Domain} = \{x|x \text{ is real, } x \neq 3, x \neq -2\} \qquad \blacktriangle$$

Function Notation

The notation $f(x)$ is read "f of x" and is defined to be the value of the function f at x. It is not to be read or interpreted as f times x. The equations $y = 2x + 3$ and $f(x) = 2x + 3$ are essentially the same; that is, $y = f(x)$.

Suppose we are given the equation $y = 2x + 3$ and we want to find the value of y when x is 4. We would have to say, "If $x = 4$, then $y = 2(4) + 3 = 11$." Using the new notation, $f(x) = 2x + 3$, we have $f(4) = 2(4) + 3 = 11$. The following table gives more examples of this concept.

Function Notation	y in Terms of x
$f(x) = 2x + 3$	$y = 2x + 3$
$f(1) = 2(1) + 3 = 5$	If $x = 1$, then $y = 2(1) + 3 = 5$
$f(0) = 2(0) + 3 = 3$	If $x = 0$, then $y = 2(0) + 3 = 3$
$f(-5) = 2(-5) + 3 = -7$	If $x = -5$, then $y = 2(-5) + 3 = -7$

Here are some additional examples of the use of function notation.

▼ **Example 3**   If $f(x) = x^2 - 4$, find $f(0), f(-1), f(3), f(5), f(a)$.

SOLUTION

$$f(x) = x^2 - 4$$
$$f(0) = 0^2 - 4 = -4$$
$$f(-1) = (-1)^2 - 4 = -3$$
$$f(3) = 3^2 - 4 = 5$$
$$f(5) = 5^2 - 4 = 21$$
$$f(a) = a^2 - 4$$

▲

▼ **Example 4**   If $f(x) = 2x + 5$ and $g(x) = 2x^2 + 1$, find $f(2), g(-3)$, $g(a), f(z)$.

SOLUTION

$$f(2) = 2(2) + 5 = 4 + 5 = 9$$
$$g(-3) = 2(-3)^2 + 1 = 2(9) + 1 = 19$$
$$g(a) = 2a^2 + 1$$
$$f(z) = 2z + 5$$

▲

Specify the domain and range of the following functions.   Problem Set 9.3

1. $\{(1, 2), (3, 4), (5, 6)\}$
2. $\{(-1, 1), (-2, 2), (-3, 3)\}$
3. $\{(1, a) ,(3, b), (4, c)\}$
4. $\{(a, -1), (b, c), (d, -2)\}$
5. $y = x, \quad x = 1, 2, 3, 4$
6. $y = 2x, \quad x = 2, 4, 6, 8$
7. $y = 3x - 1, \quad x = 1, 3, 5, 7, 9$
8. $y = 2x - 3, \quad x = 2, 4, 6, 8, 10$
9. $2x + 3y = 6, \quad x = 0, 2, 4$
10. $3x - 2y = 6, \quad y = 0, -2, -3$
11. $2x - y = 4, \quad y = 2, 4, 6, 8, 10$
12. $3x - 5y = 15, \quad x = 0, 5, 15$

Specify any restrictions on the domain of the following functions.

13. $y = \dfrac{1}{x - 2}$

14. $y = \dfrac{2}{x - 1}$

15. $y = \dfrac{3}{x + 3}$

16. $y = \dfrac{5}{x + 5}$

17. $y = \dfrac{4}{(x + 1)(x + 2)}$

18. $y = \dfrac{(x + 5)}{(x - 3)(x + 2)}$

19. $y = \dfrac{x}{(x + 5)(x - 1)}$

20. $y = \dfrac{2x}{(x - 7)(x + 6)}$

21. $y = \dfrac{3}{x^2 - 9}$

22. $y = \dfrac{2}{x^2 - 25}$

23. $y = \dfrac{3}{x^2 - 5x + 6}$

24. $y = \dfrac{2x}{x^2 + 4x + 4}$

25. $y = \dfrac{x}{2x^2 - 5x - 3}$

26. $y = \dfrac{3x}{3x^2 + x - 2}$

27. If $f(x) = 2x$, find $f(0)$, $f(1)$, $f(2)$.
28. If $f(x) = -3x$, find $f(-2)$, $f(0)$, $f(2)$.
29. If $f(x) = x - 5$, find $f(2)$, $f(0)$, $f(-1)$, $f(a)$.
30. If $f(x) = x + 4$, find $f(-3)$, $f(-2)$, $f(0)$, $f(b)$.
31. If $f(x) = 2x - 3$, find $f(1)$, $f(2)$, $f(\frac{1}{2})$, $f(z)$.
32. If $f(x) = 5x - 6$, find $f(-3)$, $f(-5)$, $f(\frac{2}{5})$, $f(c)$.
33. If $f(x) = x^2 + 1$, find $f(2)$, $f(3)$, $f(4)$, $f(a)$.
34. If $f(x) = 2x^2 - 3$, find $f(0)$, $f(-3)$, $f(2)$, $f(a + 1)$.
35. If $f(x) = 1/x$, find $f(-1)$, $f(-2)$, $f(3)$, $f(\frac{1}{2})$, $f(\frac{4}{3})$.
36. If $f(x) = \sqrt{x}$, find $f(4)$, $f(a)$, $f(16)$, $f(b)$.
37. If $f(x) = x^2 + 2x$, find $f(3)$, $f(-2)$, $f(a)$, $f(a - 2)$.
38. If $f(x) = 2x^2 + 3x$, find $f(5)$, $f(-3)$, $f(b)$, $f(a - b)$.
39. If $f(x) = x^3 - 1$, find $f(1)$, $f(2)$, $f(-3)$, $f(a)$.
40. If $f(x) = x^3 + 4$, find $f(0)$, $f(-3)$, $f(-2)$, $f(a + 1)$.

Review Problems   The problems below review material we covered in Chapter 3.

Solve the following systems of linear equations. If you do not get an ordered pair for a solution, then tell whether the lines coincide or are parallel.

41. $x + y = 4$
    $x - y = 2$

42. $3x + 2y = 1$
    $2x + \ \ y = 3$

43. $3x - \ \ y = 4$
    $6x - 2y = 2$

44. $2x + 3y = -1$
    $3x + 5y = -2$

45. $x + y = 20$
$y = 5x + 2$

46. Amy has $1.85 in dimes and quarters. If she has a total of 11 coins, how many of each coin does she have?

47. Graph the line $2x - 3y = 6$.

48. Graph the line $y = 2x - 3$.

Consider the expression $3^{1/2}$ and the expression $\sqrt{3}$. The second expression is the positive square root of 3. We have never encountered the first expression before. Assuming the properties of exponents apply to fractional exponents in the same way they apply to integer exponents, we can take a look at what happens when we square both of the above quantities:

$$(3^{1/2})^2 \qquad (\sqrt{3})^2$$
$$3^{(1/2)2} \qquad \sqrt{3} \cdot \sqrt{3}$$
$$3^1 \qquad \sqrt{9}$$
$$3 \qquad 3$$

The result is the same in both cases. This result allows us to define fractional exponents.

DEFINITION   If x represents a nonnegative real number, then:

$$\sqrt{x} = x^{1/2}$$
$$\sqrt[3]{x} = x^{1/3}$$
$$\sqrt[4]{x} = x^{1/4}$$

and, in general,

$$\sqrt[n]{x} = x^{1/n}$$

We can use this definition in many ways. We can change any expression involving radicals to an expression that involves only exponents. We can avoid radical notation altogether if we choose.

All the previous properties of exponents hold for fractional exponents. We will omit the proofs here and instead get used to the new notation by considering some examples.

▼ **Examples**   Write each of the following as a radical and then simplify.

1. $25^{1/2}$

SOLUTION   The exponent $\frac{1}{2}$ indicates that we are to find the square root of 25:

$$25^{1/2} = \sqrt{25} = 5$$

2. $9^{1/2}$

SOLUTION   Again, an exponent of $\frac{1}{2}$ indicates that we are to find the square root of 9:

$$9^{1/2} = \sqrt{9} = 3$$

3. $8^{1/3}$

SOLUTION   The exponent is $\frac{1}{3}$, indicating that we are to find the cube root of 8:

$$8^{1/3} = \sqrt[3]{8} = 2$$

4. $81^{1/4}$

SOLUTION   An exponent of $\frac{1}{4}$ indicates that we are to find the fourth root:

$$81^{1/4} = \sqrt[4]{81} = 3 \qquad\qquad \blacktriangle$$

Consider the expression $9^{3/2}$. If we assume that all our properties of exponents hold for fractional exponents, we can rewrite this expression as a power raised to another power.

$$9^{3/2} = (9^{1/2})^3$$

Since $9^{1/2}$ is equivalent to $\sqrt{9}$, we can simplify further:

$$(9^{1/2})^3 = (\sqrt{9})^3 = 3^3 = 27$$

We can summarize this discussion with the following definition.

DEFINITION If x represents a positive real number, and n and m are integers, then

$$x^{m/n} = \sqrt[n]{x^m} = (\sqrt[n]{x})^m$$

We can think of the exponent m/n as consisting of two parts. The numerator m is the power, and the denominator n is the root.

▼   **Example 5** Simplify $8^{2/3}$.

SOLUTION   Using the definition above, we have

$$8^{2/3} = (8^{1/3})^2 \quad \text{Separate exponents}$$
$$= (\sqrt[3]{8})^2 \quad \text{Write as cube root}$$
$$= 2^2 \qquad \sqrt[3]{8} = 2$$
$$= 4 \qquad 2^2 = 4$$

▲

▼ **Example 6** Simplify $27^{4/3}$.

SOLUTION

$$27^{4/3} = (27^{1/3})^4 \quad \text{Separate exponents}$$
$$= (\sqrt[3]{27})^4 \quad \text{Write as cube root}$$
$$= 3^4 \qquad \sqrt[3]{27} = 3$$
$$= 81 \qquad 3^4 = 81$$

▲

We can apply the properties of exponents to expressions that involve fractional exponents.

▼ **Examples** Use the properties of exponents to simplify each expression. Assume that all variables represent positive numbers.

7. $x^{1/3} \cdot x^{2/3}$

SOLUTION To multiply with the same base, we add exponents. The property we use is $a^r a^s = a^{r+s}$:

$$x^{1/3} \cdot x^{2/3} = x^{1/3 + 2/3}$$
$$= x^{3/3}$$
$$= x$$

8. $(x^{1/4})^8$

SOLUTION To raise a power to another power, we multiply exponents. The property we use is $(a^r)^s = a^{rs}$

$$(x^{1/4})^8 = x^{(1/4)8}$$
$$= x^2$$

9. $\dfrac{a^{5/6}}{a^{4/6}}$

SOLUTION To divide with the same base, we subtract the exponent in the denominator from the exponent in the numerator.

$$\frac{a^{5/6}}{a^{4/6}} = a^{5/6 - 4/6}$$

$$= a^{1/6}$$

10. $(8y^{12})^{1/3}$

SOLUTION   Distributing the exponent $\frac{1}{3}$ across the product, we have

$$(8y^{12})^{1/3} = 8^{1/3}(y^{12})^{1/3}$$
$$= 2y^4$$ ▲

Problem Set 9.4

Change each of the following to an expression involving roots and then simplify.

1. $4^{1/2}$	**2.** $9^{1/2}$	**3.** $16^{1/2}$
4. $25^{1/2}$	**5.** $27^{1/3}$	**6.** $8^{1/3}$
7. $125^{1/3}$	**8.** $16^{1/4}$	**9.** $81^{1/4}$
10. $36^{1/2}$	**11.** $81^{1/2}$	**12.** $144^{1/2}$
13. $8^{2/3}$	**14.** $25^{3/2}$	**15.** $125^{2/3}$
16. $36^{3/2}$	**17.** $16^{3/4}$	**18.** $9^{3/2}$
19. $16^{3/2}$	**20.** $8^{5/3}$	**21.** $4^{3/2}$
22. $4^{5/2}$	**23.** $(-8)^{2/3}$	**24.** $(-27)^{2/3}$
25. $(-32)^{1/5}$	**26.** $(-32)^{3/5}$	**27.** $4^{1/2} + 9^{1/2}$
28. $16^{1/2} + 25^{1/2}$	**29.** $16^{3/4} + 27^{2/3}$	**30.** $49^{1/2} + 64^{1/2}$
31. $4^{1/2} \cdot 27^{1/3}$	**32.** $8^{1/3} \cdot 25^{1/2}$	

Use the properties of exponents to simplify each expression below. Assume that all variables represent positive numbers.

33. $x^{1/4} \cdot x^{3/4}$	**34.** $x^{1/8} \cdot x^{3/8}$
35. $(x^{2/3})^3$	**36.** $(x^{1/4})^{12}$
37. $\dfrac{a^{3/5}}{a^{1/5}}$	**38.** $\dfrac{a^{5/7}}{a^{3/7}}$
39. $(27y^6)^{1/3}$	**40.** $(81y^8)^{1/4}$
41. $(9a^4b^2)^{1/2}$	**42.** $(25a^8b^4)^{1/2}$
43. $\dfrac{x^{3/5} \cdot x^{4/5}}{x^{2/5}}$	**44.** $\dfrac{x^{4/7} \cdot x^{6/7}}{x^{3/7}}$

Recall that negative exponents indicate reciprocals. That is, $a^{-r} = \dfrac{1}{a^r}$. Use this property of exponents, along with what you have learned in this section, to simplify the expressions below.

45. $25^{-1/2}$	**46.** $8^{-1/3}$
47. $8^{-2/3}$	**48.** $81^{-3/4}$
49. $27^{-2/3}$	**50.** $16^{-3/4}$

Review Problems   The problems below review material we covered in Chapter 4.

Simplify.

51. $\dfrac{(20x^2y^3)(5x^4y)}{(2xy^5)(10x^2y^3)}$

Perform the indicated operations.

52. $(9x^2 + 3x - 2) - (6x^2 - 5x - 7)$
53. $2x^2(3x^2 + 3x - 1)$
54. $(y + 3)(y - 5)$
55. $(2x + 3)(5x - 2)$
56. $(3x - 5)^2$
57. $(a - 3)(a^2 + 3a + 9)$
58. $(15x^{10} - 10x^8 + 25x^6) \div 5x^6$

Many branches of science require working with very large and very small numbers. In astronomy, for example, distances commonly are given in light-years. A light-year is the distance light travels in a year. It is approximately

$$5,880,000,000,000 \text{ miles}$$

In chemistry, the number used to represent the mass of a single atom of hydrogen is

$$0.000\ 000\ 000\ 000\ 000\ 000\ 000\ 016\ 7 \text{ grams}$$

Both of the numbers that we just mentioned are difficult to use in calculations because of the number of zeros they contain. It takes awhile just to write them.

Scientific notation provides a way of writing very large and very small numbers that makes them more manageable.

DEFINITION A number is in *scientific notation* when it is written as the product of a number between 1 and 10 and an integer power of 10. A number written in scientific notation has the form

$$n \times 10^r$$

where $1 \leq n < 10$ and $r =$ an integer.

9.5
Scientific Notation

The table below lists some numbers and their forms in scientific notation. Each number on the left is equal to the number on the right-hand side of the equal sign.

Number Written the Long Way		Number Written Again in Scientific Notation
376,000	=	3.76×10^5
49,500	=	4.95×10^4
3,200	=	3.2×10^3
591	=	5.91×10^2
46	=	4.6×10^1
8	=	8×10^0
0.47	=	4.7×10^{-1}
0.093	=	9.3×10^{-2}
0.00688	=	6.88×10^{-3}
0.0002	=	2×10^{-4}
0.000098	=	9.8×10^{-5}

Notice that in each case, when the number is written in scientific notation, the decimal point in the first number is placed so that the number is between 1 and 10. The exponent on 10 in the second number keeps track of the number of places we moved the decimal point in the original number to get a number between 1 and 10:

$$376{,}000 = 3.76 \times 10^5$$

Moved 5 places

Decimal point originally here

Keeps track of the 5 places we moved the decimal point

Doing arithmetic with numbers written in scientific notation is very similar to the work we did in Chapter 4 with monomials. Here are some examples.

▼ **Example 1** Multiply $(4 \times 10^7)(2 \times 10^{-4})$.

SOLUTION Since multiplication is commutative and associative, we can rearrange the order of these numbers and group them as follows:

$$(4 \times 10^7)(2 \times 10^{-4}) = (4 \times 2)(10^7 \times 10^{-4})$$
$$= 8 \times 10^3$$

Notice that we add exponents, $7 + (-4) = 3$, when we multiply with the same base. ▲

▼ **Example 2** Divide $\dfrac{9.6 \times 10^{12}}{3 \times 10^4}$.

SOLUTION We group the numbers between 1 and 10 separately from the powers of 10 and proceed as we did in Example 1:

$$\frac{9.6 \times 10^{12}}{3 \times 10^4} = \frac{9.6}{3} \times \frac{10^{12}}{10^4}$$
$$= 3.2 \times 10^8$$

Notice that the procedure we used in both of these examples is very similar to multiplication and division of monomials, for which we multiplied or divided coefficients and added or subtracted exponents. ▲

▼ **Example 3** Simplify $\dfrac{(6.8 \times 10^5)(3.9 \times 10^{-7})}{7.8 \times 10^{-4}}$.

SOLUTION We group the numbers between 1 and 10 separately from the powers of 10:

$$\frac{(6.8)(3.9)}{7.8} \times \frac{(10^5)(10^{-7})}{10^{-4}} = 3.4 \times 10^{5+(-7)-(-4)}$$

$$= 3.4 \times 10^2 \qquad \blacktriangle$$

Write each number in scientific notation.

1. 3750
2. 375,000
3. 763,000,000
4. 898,000,000
5. 25
6. 37
7. 4
8. 8
9. 0.003
10. 0.0007
11. 0.048
12. 0.099
13. 0.0000000007
14. 0.00000000007

Write each number the long way, that is, not in scientific notation.

15. 4.23×10^3
16. 4.23×10^2
17. 5.8×10^6
18. 7.3×10^7
19. 2.4×10^1
20. 9.8×10^1
21. 6×10^0
22. 8×10^0

23. 7.1×10^{-2} 24. 7.1×10^{-3}
25. 4.11×10^{-6} 26. 9.77×10^{-4}
27. 4×10^{-1} 28. 9×10^{-1}

Find each product. Write all answers in scientific notation.

29. $(3 \times 10^3)(2 \times 10^5)$ 30. $(4 \times 10^8)(1 \times 10^6)$
31. $(3.5 \times 10^4)(5 \times 10^{-6})$ 32. $(7.1 \times 10^5)(2 \times 10^{-8})$
33. $(5.5 \times 10^{-3})(2.2 \times 10^{-4})$ 34. $(3.4 \times 10^{-2})(4.5 \times 10^{-6})$
35. $(2,500,000)(36,000)$ 36. $(42,000,000)(60,000)$
37. $(5,000,000)(0.00003)$ 38. $(80,000)(0.0003)$

Find each quotient. Write all answers in scientific notation.

39. $\dfrac{8.4 \times 10^5}{2 \times 10^2}$ 40. $\dfrac{9.6 \times 10^{20}}{3 \times 10^6}$

41. $\dfrac{6 \times 10^8}{2 \times 10^{-2}}$ 42. $\dfrac{8 \times 10^{12}}{4 \times 10^{-3}}$

43. $\dfrac{2.5 \times 10^{-6}}{5 \times 10^{-4}}$ 44. $\dfrac{4.5 \times 10^{-8}}{9 \times 10^{-4}}$

45. $\dfrac{4,200,000}{70,000,000}$ 46. $\dfrac{48,000,000}{800,000,000}$

47. $\dfrac{0.00003}{50,000}$ 48. $\dfrac{0.0000028}{70,000}$

Simplify each expression and write all answers in scientific notation.

49. $\dfrac{(6 \times 10^8)(3 \times 10^5)}{9 \times 10^7}$ 50. $\dfrac{(8 \times 10^4)(5 \times 10^{10})}{2 \times 10^7}$

51. $\dfrac{(5 \times 10^3)(4 \times 10^{-5})}{2 \times 10^{-2}}$ 52. $\dfrac{(7 \times 10^6)(4 \times 10^{-4})}{1.4 \times 10^{-3}}$

53. $\dfrac{(2.8 \times 10^{-7})(3.6 \times 10^4)}{2.4 \times 10^3}$ 54. $\dfrac{(5.4 \times 10^2)(3.5 \times 10^{-9})}{4.5 \times 10^6}$

55. $\dfrac{(5.4 \times 10^{-1})(9.8 \times 10^8)}{(1.4 \times 10^6)(6.3 \times 10^{-3})}$ 56. $\dfrac{(3.5 \times 10^{-2})(4.8 \times 10^{-6})}{(1.5 \times 10^{-8})(4 \times 10^{10})}$

Review Problems   The problems below review material we covered in Chapter 5.

57. Factor 630 into the product of prime factor.

Factor each of the following completely.

58. $x^2 - 4x - 12$ **59.** $x^2 - 81$
60. $x^4 - 16$ **61.** $2x^2 + x - 21$
62. $5x^3 - 25x^2 - 30x$

Solve each equation.

63. $x^2 - 9x + 18 = 0$
64. $x^2 - 6x = 0$
65. $8x^2 = -2x + 15$

66. The product of two consecutive even integers is four more than twice their sum. Find the two integers.

We will begin this section with the definition of the ratio of two numbers.

DEFINITION   If a and b are any two numbers, $b \neq 0$, then the ratio of a to b is

$$\frac{a}{b}$$

(sometimes written $a:b$).

As you can see, ratios are another name for fractions or rational numbers. They are a way of comparing quantities. Since we can also think of a/b as the quotient of a and b, ratios are also quotients. The following table gives some ratios in words and as fractions.

Ratio	As a Fraction	In Lowest Terms	
25 to 75	$\frac{25}{75}$	$\frac{1}{3}$	
8 to 2	$\frac{8}{2}$	$\frac{4}{1}$	With ratios it is common to leave the 1 in the denominator
20 to 16	$\frac{20}{16}$	$\frac{5}{4}$	
$\frac{2}{5}$ to $\frac{3}{5}$	$\frac{\frac{2}{5}}{\frac{3}{5}}$	$\frac{2}{3}$	Found by multiplying $\frac{2}{5}$ by $\frac{5}{3}$
$3\sqrt{2}$ to $8\sqrt{2}$	$\frac{3\sqrt{2}}{8\sqrt{2}}$	$\frac{3}{8}$	

The table shows that ratios can involve all kinds of numbers—whole numbers, fractions, complex fractions, and irrational numbers.

▼   **Example 1**   A solution of hydrocloric acid (HCl) and water contains 49 milliliters of water and 21 milliliters of HCl. Find the ratio of HCl to water and of HCl to the total volume of the solution.

SOLUTION   The ratio of HCl to water is 21 to 49 or

$$\frac{21}{49} = \frac{3}{7}$$

The amount of total solution volume is $49 + 21 = 70$ milliliters. Therefore the ratio of HCl to total solution is 21 to 70, or

$$\frac{21}{70} = \frac{3}{10}$$

When two ratios are equal, we say they form a *proportion*.   ▲

DEFINITION A proportion is a statement that two ratios are equal. If *a/b* and *c/d* are two equal ratios, then the statement

$$\frac{a}{b} = \frac{c}{d}$$

is called a proportion.

Each of the four numbers in a proportion is called a *term* of the proportion. We number the terms as follows:

First term ⟶ a _ c ⟵ Third term
Second term ⟶ b = d ⟵ Fourth term

The first and fourth terms are called the *extremes,* and the second and third terms are called the *means*:

Means $\overset{\frown}{\underset{\longrightarrow}{\frac{a}{b}}} = \frac{c}{d}$ Extremes

For example, in the proportion

$$\frac{3}{8} = \frac{12}{32}$$

the extremes are 3 and 32, and the means are 8 and 12.

Means-Extremes Property

If *a, b, c,* and *d* are real numbers with $b \neq 0$ and $d \neq 0$, then

$$\text{If} \quad \frac{a}{b} = \frac{c}{d}$$

$$\text{then} \quad ad = bc$$

In words: In any proportion, the product of the extremes is equal to the product of the means.

This property of proportions comes from the multiplication property of equality. We can use it to solve for a missing term in a proportion.

▼ **Example 2**   Solve the proportion $\frac{3}{x} = \frac{6}{7}$ for x.

SOLUTION   We could solve for x by using the method developed in Section 6.5, that is, multiplying both sides by the LCD $7x$. Instead, let's use our new means-extremes property.

$$\frac{3}{x} = \frac{6}{7} \qquad \text{Extremes are 3 and 7;}$$
$$\qquad\qquad\qquad \text{means are } x \text{ and 6}$$

$$21 = 6x \qquad \text{Product of extremes} =$$
$$\qquad\qquad\qquad \text{product of means}$$

$$\frac{21}{6} = x \qquad \text{Divide both sides by 6}$$

$$x = \frac{7}{2} \qquad \text{Reduce to lowest terms} \qquad ▲$$

▼ **Example 3**   Solve for x: $\frac{x+1}{2} = \frac{3}{x}$.

SOLUTION   Again we want to point out that we could solve for x by using the method we used in Section 6.5. Using the means-extremes property is simply an alternative to the method developed in Section 6.5:

$$\frac{x+1}{2} = \frac{3}{x} \qquad \text{Extremes are } x+1 \text{ and } x;$$
$$\qquad\qquad\qquad \text{means are 2 and 3}$$

$$x^2 + x = 6 \qquad \text{Product of extremes} =$$
$$\qquad\qquad\qquad \text{product of means}$$

$$x^2 + x - 6 = 0 \qquad \text{Standard form for a quadratic equation}$$

$$(x + 3)(x - 2) = 0 \qquad \text{Factor}$$

$$x + 3 = 0 \quad \text{or} \quad x - 2 = 0 \qquad \text{Set factors to 0}$$

$$x = -3 \quad \text{or} \qquad x = 2$$

This time we have two solutions, -3 and 2. ▲

▼ **Example 4** A manufacturer knows that during a production run, 8 out of every 100 parts produced by a certain machine will be defective. If the machine produces 1450 parts, how many can be expected to be defective?

SOLUTION The ratio of defective parts to total parts produced is $\dfrac{8}{100}$. If we let x represent the number of defective parts out of the total of 1450 parts, then we can write this ratio again as $\dfrac{x}{1450}$. This gives us a proportion to solve:

Defective parts $\qquad \dfrac{x}{1450} = \dfrac{8}{100} \qquad$ Extremes are x and 100;
in numerator $\qquad\qquad\qquad\qquad\qquad$ means are 1450 and x

Total parts $\qquad\quad 100x = 11{,}600 \qquad$ Product of extremes =
in denominator $\qquad\qquad\qquad\qquad\qquad$ Product of means

$$x = 116$$

The manufacturer can expect 116 defective parts out of the total of 1450 parts if the machine usually produces 8 defective parts for every 100 parts it produces. ▲

Problem Set 9.6

Write each ratio as a fraction in lowest terms.

1. 8 to 6
2. 6 to 8
3. 200 to 250
4. 250 to 200
5. 32 to 4
6. 4 to 32
7. $\frac{3}{7}$ to $\frac{5}{7}$
8. $\frac{3}{5}$ to $\frac{7}{5}$
9. $\frac{2}{3}$ to $\frac{3}{4}$
10. $\frac{3}{4}$ to $\frac{9}{12}$
11. $5\sqrt{3}$ to $8\sqrt{3}$
12. $2\sqrt{2}$ to $6\sqrt{2}$
13. $\sqrt{8}$ to $\sqrt{18}$
14. $\sqrt{12}$ to $\sqrt{27}$

The table below gives the number of kilocalories of energy contained in an average serving of a number of foods.

Food	Kcal of Energy
Whole wheat bread	55
White bread	60
Ice cream	205
Cherry pie	420

15. Find the ratio of the number of kilocalories of energy in white bread to the number of kilocalories of energy in whole wheat bread.

16. Give the ratio of the number of kilocalories of energy in ice cream to the number of kilocalories of energy in cherry pie.

17. If a person has two servings of whole wheat bread for breakfast and that night has a midnight snack of cherry pie and ice cream, what is the ratio of breakfast kilocalories consumed to the number of kilocalories in the midnight snack?

18. How many servings of white bread equal one serving of cherry pie in terms of kilocalories of energy? (Find the ratio of the number of kilocalories of energy in cherry pie to the number of kilocalories of energy in white bread.)

19. If there are 60 minutes in 1 hour, what is the ratio of 20 minutes to 2 hours.

20. If there are 3 feet in 1 yard, what is the ratio of 2 feet to 3 yards?

Solve each of the following proportions.

21. $\dfrac{x}{2} = \dfrac{6}{12}$

22. $\dfrac{x}{4} = \dfrac{6}{8}$

23. $\dfrac{2}{5} = \dfrac{4}{x}$

24. $\dfrac{3}{8} = \dfrac{9}{x}$

25. $\dfrac{10}{20} = \dfrac{20}{x}$

26. $\dfrac{15}{60} = \dfrac{60}{x}$

27. $\dfrac{a}{3} = \dfrac{5}{12}$

28. $\dfrac{a}{5} = \dfrac{7}{20}$

29. $\dfrac{2}{x} = \dfrac{6}{7}$

30. $\dfrac{4}{x} = \dfrac{6}{7}$

31. $\dfrac{3}{11} = \dfrac{y}{5}$

32. $\dfrac{7}{12} = \dfrac{y}{5}$

33. $\dfrac{x+1}{3} = \dfrac{4}{x}$

34. $\dfrac{x+1}{6} = \dfrac{7}{x}$

35. $\dfrac{x}{2} = \dfrac{8}{x}$

36. $\dfrac{x}{9} = \dfrac{4}{x}$

37. $\dfrac{4}{a+2} = \dfrac{a}{2}$

38. $\dfrac{3}{a+2} = \dfrac{a}{5}$

39. $\dfrac{1}{x} = \dfrac{x-5}{6}$

40. $\dfrac{1}{x} = \dfrac{x-6}{7}$

41. A baseball player gets 6 hits in the first 18 games of the season. If he continues hitting at the same rate, how many hits will he get in the first 45 games?

42. A basketball player makes 8 of 12 free throws in the first game of the season. If she shoots with the same accuracy in the second game, how many of the 15 free throws she attempts will she make?

43. A solution contains 12 milliliters of alcohol and 16 milliliters of water. If another solution is to have the same concentration of alcohol in water but is to contain 28 milliliters of water, how much alcohol must it contain?

44. A solution contains 15 milliliters of HCl and 42 milliliters of water. If another solution is to have the same concentration of HCl in water but is to contain 140 milliliters of water, how much HCl must it contain?

45. If 100 grams of ice cream contain 13 grams of fat, how much fat is in 350 grams of ice cream.

46. A 6-ounce serving of grapefruit juice contains 159 grams of water. How many grams of water are in 20 ounces of grapefruit juice?

47. A map is drawn so that every 3.5 inches on the map corresponds to an actual distance of 100 miles. If the actual distance between two cities is 420 miles, how far apart are they on the map?

48. The scale on a map indicates that 1 inch on the map corresponds to an actual distance of 105 miles. Two cities are 4.5 inches apart on the map. What is the actual distance between the two cities?

49. A man drives his car 245 miles in 5 hours. At this rate, how far will he travel in 7 hours?

50. An airplane flies 1380 miles in 3 hours. How far will it fly in 5 hours.

Review Problems   The problems below review material we covered in Chapter 6.

51. Reduce to lowest terms: $\dfrac{x^2 - x - 6}{x^2 - 9}$.

52. Divide, using long division for polynomials:

$$\frac{x^2 - 2x + 6}{x - 4}$$

Perform the indicated operations.

53. $\dfrac{x^2 - 25}{x + 4} \cdot \dfrac{2x + 8}{x^2 - 9x + 20}$

54. $\dfrac{3x + 6}{x^2 + 4x + 3} \div \dfrac{x^2 + x - 2}{x^2 + 2x - 3}$

55. $\dfrac{x}{x^2 - 16} + \dfrac{4}{x^2 - 16}$

56. $\dfrac{5}{x^2 - 1} - \dfrac{4}{x^2 + 3x - 4}$

57. $\dfrac{1 - \dfrac{25}{x^2}}{1 - \dfrac{8}{x} + \dfrac{15}{x^2}}$

58. Solve for x: $\dfrac{x}{2} - \dfrac{5}{x} = -\dfrac{3}{2}$.

Two variables are said to *vary directly* if one is a constant multiple of the other. For instance, y varies directly as x if $y = Kx$, where K is a constant. The constant K is called the constant of variation. The following table gives the relation between direct variation statements and their equivalent algebraic equations.

**9.7
Variation**

Statement	Equation ($K = $ constant of variation)
y varies directly as x	$y = Kx$
y varies directly as the square of x	$y = K \cdot x^2$
s varies directly as the square root of t	$s = K\sqrt{t}$
r varies directly as the cube of s	$r = Ks^3$

Any time we run across a statement similar to the ones above, we can immediately write an equivalent expression involving variables and a constant of variation K.

▼ **Example 1** Suppose y varies directly as x. When y is 15, x is 3. Find y when x is 4.

SOLUTION   From the first sentence we can write the relationship between x and y as

$$y = Kx$$

We now use the second sentence to find the value of K. Since y is 15 when x is 3, we have

$$15 = K(3) \quad \text{or} \quad K = 5$$

Now we can rewrite the relationship between x and y more specifically as

$$y = 5x$$

To find the value of y when x is 4 we simply substitute $x = 4$ into our last equation. Substituting

$$x = 4$$

into

$$y = 5x$$

we have

$$y = 5(4)$$
$$y = 20 \qquad\qquad \blacktriangle$$

▼ **Example 2**   Suppose y varies directly as the square of x. When x is 4, y is 32. Find x when y is 50.

SOLUTION   The first sentence gives us

$$y = Kx^2$$

Since y is 32 when x is 4, we have

$$32 = K(4)^2$$
$$32 = 16K$$
$$K = 2$$

The equation now becomes

$$y = 2x^2$$

When y is 50, we have

$$50 = 2x^2$$
$$25 = x^2$$
$$x = \pm 5$$

There are two possible solutions, $x = 5$ or $x = -5$.   ▲

Two variables are said to *vary inversely* if one is a constant multiple of the reciprocal of the other. For example, y varies inversely as x if $y = \dfrac{K}{x}$, where K is a real number constant. Again, K is called the constant of variation. The table that follows gives some examples of inverse variation statements and their associated algebraic equations.

Statement	Equation (K = constant of variation)
y varies inversely as x	$y = \dfrac{K}{x}$
y varies inversely as the square of x	$y = \dfrac{K}{x^2}$
F varies inversely as the square root of t	$F = \dfrac{K}{\sqrt{t}}$
r varies inversely as the cube of s	$r = \dfrac{K}{s^3}$

Every inverse variation statement has an associated inverse variation equation.

▼ **Example 3** Suppose y varies inversely as x. When y is 4, x is 5. Find y when x is 10.

SOLUTION The first sentence gives us the relationship between x and y:

$$y = \frac{K}{x}$$

We use the second sentence to find the value of the constant K:

$$4 = \frac{K}{5}$$

or

$$K = 20$$

We can now write the relationship between x and y more specifically as

$$y = \frac{20}{x}$$

We use this equation to find the value of y when x is 10. Substituting

$$x = 10$$

into

$$y = \frac{20}{x}$$

we have

$$y = \frac{20}{10}$$

$$y = 2 \qquad \blacktriangle$$

▼ **Example 4** The intensity (I) of light from a source varies inversely as the square of the distance (d) from the source. Ten feet away from the source the intensity is 200 candlepower. What is the intensity 5 feet from the source?

SOLUTION

$$I = \frac{K}{d^2}$$

Since $I = 200$ when $d = 10$, we have

$$200 = \frac{K}{10^2}$$

$$200 = \frac{K}{100}$$

$$K = 20{,}000$$

The equation becomes

$$I = \frac{20{,}000}{d^2}$$

When $d = 5$, we have

$$I = \frac{20{,}000}{5^2}$$

$$= \frac{20,000}{25}$$

$$= 800 \text{ candlepower} \qquad \blacktriangle$$

For each of the following problems, y varies directly as x.

1. If $y = 10$ when $x = 5$, find y when x is 4.
2. If $y = 20$ when $x = 4$, find y when x is 11.
3. If $y = 39$ when $x = 3$, find y when x is 10.
4. If $y = -18$ when $x = 6$, find y when x is 3.
5. If $y = -24$ when $x = 4$, find x when y is -30.
6. If $y = 30$ when $x = -15$, find x when y is 8.
7. If $y = -7$ when $x = -1$, find x when y is -21.
8. If $y = 30$ when $x = 4$, find y when x is 7.

For each of the following, y varies directly as the square of x.

9. If $y = 75$ when $x = 5$, find y when x is 1.
10. If $y = -72$ when $x = 6$, find y when x is 3.
11. If $y = 48$ when $x = 4$, find y when x is 9.
12. If $y = 27$ when $x = 3$, find x when y is 75.

For each of the following problems, y varies inversely with x.

13. If $y = 5$ when $x = 2$, find y when x is 5.
14. If $y = 2$ when $x = 10$, find y when x is 4.
15. If $y = 2$ when $x = 1$, find y when x is 4.
16. If $y = 4$ when $x = 3$, find y when x is 6.
17. If $y = 5$ when $x = 3$, find x when y is 15.
18. If $y = 12$ when $x = 10$, find x when y is 60.
19. If $y = 10$ when $x = 10$, find x when y is 20.
20. If $y = 15$ when $x = 2$, find x when y is 6.

For each of the following, y varies inversely as the square of x.

21. If $y = 4$ when $x = 5$, find y when x is 2.
22. If $y = 5$ when $x = 2$, find y when x is 6.
23. If $y = 4$ when $x = 3$, find y wnen x is 2.
24. If $y = 9$ when $x = 4$, find y when x is 3.

Solve the following problems.

25. The tension t in a spring varies directly with the distance d the spring is stretched. If the tension is 42 pounds when the spring is stretched 2 inches, find the tension when the spring is stretched twice as far.
26. The time t it takes to fill a bucket varies directly with the volume g of the

bucket. If it takes 1 minute to fill a 4-gallon bucket, how long will it take to fill a 6-gallon bucket?

27. The power P in an electric circuit varies directly with the square of the current I. If $P = 30$ when $I = 2$, find P when $I = 7$.

28. The resistance R in an electric circuit varies directly with the voltage V. If $R = 20$ when $V = 120$, find R when $V = 240$.

29. The amount of money M a person makes per week varies directly with the number of hours h he works per week. If he works 20 hours and earns $157, how much does he make if he works 30 hours?

30. The volume V of a gas varies directly as the temperature T. If $V = 3$ when $T = 150$, find V when T is 200.

31. The weight F of a body varies inversely with the square of the distance d between the body and the center of the earth. If a man weighs 150 pounds 4000 miles from the center of the earth, how much will he weigh at a distance of 5000 miles from the center of the earth?

32. The intensity I of a light source varies inversely with the square of the distance d from the source. Four feet from the source the intensity is 9 footcandles. What is the intensity 3 feet from the source?

33. The current I in an electric circuit varies inversely with the resistance R. If a current of 30 amps is produced by a resistance of 2 ohms, what current will be produced by a resistance of 5 ohms?

34. The pressure exerted by a gas on the container in which it is held varies inversely with the volume of the container. A pressure of 40 pounds per square inch is exerted on a container of volume 2 cubic feet. What is the pressure on a container whose volume is 8 cubic feet?

Review Problems   The problems below review material we covered in Chapter 7.

Find each root.

35. $\sqrt{49}$

36. $\sqrt[3]{-8}$

Write in simplified form for radicals.

37. $\sqrt{50}$

38. $2\sqrt{18x^2y^3}$

Perform the indicated operations.

39. $3\sqrt{12} + 5\sqrt{27}$

40. $\sqrt{3}(\sqrt{3} - 4)$

41. $(\sqrt{6} + 2)(\sqrt{6} - 5)$

42. $(\sqrt{3} - \sqrt{2})^2$

Rationalize the denominator.

43. $\dfrac{8}{\sqrt{5} - \sqrt{3}}$

Solve for x

44. $\sqrt{2x - 5} = 3$

Formulas are very common in science and industry. Almost all disciplines that use mathematics use a number of formulas. Formulas almost always contain more than one variable.

In this section we will learn how to "solve" a formula for one of its variables without the need for new properties or definitions. We need only apply the material we have already learned to this new situation.

Consider the formula for the area of a triangle:

$$A = \tfrac{1}{2} bh$$

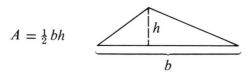

where A = area, b = length of the base, and h = the height of the triangle.

Suppose we want to solve this formula for h. What we must do is isolate the variable h on one side of the equal sign. We begin by multiplying both sides by 2:

$$2 \cdot A = 2 \cdot \frac{1}{2} bh$$

$$2A = bh$$

Then we divide both sides by b:

$$\frac{2A}{b} = \frac{bh}{b}$$

$$h = \frac{2A}{b}$$

The original formula $A = \dfrac{1}{2} bh$ and the final formula $h = \dfrac{2A}{b}$ both give the same relationship among A, b, and h. The first one has been solved for A and the second one has been solved for h.

RULE   To solve a formula for one of its variables, we must isolate that variable on either side of the equal sign. All other variables and constants will appear on the other.

▼ **Example 1**   Solve $C = \frac{5}{9}(F - 32)$ for F.

SOLUTION   This formula gives the relationship between the Fahrenheit temperature scale and the Celsius temperature scale. We begin by multiplying both sides by $\frac{9}{5}$.

$$\frac{9}{5}C = \frac{9}{5} \cdot \frac{5}{9}(F - 32)$$
$$\frac{9}{5}C = F - 32$$

We finish the problem by adding 32 to both sides:

$$\frac{9}{5}C + 32 = F - 32 + 32$$
$$\frac{9}{5}C + 32 = F$$

or

$$F = \frac{9}{5}C + 32$$ ▲

▼ **Example 2**   Solve $A = \frac{1}{2}(b + B)h$ for B.

SOLUTION   Since multiplication is both commutative and associative, we can rewrite the formula as follows:

$$A = \frac{h}{2}(b + B)$$

Multiplying both sides by $\dfrac{2}{h}$, we have

$$\frac{2}{h}A = \frac{2}{h} \cdot \frac{h}{2}(b + B)$$

$$\frac{2A}{h} = b + B$$

Now, adding $-b$ to both sides completes the problem.

$$\frac{2A}{h} + (-b) = b + (-b) + B$$

$$\frac{2A}{h} - b = B$$

or

$$B = \frac{2A}{h} - b$$ ▲

There are a number of other ways to go about solving for B. This is only one of them. The problem is complete when the variable we are solving for appears alone on only one side of the equal sign.

Solve each of the following formulas for the indicated variable.

1. $A = lw$ for l
2. $A = lw$ for w
3. $d = rt$ for r
4. $d = rt$ for t
5. $V = lwh$ for h
6. $V = lwh$ for l
7. $V = IR$ for R
8. $V = IR$ for I
9. $PV = nRT$ for P
10. $PV = nRT$ for T
11. $PV = nRT$ for R
12. $A = \frac{1}{2}bh$ for b
13. $c^2 = a^2 + b^2$ for a^2
14. $c^2 = a^2 + b^2$ for b^2
15. $p = a + b + c$ for a
16. $p = a + b + c$ for c
17. $p = 2l + 2w$ for l
18. $p = 2l + 2w$ for w
19. $F = \frac{9}{5}C + 32$ for C
20. $A = \frac{1}{2}(b + B)h$ for B
21. $A = \frac{1}{2}(b + B)h$ for b
22. $A = \pi r^2 + 2\pi rh$ for h

23. The formula $C = \frac{5}{9}(F - 32)$ gives the relationship between temperatures measured in degrees Fahrenheit and the corresponding temperature in degrees Celsius. Find the temperature in degrees Celsius by letting $F = 77$ in the formula.
24. If the temperature is 86° Fahrenheit, what is it in degrees Celsius.
25. The formula $A = \frac{1}{2}(b + B)h$ gives the area A of a trapezoid in terms of the height h and the two bases, b and B. If $h = 6$ inches, $b = 3$ inches, and $B = 5$ inches, find the area.
26. Find the area of a trapezoid with a height of 8 inches, if the bases are 4 inches and 7 inches.

Review Problems   The problems below review material we covered in Chapter 8.

Solve each equation. Use any convenient method.

27. $(2x + 1)^2 = 25$
28. $x^2 - 2x - 8 = 0$
29. $3x^2 = 4x + 2$
30. $x^2 - 5x - 5 = 0$

Perform the indicated operations.

31. $(4 + 3i) - (2 - 5i)$
32. $2i(1 - 3i)$

33. $(4 + 5i)(2 - 3i)$
34. $\dfrac{8}{2 - 3i}$

Graph each equation.

35. $y = x^2 - 3$
36. $y = (x - 3)^2$

Chapter 9 Summary and Review

Examples

1. The slope of the line through $(3, -5)$ and $(-2, 1)$ is

$$m = \frac{-5 - 1}{3 - (-2)} = \frac{-6}{5}$$

SLOPE OF A LINE [9.1]

The *slope* of the line containing the points (x_1, y_1) and (x_2, y_2) is given by

$$\text{Slope} = m = \frac{y_2 - y_1}{x_2 - x_1} = \frac{\text{rise}}{\text{run}}$$

2. The x-intercept for the line $3x - 2y = 6$ is 2. We find it by letting $y = 0$ and solving for x. The y-intercept is -3, which is found by letting $x = 0$.

INTERCEPTS [9.1]

The *x-intercept* of a graph is the x-coordinate of the point where the graph crosses the x-axis. The *y-intercept* of a graph is the y-coordinate of the point where the graph crosses the y-axis.

If a graph has an x-intercept a, the point $(a, 0)$ is on the graph. If a graph has a y-intercept b, the point $(0, b)$ is on the graph.

3. The equation of the line through $(1, 2)$ with a slope of 3 is

$$y - 2 = 3(x - 1)$$
$$y - 2 = 3x - 3$$
$$y = 3x - 1$$

POINT-SLOPE FORM OF A STRAIGHT LINE [9.2]

If a line has a slope of m and contains the point (x_1, y_1), the equation can be written as

$$y - y_1 = m(x - x_1)$$

4. The equation of the line with a slope of 2 and a y-intercept 5 is

$$y = 2x + 5$$

SLOPE-INTERCEPT FORM OF A STRAIGHT LINE [9.2]

The equation of the line with a slope of m and a y-intercept b is

$$y = mx + b$$

5. If $f(x) = 3x - 5$
then $f(0) = 3(0) - 5 = -5$
$f(2) = 3(2) - 5 = 1$
$f(3) = 3(3) - 5 = 4$

FUNCTIONS [9.3]

A *function* is any set of ordered pairs in which no two different ordered pairs have the same first coordinates. The *domain* of the function is the set of all first coordinates. The *range* is the set of all second coordinates.

The notation $f(x)$ is read "f of x" and is defined to be the value of the function f at x.

FRACTIONAL EXPONENTS [9.4]

Fractional exponents are used to specify roots and avoid radical notation. For any nonnegative number a, we have

$$a^{m/n} = \sqrt[n]{a^m}$$

where m is the power and n represents the root.

6. $8^{2/3} = (\sqrt[3]{8})^2 = 2^2 = 4$
$9^{3/2} = (\sqrt{9})^3 = 3^3 = 27$

SCIENTIFIC NOTATION [9.5]

A number is in scientific notation when it is written as the product of a number between 1 and 10 and an integer power of 10.

7. $768,000 = 7.68 \times 10^5$
$0.00039 = 3.9 \times 10^{-4}$

RATIO AND PROPORTION [9.6]

The ratio of a to b is

$$\frac{a}{b}$$

Two equal ratios form a proportion. In the proportion

$$\frac{a}{b} = \frac{c}{d}$$

a and d are the *extremes*, and b and c are the *means*. In any proportion, the product of the extremes is equal to the product of the means.

8. Solve for x: $\dfrac{3}{x} = \dfrac{5}{20}$

$3 \cdot 20 = 5 \cdot x$
$60 = 5x$
$x = 12$

DIRECT VARIATION [9.7]

The variable y is said to vary directly with the variable x if $y = Kx$, where K is a real number.

9. If y varies directly with the square of x, then

$$y = Kx^2$$

INVERSE VARIATION [9.7]

The variable y is said to vary inversely with the variable x if $y = K/x$, where K is a real number constant.

10. If y varies inversely with the cube of x, then

$$y = \frac{K}{x^3}$$

FORMULAS [9.8]

A formula is an equation with more than one variable. To solve a formula for one of its variables we use the addition and multiplication properties of equality to move everything except the variable in question to one side of the equal sign so the variable in question is alone on the other side.

11. Solving $P = 2l + 2w$ for l, we have

$$P - 2w = 2l$$
$$\frac{P - 2w}{2} = l$$

Give the slope, x-intercept, and y-intercept of the following straight lines and sketch their graphs.

1. $3x - 2y = 6$ 2. $y = 3x - 2$
3. $2x - y = 5$ 4. $y = -1$

5. Find the equation of the line through $(-2, 5)$ with a slope of 3.
6. Find the equation of the line with a slope of 4 and y-intercept 8.
7. Line l passes through the points $(3, 1)$ and $(-2, 4)$. Find the equation of line l.
8. A straight line has an x-intercept 1 and contains the point $(3, 4)$. Find its equation.
9. If $f(x) = 3x^2 - 1$ and $g(x) = 5x + 2$, find the following.

   (a) $f(3)$ (b) $f(-2)$ (c) $f(a)$
   (d) $g(0)$ (e) $g(-5)$ (f) $g(b)$

10. Simplify each of the following expressions.

    (a) $8^{2/3}$ (b) $9^{3/2} - 16^{3/4}$ (c) $\dfrac{3^{1/2} \cdot 3^{1/3}}{3}$

11. Write in scientific notation.

    (a) 78,000   (b) 0.00397

12. Simplify and write your answer in scientific notation.

    (a) $(4 \times 10^5)(2 \times 10^{-3})$ (b) $\dfrac{(3 \times 10^6)(8 \times 10^4)}{4 \times 10^{-2}}$

13. Solve each proportion.

    (a) $\dfrac{x}{5} = \dfrac{10}{20}$ (b) $\dfrac{3}{7} = \dfrac{2}{x}$

14. Suppose y varies directly as the square of x. When x is 3, y is 36. Find y when x is 5.
15. Suppose y varies inversely with x. When x is 3, y is 6. Find y when x is 9.
16. In the formula $C = \frac{5}{9}(F - 32)$, solve for F.

Answers to Odd-Numbered Exercises and Chapter Tests

CHAPTER 1

PROBLEM SET 1.1,　P. 4

1. The sum of 7 and 8 is 15.　**3.** 7 is less than 10.　**5.** The sum of 8 and 2 is not 6.
7. 21 is greater than 20.　**9.** The sum of x and 1 is 5.　**11.** x is less than y.
13. The sum of x and 2 is equal to the sum of y and 3.　**15.** The difference of x and 1 is less than the product of 2 and x.　**17.** 1 more than the product of 2 and x is 10.　**19.** 4 times the sum of x and 1 is 6.　**21.** True　**23.** False　**25.** False　**27.** True　**29.** False
31. 11　**33.** 17　**35.** 31　**37.** 20　**39.** 17　**41.** 20　**43.** 27　**45.** 35
47. 13　**49.** 13　**51.** $2(10 + 4)$　**53.** $3 \cdot 50 - 14$

PROBLEM SET 1.2,　P. 10

1.–11.

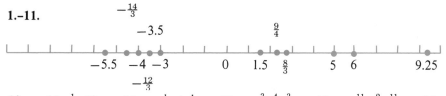

13. $-10, \frac{1}{10}, 10$　**15.** $-\frac{1}{4}, 4, \frac{1}{4}$　**17.** $-\frac{3}{4}, \frac{4}{3}, \frac{3}{4}$　**19.** $-\frac{11}{2}, \frac{2}{11}, \frac{11}{2}$　**21.** $3, -\frac{1}{3}, 3$　**23.** $\frac{1}{4},$
$-4, \frac{1}{4}$　**25.** $\frac{2}{5}, -\frac{5}{2}, \frac{2}{5}$　**27.** $-x, \frac{1}{x}$, the distance from x to zero on the number line　**29.** $<$
31. $>$　**33.** $<$　**35.** $>$　**37.** $>$　**39.** $=$　**41.** $<$　**43.** $<$　**45.** $<$
47. True　**49.** False　**51.** False　**53.** True　**55.** False　**57.** True
59. False　**61.** $\frac{8}{15}$　**63.** $\frac{20}{27}$　**65.** 1　**67.** 1　**69.** 1　**71.** 1　**73.** Opposite 7,
reciprocal $-\frac{1}{7}$　**75.** $+16{,}732, -1299$　**77.** -2 and 8

PROBLEM SET 1.3,   P. 15

1. $8, -8, 2, -2$ **3.** $35, -35, 5, -5$ **5.** 3 **7.** 5 **9.** 6 **11.** -7 **13.** -14
15. -3 **17.** -8 **19.** -25 **21.** -12 **23.** -19 **25.** -25 **27.** -25
29. -15 **31.** -8 **33.** -4 **35.** 6 **37.** -30 **39.** 6 **41.** 8 **43.** -4
45. -20 **47.** -7 **49.** -7 **51.** -10 **53.** -10 **55.** -25 **57.** -25 [A rule
that summarizes these results is $-(a + b) = -a + (-b)$.]   **59.** $5 + 9 = 14$
61. $4 + [-7 + (-5)] = -8$ **63.** $[-2 + (-3)] + 10 = 5$ **65.** 3 **67.** -3
69. $-12 + 4$ **71.** $10 + (-6) + (-8) = -4$

PROBLEM SET 1.4,   P. 18

1. $6, -6, 10, -10$ **3.** $-8, 8, 32, -32$ **5.** -3 **7.** 21 **9.** 41 **11.** -7
13. -5 **15.** -20 **17.** -4 **19.** 4 **21.** -1 **23.** -28 **25.** 25 **27.** -11
29. 192 **31.** 4 **33.** -7 **35.** 18 **37.** 10 **39.** 17 **41.** -11 **43.** -48
45. 20 **47.** 2 **49.** -18 **51.** $8 - 5 = 3$ **53.** $-8 - 5 = -13$
55. $8 - (-5) = 13$ **57.** $[3 - (-5)] + 7 = 15$ **59.** 10 **61.** -2 **63.** $1500 - 730$
65. $-35 + 15 - 20 = -40$

PROBLEM SET 1.5,   P. 25

1. Commutative   **3.** Multiplicative inverse   **5.** Commutative   **7.** Distributive
9. Associative, commutative   **11.** Commutative, associative   **13.** Commutative
15. Commutative, associative   **17.** Commutative   **19.** Additive inverse   **21.** $3x - 6$
23. $9a + 9b$ **25.** 0 **27.** 0 **29.** 10 **31.** $(4 + 2) + x = 6 + x$
33. $x + (2 + 7) = x + 9$ **35.** $(3 \cdot 5)x = 15x$ **37.** $(9 \cdot 6)y = 54y$ **39.** $(2 \cdot 3)a = 6a$
41. $(\frac{1}{3} \cdot 3)x = x$ **43.** $(\frac{1}{2} \cdot 2)y = y$ **45.** $(\frac{3}{4} \cdot \frac{4}{3})x = x$ **47.** $(\frac{6}{5} \cdot \frac{5}{6})a = a$
49. $8(x) + 8(2) = 8x + 16$ **51.** $8(x) - 8(2) = 8x - 16$ **53.** $4(y) + 4(1) = 4y + 4$
55. $3(6x) + 3(5) = 18x + 15$ **57.** $2(3a) + 2(7) = 6a + 14$ **59.** $9(6y) - 9(8) = 54y - 72$
61. $3(4) + 3(2x) = 12 + 6x$ **63.** $7(3) - 7(a) = 21 - 7a$ **65.** $3x + 3y$ **67.** $8a - 8b$
69. No   **71.** No

PROBLEM SET 1.6,   P. 30

1. -42 **3.** -21 **5.** -16 **7.** 3 **9.** 121 **11.** 6 **13.** -60 **15.** 12
17. -48 **19.** 24 **21.** 3 **23.** 6 **25.** 10 **27.** -31 **29.** 1 **31.** 14
33. 54 **35.** -2 **37.** -35 **39.** 9 **41.** $-\frac{10}{21}$ **43.** -4 **45.** 1 **47.** 1
49. $-8x$ **51.** $-45y$ **53.** $42x$ **55.** x **57.** x **59.** $-4(a) + (-4)(2) = -4a - 8$
61. $-2(4x) + (-2)(1) = -8x - 2$ **63.** $-5(3x) + (-5)(4) = -15x - 20$
65. $-7(6x) - (-7)(3) = -42x + 21$ **67.** $-1(3x) - (-1)(4) = -3x + 4$ **69.** -25
71. $2(-4x) = -8x$ **73.** -26

PROBLEM SET 1.7,   P. 34

1. -2 **3.** -5 **5.** $-\frac{1}{3}$ **7.** -3 **9.** -9 **11.** $-\frac{1}{3}$ **13.** 3 **15.** 3 **17.** $\frac{1}{5}$
19. 0 **21.** $\frac{3}{5}$ **23.** $-\frac{5}{3}$ **25.** -2 **27.** -3 **29.** 10 **31.** -6 **33.** 5
35. $-\frac{35}{9}$ **37.** -1 **39.** $\frac{9}{7}$ **41.** $\frac{16}{11}$ **43.** $\frac{16}{15}$ **45.** $\frac{4}{3}$ **47.** $-\frac{8}{13}$ **49.** -1
51. 3 **53.** -10 **55.** -8

PROBLEM SET 1.8, P. 39

1. $\{0, 1\}$ **3.** $\{-3, -2.5, 0, 1, \frac{3}{2}\}$ **5.** All   **7.** $\{-10, -8, -2, 9\}$ **9.** π **11.** True
13. False   **15.** False   **17.** True   **19.** Whole/counting/integer/rational/real
21. Integer/rational/real   **23.** Rational/real   **25.** Irrational/real
27. Whole/counting/integer/rational/real   **29.** Rational/real   **31.** Rational/real
33. Whole/counting/integer/rational/real   **35.** Irrational/real   **37.** Rational/real

CHAPTER 1 TEST, P. 43

1. -4 **2.** -7 **3.** -7 **4.** -15 **5.** -18 **6.** -30 **7.** 15 **8.** -7
9. 6 **10.** 1 **11.** -5 **12.** $-\frac{11}{3}$ **13.** 22 **14.** 2 **15.** 6 **16.** False
17. True   **18.** False   **19.** False   **20.** False   **21.** False   **22.** $\{1, -8\}$
23. $\{1, 1.5, \frac{3}{4}, -8\}$ **24.** $\{\sqrt{2}\}$ **25.** All   **26.** A, C   **27.** E   **28.** B, D   **29.** A
30. E   **31.** B   **32.** $8 + (-3) = 5$ **33.** $-24 - 2 = -26$ **34.** $-5(-4) = 20$
35. $\dfrac{-24}{-2} = 12$ **36.** $5 + (6)(-1) = -1$ **37.** $2(-9 + 2) = -14$ **38.** $2(3) + (-4) = 2$
39. $(3 + 5) + x = 8 + x$ **40.** $(-4 + 2) + x = -2 + x$ **41.** $[-\frac{1}{3}(-3)]x = x$
42. $[5(-4)]x = -20x$ **43.** $2(3x) + 2(5) = 6x + 10$ **44.** $-3(x) + (-3)(4) = -3x - 12$
45. $-5(2x) - (-5)(1) = -10x + 5$ **46.** $-1(5x) - (-1)(9) = -5x + 9$

CHAPTER 2

PROBLEM SET 2.1, P. 48

1. $-3x$ **3.** $-a$ **5.** $12x$ **7.** $6a$ **9.** $6x - 3$ **11.** $7a + 5$ **13.** $5x - 5$
15. $4a + 2$ **17.** $-9x - 2$ **19.** $12a + 3$ **21.** $10x - 1$ **23.** $21y + 6$
25. $-6x + 8$ **27.** $-2a + 3$ **29.** $-4x + 26$ **31.** $4y - 16$ **33.** $-6x - 1$
35. $5x - 12$ **37.** $7y - 39$ **39.** $-21x - 24$ **41.** 5 **43.** 6 **45.** 11 **47.** -9
49. 8 **51.** 16 **53.** -28 **55.** -37 **57.** -17 **59.** -41 **61.** Three times the
quantity $5x$ plus 1, minus 6   **63.** Four times the quantity $2x$ plus $3y$, minus 1   **65.** Four
times the quantity $2x$ plus $3y$ minus 1   **67.** Two times the quantity $4x$ plus 1, plus 3 times the
quantity $2x$ plus 1   **69.** 10 **71.** 17 **73.** -17 **75.** -4 **77.** 2

PROBLEM SET 2.2, P. 54

1. 11 **3.** 4 **5.** -9 **7.** -8 **9.** -17 **11.** 3 **13.** -4 **15.** -1 **17.** 1
19. 5 **21.** 3 **23.** 13 **25.** 21 **27.** 7 **29.** 6 **31.** -22 **33.** 6 **35.** 0
37. -2 **39.** -16 **41.** -3 **43.** 10 **45.** -12 **47.** -1 **49.** 4 **51.** 2
53. -5 **55.** -1 **57.** -3 **59.** 8 **61.** -8 **63.** 5 **65.** -11 **67.** $18x$
69. x **71.** y **73.** x

PROBLEM SET 2.3, P. 59

1. 2 **3.** 4 **5.** -4 **7.** $-\frac{1}{2}$ **9.** -2 **11.** 3 **13.** 4 **15.** 1 **17.** 0
19. 0 **21.** 6 **23.** 50 **25.** -6 **27.** 15 **29.** -55 **31.** -2 **33.** -2
35. -8 **37.** $\frac{1}{2}$ **39.** 4 **41.** 3 **43.** -4 **45.** 4 **47.** -15 **49.** $-\frac{1}{2}$
51. 3 **53.** 7 **55.** 1 **57.** $\frac{1}{4}$ **59.** -3 **61.** 3 **63.** 2 **65.** -9 **67.** 5
69. -6 **71.** $10x - 43$ **73.** $-3x - 13$ **75.** $-6y + 4$ **77.** $-5x + 7$

PROBLEM SET 2.4,   P. 63

1. 3   **3.** −2   **5.** −1   **7.** 2   **9.** −4   **11.** −2   **13.** 0   **15.** 1   **17.** $\frac{1}{2}$
19. 7   **21.** 8   **23.** −3   **25.** −$\frac{1}{3}$ **27.** 0   **29.** 2   **31.** 6   **33.** 8   **35.** 0
37. All real numbers   **39.** No solution   **41.** All real numbers   **43.** No solution
45. The sum of 4 and 1 is 5.   **47.** The difference of 6 and 2 is 4.   **49.** 2(6 + 3)
51. 2·5 + 3 = 13

PROBLEM SET 2.5,   P. 68

Along with the answers to the odd-numbered problems in this problem set we are including the equations used to solve each problem. Be sure that you try the problems on your own before looking here to see what the correct equations are.

1. $x + 5 = 13$; 8   **3.** $2x + 4 = 14$; 5   **5.** $5(x + 7) = 30$; −1   **7.** The two numbers are x and $x + 2$; $x + (x + 2) = 8$; 3 and 5   **9.** The two numbers are x and $3x − 4$; $x + (3x − 4) + 5 = 25$; 6 and 14   **11.** The width is x, the length is $x + 5$; $2x + 2(x + 5) = 34$; 6 inches and 11 inches   **13.** The length of a side is x; $4x = 48$; 12 meters   **15.** The width is x, the length is $2x − 3$; $2x + 2(2x − 3) = 54$; 10 inches and 17 inches   **17.** Barney's age is x, Fred's age is $x + 4$; $(x − 5) + (x − 1) = 48$; Barney is 27, Fred 31.   **19.** Lacy's age is x, Jack's age is $2x$; $(x + 3) + (2x + 3) = 54$; Lacy is 16, Jack is 32.   **21.** 4 is less than 10.   **23.** 9 is greater than or equal to −5.   **25.** <   **27.** >   **29.** <

PROBLEM SET 2.6,   P. 73

1. $x < 12$

3. $a \leq 12$

5. $x > 13$

7. $y \geq 4$

9. $x > 9$

11. $x < 2$

13. $a \leq 5$

15. $x > 15$

17. $y \geq 10$

19. $x < -3$

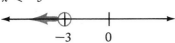

21. $a > -7$

23. $x < 6$

25. $x \geq -50$

27. $x < 6$

29. $y \geq -5$

31. $x < 3$

33. $a \leq 3$

35. $x > \frac{15}{2}$

37. $x < -1$

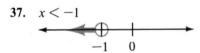

39. $y \geq -2$

41. $x < -1$

43. $m \leq -6$

45. $x \leq -5$

47. $2x + 6 < 10; \; x < 2$ **49.** $4x > x - 8; \; x > -\frac{8}{3}$ **51.** $2(x + 5) \leq 12; \; x \leq 1$
53. $3x - 5 < x + 7; \; x < 6$ **55.** B **57.** A **59.** B and C

PROBLEM SET 2.7, P. 77

1.

3.

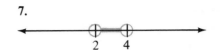

5.

7.

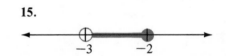

9.

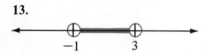

11.

13.

15.

17.

19.

21.

23.

25.

27.

29.

31.

33.

35.

37.

39.

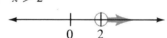

41. $\{3\}$ **43.** $\{-10, 0, 3\}$ **45.** $\{-\sqrt{5}, \sqrt{15}\}$

CHAPTER 2 TEST,   P. 81
1. $-4x + 5$ **2.** $10x + 2$ **3.** $3a - 4$ **4.** $5x + 6$ **5.** $-3y - 12$ **6.** $11x + 28$
7. 2   **8.** $5y + 3$ **9.** 6   **10.** 5   **11.** 4   **12.** 2   **13.** 2   **14.** -3 **15.** 2
16. -3 **17.** -1 **18.** -2
19. $x < 1$

20. $x \leq -3$

21. $a < -4$

22. $x > 2$

23. $x \leq 8$

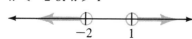

24. $m > -2$

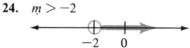

25. $x < -2$ or $x > 1$

26. $-2 \leq x \leq 4$

27. 3   **28.** 3, 15   **29.** 20, 40   **30.** 10, 20 inches

CHAPTER 3

PROBLEM SET 3.1,   P. 88

1. (0, 6)(3, 0)(6, −6)   **3.** (0, 3)(4, 0)(−4, 6)   **5.** (1, 1)($\frac{3}{4}$, 0)(5, 17)
7. (2, 13)(1, 6)(0, −1)   **9.** (−5, 4)(−5, −3)(−5, 0)

11.

x	y
1	3
−3	−9
4	12
6	18

13.

x	y
0	0
−$\frac{1}{2}$	−2
−3	−12
3	12

15.

x	y
2	3
3	2
5	0
9	−4

17.

x	y
2	0
3	2
3	2
−3	−10

19.

x	y
0	−1
−1	−7
−3	−19
$\frac{3}{2}$	8

21. (0, −2)   **23.** (1, 5)(0, −2)(−2, −16)   **25.** (1, 6)(−2, −12)(0, 0)   **27.** (2, −2)
29. (3, 0)(3, −3)

41. 12 inches   **43.** (5, 10)   **45.** (3, 3), (−3, 3), (5, 5), (−5, 5)   **47.** −12   **49.** −4
51. −5   **53.** 3

PROBLEM SET 3.2,   P. 97

1–19.

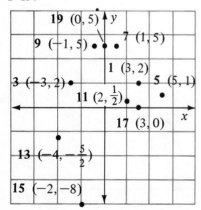

21.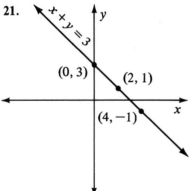

23.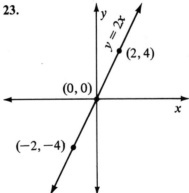

25.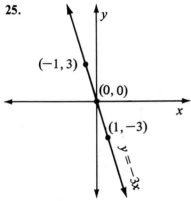

27.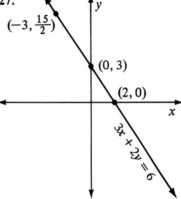

29.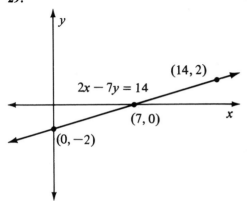

31.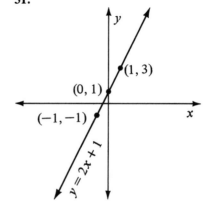

33.

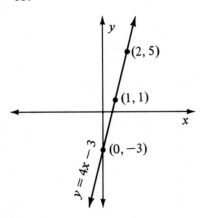

35.

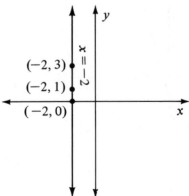

37.

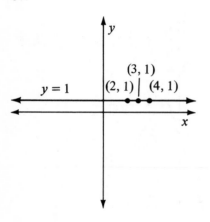

39.

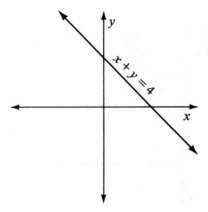

41.

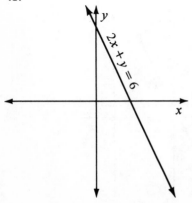

43.

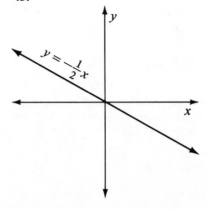

45.

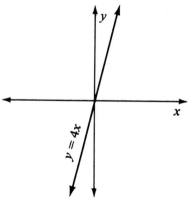

47.

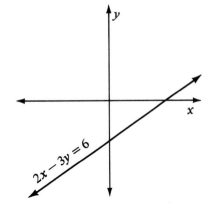

49.

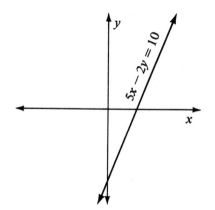

51.

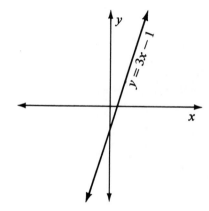

53.

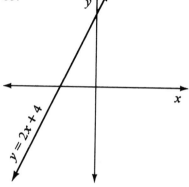

55.

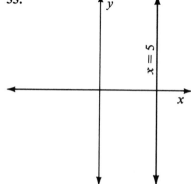

57.

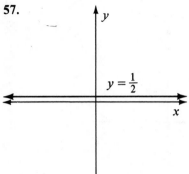

59.

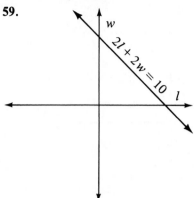

61.

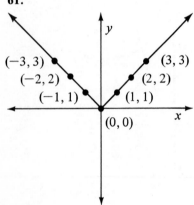

63.

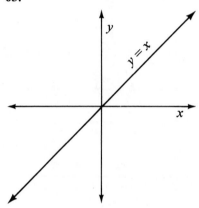

65.

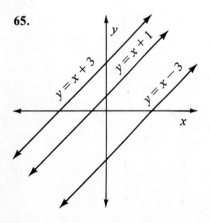

67. 5   **69.** −2   **71.** 3

PROBLEM SET 3.3, P. 106
1. $(2, 1)$ **3.** $(-1, 2)$ **5.** $(3, 7)$ **7.** $(4, 3)$ **9.** $(0, -6)$ **11.** $(1, 0)$ **13.** $(0, 0)$
15. $(-5, -6)$ **17.** $(-1, -1)$ **19.** $(-3, 2)$ **21.** $(2, 1)$ **23.** $(-3, 5)$ **25.** $(-4, 6)$
27. Lines are parallel; there is no solution to the system; $\emptyset$ **29.** Lines coincide; any solution to one equation is a solution to the other   **31.** -15 **33.** 27 **35.** -6 **37.** -27

PROBLEM SET 3.4, P. 112
1. $(2, 1)$ **3.** $(3, 7)$ **5.** $(2, -5)$ **7.** $(-1, 0)$ **9.** Lines coincide.   **11.** $(4, 8)$
13. $(2, 3)$ **15.** $(1, 0)$ **17.** $(-1, -2)$ **19.** $(5, -1)$ **21.** $(-4, 5)$ **23.** $(-3, -10)$
25. $(3, 2)$ **27.** $(6, 5)$ **29.** $(4, -3)$ **31.** $(2, 2)$ **33.** Lines are parallel.   **35.** $(1, 1)$
37. Lines are parallel.   **39.** $(7, 5)$ **41.** $(10, 12)$ **43.** $x < 5$ **45.** $x \leq -4$
47. $x \geq 3$ **49.** $x > -4$

PROBLEM SET 3.5, P. 117
1. $(4, 7)$ **3.** $(3, 17)$ **5.** $(-1, -2)$ **7.** $(2, 4)$ **9.** $(0, 4)$ **11.** $(-1, 3)$ **13.** $(1, 1)$
15. $(2, -3)$ **17.** $(8, -2)$ **19.** $(-3, 5)$ **21.** Lines are parallel.   **23.** $(3, 1)$
25. $(3, -3)$ **27.** $(2, 6)$ **29.** $(4, 4)$ **31.** $(5, -2)$ **33.** $(18, 10)$ **35.** Lines coincide.
37. $x + 4 = 12; 8$ **39.** $6x + 2 = 20; 3$

PROBLEM SET 3.6, P. 123
As you can see, in addition to the answers to the problems we have sometimes included the system of equations used to solve the problems. Remember, you should attempt the problem on your own before looking here to check your answers or equations.

1. $x + y = 25$ The two numbers   **3.** 3, 12
      $y = x + 5$ are 10 and 15.
5. $x - y = 5$ The two numbers   **7.** 6, 29   **9.** 8 quarters, 6 nickels
       $x = 2y + 1$ are 9 and 4.
11. Let $x =$ the number of dimes and $y =$ the number of quarters.

$$x + y = 21 \qquad \text{He has 12 dimes}$$
$$0.10x + 0.25y = 3.45 \qquad \text{and 9 quarters.}$$

13. Let $x =$ the amount invested at 6% and $y =$ the amount invested at 8%.

$$x + y = 20,000 \qquad \text{He has \$9,000 at 8\%}$$
$$0.06x + 0.08y = 1380 \qquad \text{and \$11,000 at 6\%.}$$

15. \$8000 at 5%, \$2000 at 6%   **17.** 4 pounds at 85¢ per pound, 6 pounds at 90¢
19. Let $x =$ the number of adult tickets and $y =$ the number of children's tickets.

$$x + y = 275 \qquad \text{150 adults,}$$
$$1.5x + 1.0y = 350 \qquad \text{125 children}$$

21. $x < -4$ **23.** $x \leq -3$ **25.** $x \leq -1$ **27.** $x > 1$

PROBLEM SET 3.7, P. 129

1.

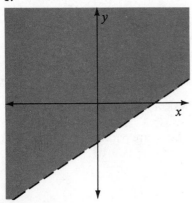

3.

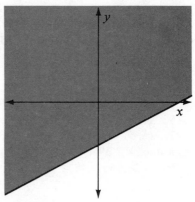

5.

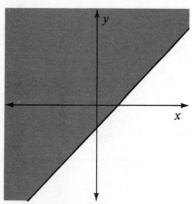

7.

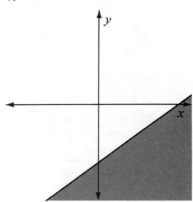

9.

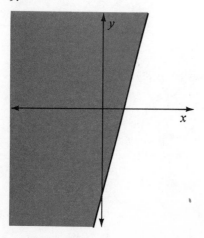

11.

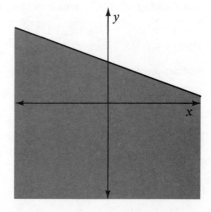

13.

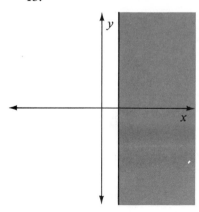

15.

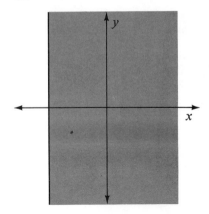

17.

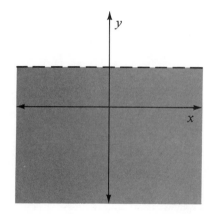

19.

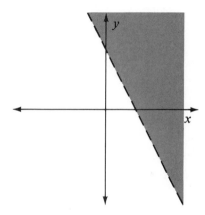

21.

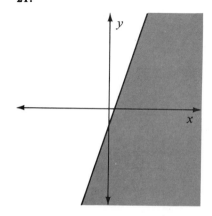

23.

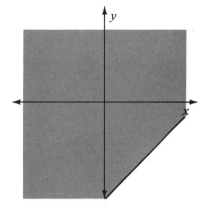

25.

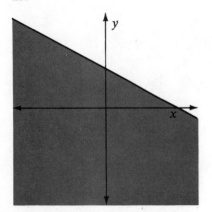

27.

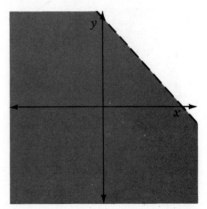

29.

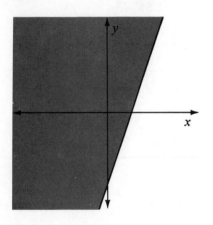

31.

33.

35.

CHAPTER 3 TEST,   P. 132

1.

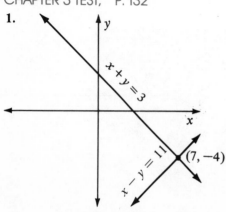

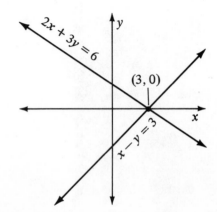

3.

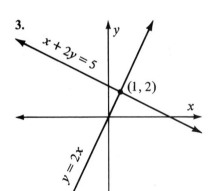

4.

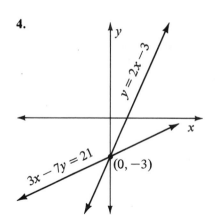

5.

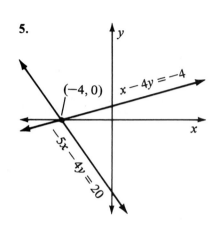

6.

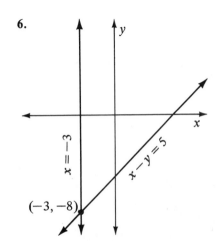

7. (5, 3)   **8.** (2, 7)   **9.** Lines are parallel.   **10.** (4, 3)   **11.** (2, 9)   **12.** (5, 11)

13. (−3, −4)   **14.** (−5, −1)   **15.** (5, −3)   **16.** Lines are parallel.   **17.** Lines coincide.

18. (2, 2)   **19.** (2, −2)   **20.** (−3, −2)   **21.** 5, 7   **22.** 3, 12   **23.** $6000

24. 7 nickels, 5 quarters

25.

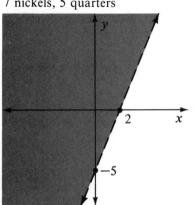

CHAPTER 4

PROBLEM SET 4.1,   P. 139

1. Base 4; exponent 2; 16   **3.** Base 7; exponent 2; 49   **5.** Base 4; exponent 3; 64   **7.** Base −5; exponent 2; 25   **9.** Base 2; exponent 3; −8   **11.** Base 3; exponent 4; 81   **13.** Base $\frac{2}{3}$, exponent 2; $\frac{4}{9}$ **15.** Base $\frac{1}{2}$, exponent 4; $\frac{1}{16}$ **17.** x^9 **19.** 7^7 **21.** y^{30} **23.** 2^{12} **25.** x^{28} **27.** x^{10} **29.** 5^{12} **31.** y^9 **33.** 2^{50} **35.** a^{3x} **37.** b^{xy} **39.** $16x^2$ **41.** $32y^5$ **43.** $81x^4$ **45.** $49a^2b^2$ **47.** $64x^3y^3z^3$ **49.** $8x^{12}$ **51.** $16a^6$ **53.** x^{14} **55.** a^{11} **57.** $128x^7$ **59.** $432x^{10}$ **61.** $16x^4y^6$ **63.** $8a^{12}b^3$ **65.** $30x^9$ **67.** $256x^8y^5$ **69.** 100,000; the exponent and the number of zeros are the same   **71.** $(2^3)^2 = 2^6 = 64$ $2^{32} = 2^9 = 512$ **73.** b **75.** −3   **77.** 11   **79.** −5   **81.** 5

PROBLEM SET 4.2,   P. 146

1. $\frac{1}{9}$ **3.** $\frac{1}{36}$ **5.** $\frac{1}{64}$ **7.** $\frac{1}{125}$ **9.** $\dfrac{2}{x^3}$ **11.** $\dfrac{1}{8x^3}$ **13.** $\dfrac{1}{25y^2}$ **15.** $\frac{1}{100}$ **17.** 25

19. $\frac{1}{25}$ **21.** x^6 **23.** 64   **25.** $8x^3$ **27.** 6^{10} **29.** $\dfrac{1}{6^{10}}$ **31.** $\dfrac{1}{2^8}$ **33.** 2^8

35. $(ab)^3$ **37.** $\dfrac{16}{x^{12}}$ **39.** 1   **41.** $2a^2b$ **43.** $\dfrac{1}{49y^6}$ **45.** $\dfrac{1}{x^8}$ **47.** $\dfrac{1}{y^3}$ **49.** x^2

51. a^6 **53.** $\dfrac{1}{y^9}$ **55.** $\dfrac{1}{x^3}$ **57.** $\dfrac{1}{x}$ **59.** x^{18} **61.** a^{16} **63.** $\dfrac{1}{a^4}$ **65.** x

67. 3   **69.** 9   **71.** 49   **73.** 25   **75.** $7x$ **77.** $2a$ **79.** $10y$

PROBLEM SET 4.3,   P. 152

1. $12x^7$ **3.** $-16y^{11}$ **5.** $32x^2$ **7.** $200a^6$ **9.** $-24a^3b^3$ **11.** $-6x^{11}y^{10}$ **13.** $24x^6y^8$

15. $3x$ **17.** $\dfrac{6}{y^3}$ **19.** $\dfrac{1}{2a}$ **21.** $-\dfrac{3a}{b^2}$ **23.** $\dfrac{9x^2}{z^2}$ **25.** $-\dfrac{12}{xy^6}$ **27.** $\dfrac{6x^6}{y^2}$

29. $8x^2$ **31.** $-11x^5$ **33.** $46a$ **35.** 0   **37.** $-x^2$ **39.** $4x^3$ **41.** $31ab^2$

43. $-14abc$ **45.** $\frac{7}{5}$ **47.** $\frac{3}{5}$ **49.** −1   **51.** $\frac{2}{3}$ **53.** $\frac{4}{9}$ **55.** $4x^3$ **57.** $\dfrac{1}{b^2}$

59. $\dfrac{6y^{10}}{x^4}$ **61.** $9x^3$ **63.** $-20a^2$ **65.** $6x^5y^2$ **67.** 2   **69.** 4

71. $(4 + 5)^2 = 9^2 = 81$ $4^2 + 5^2 = 16 + 25 = 41$ **73.** −8   **75.** 1   **77.** −13   **79.** −15

PROBLEM SET 4.4,   P. 157

1. Trinomial, 3   **3.** Trinomial, 3   **5.** Binomial, 1   **7.** Binomial, 2   **9.** Monomial, 2   **11.** Monomial, 0   **13.** $5x^2 + 5x + 9$ **15.** $5a^2 - 9a + 7$ **17.** $x^2 + 6x + 8$

19. $6x^2 - 13x + 5$ **21.** $x^2 - 9$ **23.** $3y^2 - 11y + 10$ **25.** $6x^3 + 5x^2 - 4x + 3$
27. $8x^2 - 2$ **29.** $2a^2 - 2a - 2$ **31.** $-x^2 - 26x + 4$ **33.** $-4y^2 + 15y - 22$
35. $x^2 - 33x + .63$ **37.** $8y^2 + 4y + 26$ **39.** $75x^2 - 150x - 75$ **41.** $12x + 2$
43. 4 **45.** 25 **47.** 16 **49.** $10x^2$ **51.** $-15x^2$ **53.** $6x^3$ **55.** $6x^4$

PROBLEM SET 4.5, P. 161

1. $6x^2 + 2x$ **3.** $6x^4 - 4x^3 + 2x^2$ **5.** $2a^3b - 2a^2b^2 + 2ab$ **7.** $3y^4 + 9y^3 + 12y^2$
9. $8x^5y^2 + 12x^4y^3 + 32x^2y^4$ **11.** $x^2 + 7x + 12$ **13.** $x^2 + 7x + 6$ **15.** $x^2 - 2x - 3$
17. $a^2 + 2a - 15$ **19.** $x^2 + 8x + 16$ **21.** $x^2 - 36$ **23.** $y^2 - 4$ **25.** $2x^2 - 11x + 12$
27. $2a^2 + 3a - 2$ **29.** $6x^2 - 19x + 10$ **31.** $9x^2 - 12x + 4$ **33.** $25x^2 - 16$
35. $8x^2 - 13x - 6$ **37.** $3 - 10a + 8a^2$ **39.** $56 - 83x + 30x^2$ **41.** $x^3 + 4x^2 - x - 4$
43. $a^3 - 6a^2 + 11a - 6$ **45.** $x^3 + 8$ **47.** $2x^3 + 17x^2 + 26x + 9$
49. $5x^4 - 13x^3 + 20x^2 + 7x + 5$ **51.** $2x^2 + 10x - 28$ **53.** $3x^3 + 3x^2 - 18x$
55. $x^3 + 6x^2 + 11x + 6$
57. $(x + 4)(x + 2) = x^2 + 2x + 4x + 8$
$\qquad\qquad\qquad\quad = x^2 + 6x + 8$

59.

	x	10
x	x^2	$10x$
5	$5x$	50

61. $(6, 2)(-4, 12)(0, 8)$ **63.** $(0, 4)(6, 0)(3, 2)$

PROBLEM SET 4.6, P. 165

1. $x^2 - 4x + 4$ **3.** $a^2 + 6a + 9$ **5.** $x^2 - 10x + 25$ **7.** $a^2 - 14a + 49$
9. $x^2 + 20x + 100$ **11.** $a^2 + 2ab + b^2$ **13.** $4x^2 - 4x + 1$ **15.** $16a^2 + 40a + 25$
17. $9x^2 - 12x + 4$ **19.** $9a^2 + 30ab + 25b^2$ **21.** $16x^2 - 40xy + 25y^2$
23. $49m^2 + 28mn + 4n^2$ **25.** $36x^2 - 120xy + 100y^2$ **27.** $x^4 + 10x^2 + 25$
29. $a^4 + 2a^2 + 1$ **31.** $x^6 - 14x^3 + 49$ **33.** $x^2 - 9$ **35.** $a^2 - 25$ **37.** $y^2 - 1$
39. $81 - x^2$ **41.** $4x^2 - 25$ **43.** $16x^2 - 1$ **45.** $4a^2 - 49$ **47.** $36 - 49x^2$
49. $x^4 - 9$ **51.** $a^4 - 16$ **53.** $(50 - 1)(50 + 1) = 2500 - 1 = 2499$ **55.** Both equal 25.
59. $2x^2$ **61.** $5x$ **63.** $\dfrac{a^4}{2b^2}$

PROBLEM SET 4.7, P. 168

1. $x - 2$ **3.** $3 - 2x^2$ **5.** $5xy - 2y$ **7.** $7x^4 - 6x^3 + 5x^2$ **9.** $10x^4 - 5x^2 + 1$
11. $-4a + 2$ **13.** $-8a^4 - 12a^3$ **15.** $-4b - 5a$ **17.** $-6a^2b + 3ab^2 - 7b^3$
19. $-\dfrac{1}{2}a - b - \dfrac{b^2}{2a}$ **21.** $3x + 4y$ **23.** $-y + 3$ **25.** $5y - 4$ **27.** $xy - x^2y^2$
29. $-1 + xy$ **31.** $-a + 1$ **33.** $x^2 - 3xy + y^2$ **35.** $2 - 3b + 5b^2$ **37.** $-2xy + 1$
39. $xy - \frac{1}{2}$ **41.** $\dfrac{1}{4x} - \dfrac{1}{2a} + \dfrac{3}{4}$ **43.** $\dfrac{4x^2}{3} + \dfrac{2}{3x} + \dfrac{1}{x^2}$ **45.** Both equal 7.
47. $\dfrac{3(10) + 8}{2} = \dfrac{38}{2} = 19$ **49.** $(7, -1)$ **51.** $(2, 3)$

$\qquad 3(10) + 4 = 34$

CHAPTER 4 TEST,   P. 171

1. 5^5 **2.** 10^7 **3.** $\frac{1}{27}$ **4.** $\frac{1}{16}$ **5.** $\frac{1}{25}$ **6.** $\frac{1}{32}$ **7.** $\frac{9}{16}$ **8.** $\frac{25}{36}$ **9.** 10

10. 1 **11.** $27x^6$ **12.** $8x^6y^9$ **13.** $\frac{1}{x^2}$ **14.** $\frac{1}{x^3}$ **15.** a^4 **16.** $\frac{3}{a^7}$ **17.** $2x^4$

18. $\frac{8x^2}{y^6}$ **19.** $7x^2 - 6x - 3$ **20.** $8x^2 - 3x + 3$ **21.** $x^3 - 3x - 8$

22. $2a^3 - a^2 + a + 2$ **23.** $6x^2 - 7x - 5$ **24.** $21x^2 + 11x - 2$ **25.** $a^3 - 7a - 6$
26. $2a^3 - a^2 - 32a + 16$ **27.** $6x^3 - 19x^2 + 19x - 6$ **28.** $12x^3 - 7x^2 + 4x - 1$
29. $a^3 - 1$ **30.** $a^3 - 8$ **31.** $4x^2 - 20x + 25$ **32.** $9x^2 - 24x + 16$
33. $4a^2 + 12a + 9$ **34.** $9a^2 - 12a + 4$ **35.** $2x^2 - 4x + 1$ **36.** $-3x^3 + 2x^2 - 5x$

37. $2x^2 + \frac{3}{2} - \frac{1}{x^2}$ **38.** $-2x^2 + 1 - \frac{3}{2x^2}$

CHAPTER 5

PROBLEM SET 5.1,   P. 176

1. Not prime   **3.** Prime   **5.** Not prime   **7.** Not prime   **9.** Prime
11. Not prime   **13.** Prime   **15.** Not prime   **17.** Not prime   **19.** Not prime
21. $5 \cdot 7$ **23.** 2^7 **25.** $2^4 \cdot 3^2$ **27.** $2 \cdot 19$ **29.** $3 \cdot 5 \cdot 7$ **31.** $2^2 \cdot 3^2 \cdot 5$
33. $5 \cdot 7 \cdot 11$ **35.** 11^2 **37.** $2^2 \cdot 3 \cdot 5 \cdot 7$ **39.** $2^2 \cdot 5 \cdot 31$ **41.** $2^2 \cdot 3^2$ **43.** $2^8 \cdot 3^8$
45. $3x - 15$ **47.** $5x^3 - 15x^2$ **49.** $8x^3y^3 - 12x^2y^4$ **51.** $6x^4 - 12x^3 + 6x^2$

PROBLEM SET 5.2,   P. 178

1. $5(3x + 5)$ **3.** $3(2a + 3)$ **5.** $4(x - 2y)$ **7.** $3(x^2 - 2x - 3)$ **9.** $3(a^2 - a - 20)$
11. $4(6y^2 - 13y + 6)$ **13.** $x^2(9 - 8x)$ **15.** $13a^2(1 - 2a)$ **17.** $7xy(3x - 4y)$
19. $11ab^2(2a - 1)$ **21.** $7x(x^2 + 3x - 4)$ **23.** $11(11y^4 - x^4)$ **25.** $25x^2(4x^2 - 2x + 1)$
27. $8(a^2 + 2b^2 + 4c^2)$ **29.** $-x^3(31x^2 + 28x + 22)$ **31.** $4ab(a - 4b + 8ab)$
33. $11a^2b^2(11a - 2b + 3ab)$ **35.** $9rst(3s + rs - 2t)$ **37.** $12x^2y^3(1 - 6x^3 - 3x^2y)$
39. 6 **41.** $3(4x^2 + 2x + 1)$ **43.** $x^2 + 7x + 12$ **45.** $x^2 + 5x - 14$
47. $x^2 - 5x - 14$ **49.** $x^2 - x - 6$

PROBLEM SET 5.3,   P. 182

1. $(x + 3)(x + 4)$ **3.** $(x + 1)(x + 2)$ **5.** $(a + 3)(a + 7)$ **7.** $(x - 2)(x - 5)$
9. $(y - 3)(y - 7)$ **11.** $(x - 4)(x + 3)$ **13.** $(y + 4)(y - 3)$ **15.** $(x + 7)(x - 2)$
17. $(r - 9)(r + 1)$ **19.** $(x - 6)(x + 5)$ **21.** $(a + 7)(a + 8)$ **23.** $(y - 7)(y + 6)$
25. $(x + 7)(x + 6)$ **27.** $2(x + 1)(x + 2)$ **29.** $3(a - 5)(a + 4)$ **31.** $4x(x - 2)(x + 6)$
33. $a(a - 2)(a - 3)$ **35.** $x^2(x - 4)(x + 3)$ **37.** $2r(r - 3)(r + 5)$
39. $2y^2(y - 4)(y + 1)$ **41.** $x^3(x + 2)(x + 2)$ **43.** $3y^2(y - 5)(y + 1)$
45. $4x^2(x - 4)(x - 9)$ **47.** $(x + 2y)(x + 3y)$ **49.** $(x - 4y)(x - 5y)$
51. $(a + 4b)(a - 2b)$ **53.** $(a - 5b)(a - 5b)$ **55.** $(a + 5b)(a + 5b)$ **57.** $(x - 6a)(x + 8a)$

59. $(x - 9b)(x + 4b)$ **61.** $(x + 16)$ **63.** $4x^2 - x - 3$ **65.** $6a^2 + 13a + 2$
67. $6a^2 + 7a + 2$ **69.** $6a^2 + 8a + 2$

PROBLEM SET 5.4, P. 186

1. $(2x + 1)(x + 3)$ **3.** $(2a - 3)(a + 1)$ **5.** $(3x + 5)(x - 1)$ **7.** $(3y + 1)(y - 5)$
9. $(2x + 3)(3x + 2)$ **11.** $(2x - 3y)(2x - 3y)$ **13.** $(4y + 1)(y - 3)$ **15.** $(4x - 5)(5x - 4)$
17. $(10a - b)(2a + 5b)$ **19.** $(4x - 5)(5x + 1)$ **21.** $(2m + 3)(6m - 1)$
23. $(4x + 5)(5x + 3)$ **25.** $(3a - 4b)(4a - 3b)$ **27.** $(3x - 7y)(x + 2y)$
29. $(2x + 5)(7x - 3)$ **31.** $2(2x + 3)(x - 1)$ **33.** $2(4a - 3)(3a - 4)$ **35.** $x(5x - 4)(2x - 3)$
37. $x^2(2x - 5)(3x + 2)$ **39.** $2a(5a + 2)(a - 1)$ **41.** $3x(5x + 1)(x - 7)$
43. $5y(7y + 2)(y - 2)$ **45.** $a^2(3a - 1)(5a + 1)$ **47.** $3y(4x + 5)(2x - 3)$
49. $2y(3x - 7y)(2x - y)$ **51.** Both equal 25. **53.** $4x^2 - 9$ **55.** $x^2 - 9$ **57.** $36x^2 - 1$
59. $x^2 + 8x + 16$ **61.** $4x^2 + 12x + 9$

PROBLEM SET 5.5, P. 189

1. $(x - 3)(x + 3)$ **3.** $(a - 6)(a + 6)$ **5.** $(x - 7)(x + 7)$ **7.** $4(a - 2)(a + 2)$
9. No factors **11.** $(5x - 13)(5x + 13)$ **13.** $(3a - 4b)(3a + 4b)$ **15.** $(3 - m)(3 + m)$
17. $(5 - 2x)(5 + 2x)$ **19.** $2(x - 3)(x + 3)$ **21.** $32(a - 2)(a + 2)$
23. $2y(2x - 3)(2x + 3)$ **25.** $(a^2 + b^2)(a - b)(a + b)$ **27.** $(4m^2 + 9)(2m - 3)(2m + 3)$
29. $3xy(x - 5y)(x + 5y)$ **31.** $(x - 1)^2$ **33.** $(x + 1)^2$ **35.** $(a - 5)^2$ **37.** $(y + 2)^2$
39. $(x - 2)^2$ **41.** $(m - 6)^2$ **43.** $(2a + 3)^2$ **45.** $(7x - 1)^2$ **47.** $(3y - 5)^2$
49. $(x + 5y)^2$ **51.** $(3a + b)^2$ **53.** $3(a + 3)^2$ **55.** $2(x + 5y)^2$ **57.** $5x(x + 3y)^2$
59. 14 **61.** 25 **63.** 3 **65.** -2 **67.** 5 **69.** $-\frac{3}{2}$

PROBLEM SET 5.6, P. 193

1. $-2, 1$ **3.** $4, 5$ **5.** $0, -1, 3$ **7.** $-\frac{2}{3}, -\frac{3}{2}$ **9.** $0, -\frac{4}{3}, \frac{4}{3}$ **11.** $0, -\frac{1}{3}, -\frac{3}{5}$
13. $-1, -2$ **15.** $4, 5$ **17.** $6, -4$ **19.** $9, 1$ **21.** -3 **23.** $4, -4$ **25.** $-4, \frac{3}{2}$
27. $-\frac{2}{3}$ **29.** $-\frac{5}{2}, 4$ **31.** $\frac{5}{3}, -4$ **33.** $\frac{7}{2}, -\frac{7}{2}$ **35.** $0, 3$ **37.** $\frac{5}{3}, -\frac{5}{3}$ **39.** $\frac{1}{2}, -\frac{4}{3}$
41. $0, 2$ **43.** $3, -\frac{5}{2}$ **45.** $0, \frac{1}{5}, -7$ **47.** $\frac{1}{2}, -\frac{5}{2}$ **49.** $0, \frac{3}{5}, -\frac{3}{2}$ **51.** $x + 5x = 90$;
cost of suit $= \$15$, cost of bicycle $= \$75$ **53.** $x + 4x = 3000$; cost of lot $= \$600$,
cost of house $= \$2400$

PROBLEM SET 5.7, P. 198

1. Two consecutive even integers are x and $x + 2$; $x(x + 2) = 80$; 8, 10 and $-10, -8$ **3.** 9, 11
and $-11, -9$ **5.** $x(x + 2) = 5(x + x + 2) - 10$; 8, 10 and 0, 2 **7.** $-8, -6$ **9.** The numbers
are x and $5x + 2$; $x(5x + 2) = 24$; 2, 12 and $-\dfrac{12}{5}, -10$ **11.** 20, 5 and 0, 0 **13.** Let $x = $ the
width; $x(x + 1) = 12$; width 3 inches, length 4 inches **15.** 5 **17.** Let $x = $ the base;
$\frac{1}{2}(x)(2x) = 9$; base 3 inches, height 6 inches **19.** 2, 7 **21.** $(x + 4)^2 = (x + 2)^2 + x^2$; 6, 8, 10
23. 12 meters **25.** $(2, 3)$ **27.** $(1, 4)$ **29.** $(5, 7)$

CHAPTER 5 TEST,   P. 201

1. $5 \cdot 7$ **2.** $7 \cdot 11$ **3.** Prime   **4.** $2^2 \cdot 3 \cdot 7^2$ **5.** 2^7 **6.** $2^2 \cdot 3^4$ **7.** $(x - 2)(x - 3)$
8. $(x - 3)(x + 2)$ **9.** $(a - 4)(a + 4)$ **10.** $x^2 + 25$ **11.** $(x^2 + 9)(x - 3)(x + 3)$
12. $3(3x - 5y)(3x + 5y)$ **13.** $2(2a + 1)(a + 5)$ **14.** $3(m - 3)(m + 2)$
15. $(2y - 1)(3y + 5)$ **16.** $2x(2x + 1)(3x - 5)$ **17.** $-3, -4$ **18.** 2 **19.** $-6, 6$
20. $5, -4$ **21.** $5, 6$ **22.** $0, 4, -4$ **23.** $3, -\frac{5}{2}$ **24.** $0, 1, -\frac{1}{3}$ **25.** $4, 16$
26. $3, 5$ and $-3, -1$ **27.** $3, 14$ feet

CHAPTER 6

PROBLEM SET 6.1,   P. 207

1. $\dfrac{1}{x - 2}, x \neq 2$ **3.** $\dfrac{-1}{x + 2}, x \neq -2$ **5.** $\dfrac{1}{a + 3}, a \neq -3$ **7.** $\dfrac{1}{x - 5}, x \neq 5$

9. $\dfrac{1}{5a}, a \neq 0$ **11.** $\dfrac{x^2 - 4}{2}$ **13.** $\dfrac{5}{m^2}, m \neq 0$ **15.** $\dfrac{2x - 10}{3x - 6}, x \neq 2$ **17.** 2

19. $\dfrac{5(x - 1)}{4}$ **21.** $\dfrac{1}{x - 3}, x \neq 3$ **23.** $\dfrac{1}{y - 3}, y \neq 3$ **25.** $\dfrac{1}{x + 3}, x \neq -3$

27. $\dfrac{1}{a - 5}, a \neq 5$ **29.** $\dfrac{1}{3x + 2}, x \neq -\dfrac{2}{3}$ **31.** $\dfrac{x + 5}{x + 2}, x \neq -2$ **33.** $\dfrac{m + 2}{m - 2}, m \neq 2$

35. $\dfrac{3x}{2}$ **37.** $\dfrac{x - 1}{x - 4}, x \neq 4$ **39.** $\dfrac{a - 2}{3a + 2}, a \neq -\dfrac{2}{3}$ **41.** $\dfrac{2x - 3}{2x + 3}, x \neq -\dfrac{3}{2}$

43. $\dfrac{1}{(x - 3)(x^2 + 9)}, x \neq 3$ **45.** $\dfrac{1}{(x + 2)(x^2 + 4)}, x \neq -2$ **47.** $5 + 4 = 9$ **49.** $2x - 4$

51. $-3x^2 + 2x - 3$ **53.** $-2x$ **55.** $10x^2$

PROBLEM SET 6.2,   P. 213

1. $x - 2$ **3.** $x + 1$ **5.** $a + 4$ **7.** $x - 3$ **9.** $a - 6$ **11.** $x + 3$ **13.** $a - 5$

15. $x + 7$ **17.** $x - 8$ **19.** $x + 2 + \dfrac{2}{x + 3}$ **21.** $a - 2 + \dfrac{12}{a + 5}$ **23.** $x + 4 + \dfrac{9}{x - 2}$

25. $x + 4 + \dfrac{-10}{x + 1}$ **27.** $a + 1 + \dfrac{-1}{a + 2}$ **29.** $x - 3 + \dfrac{17}{2x + 4}$ **31.** $3a - 2 + \dfrac{7}{2a + 3}$

33. $2a^2 - a - 3$ **35.** $x^2 - x + 5$ **37.** $x^2 + x + 1$ **39.** $x^2 + 2x + 4$ **41.** $\frac{15}{16}$
43. $\frac{1}{6}$ **45.** 2 **47.** $\frac{8}{9}$

PROBLEM SET 6.3,   P. 218

1. 2 **3.** 1 **5.** $\frac{3}{2}$ **7.** $\dfrac{1}{2(x - 3)}$ **9.** $\dfrac{4(a + 5)}{7(a + 4)}$ **11.** $\dfrac{y - 2}{4}$ **13.** $\dfrac{2(x + 4)}{x - 2}$

15. $\dfrac{x + 3}{(x - 3)(x + 1)}$ **17.** 1 **19.** $\dfrac{y - 5}{(y + 2)(y - 2)}$ **21.** $\dfrac{x + 5}{x - 5}$ **23.** 1 **25.** $\dfrac{a + 3}{a + 4}$

27. $\dfrac{2y - 3}{y - 6}$ **29.** 1 **31.** 15 **33.** $2(x - 3)$ **35.** $2a$ **37.** $(x + 2)(x + 1)$

39. $-2x(x - 5)$ **41.** $\frac{9}{2}$ **43.** 1 **45.** $\frac{2}{3}$ **47.** $\frac{4}{5}$

PROBLEM SET 6.4, P. 225

1. $\dfrac{7}{x}$ **3.** $\dfrac{4}{a}$ **5.** 1 **7.** $y + 1$ **9.** $x + 2$ **11.** $\dfrac{15 + x}{3x}$ **13.** $\dfrac{18 - 5a}{6a}$

15. $\dfrac{(y + 2)(y - 2)}{2y}$ **17.** $\dfrac{3 + 2a}{6}$ **19.** $\dfrac{7x + 3}{4(x + 1)}$ **21.** $\dfrac{17}{5(x - 2)}$ **23.** $\dfrac{4 - 3x}{3(x + 4)}$

25. $\dfrac{3}{x - 2}$ **27.** $\dfrac{4}{a + 3}$ **29.** $\dfrac{2x - 9}{(x + 5)(x - 5)}$ **31.** $\dfrac{1}{x + 3}$ **33.** $\dfrac{3a - 4}{(a - 3)(a + 2)}$

35. $\dfrac{x - 2}{(x - 3)(x + 3)}$ **37.** $\dfrac{7a + 5}{a(a + 1)}$ **39.** $\dfrac{1}{(x + 4)(x + 3)}$ **41.** $\dfrac{y}{(y + 4)(y + 5)}$

43. $-3, -2$ **45.** $-2, 3$ **47.** $-5, 4$ **49.** 4, 5

PROBLEM SET 6.5, P. 231

1. -3 **3.** 20 **5.** -1 **7.** 5 **9.** -2 **11.** 4 **13.** 3, 5 **15.** $-8, 1$
17. 2 **19.** 1 **21.** 8 **23.** 5 **25.** Possible solution 2, which does not check; Ø
27. 3 **29.** Possible solution -3, which does not check; Ø **31.** 0
33. Possible solutions 2 and 3, but only 2 checks; 2 **35.** -4 **37.** -1 **39.** $-6, -7$
41. Possible solutions -3 and -1, but only -1 checks; -1

43.

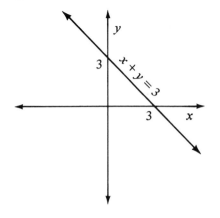

45.

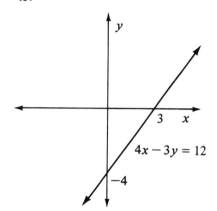

47.

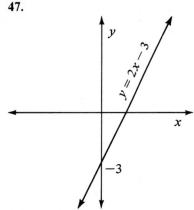

49.

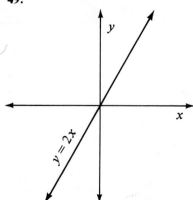

PROBLEM SET 6.6,   P. 235

1. $\dfrac{1}{x} + \dfrac{1}{3x} = \dfrac{16}{3}$; $\frac{1}{4}$ and $\frac{3}{4}$ **3.** 5   **5.** $x + \dfrac{10}{x} = 7$; 2 or 5   **7.** $\frac{5}{3}$

9. $\dfrac{1}{x} + \dfrac{1}{x+1} = \dfrac{7}{12}$; 3 and 4   **11.** 20 mph

13. Let x = plane speed in still air; $\dfrac{600}{x+50} = \dfrac{500}{x-50}$; 550 mph   **15.** 60 mph, 55 mph

17. $\dfrac{1}{12} - \dfrac{1}{15} = \dfrac{1}{x}$; 60 hours   **19.** $\frac{60}{11}$ minutes   **21.** $(2, -1)$ **23.** $(4, 3)$ **25.** $(1, 1)$

27. $(5, 2)$

PROBLEM SET 6.7,   P. 239

1. 6   **3.** $\frac{1}{6}$ **5.** xy^2 **7.** $\dfrac{xy}{2}$ **9.** $\dfrac{y}{x}$ **11.** $\dfrac{a+1}{a-1}$ **13.** $\dfrac{x+1}{2(x-3)}$ **15.** $a - 3$

17. $\dfrac{y+3}{y+2}$ **19.** $x + y$ **21.** $\dfrac{a+1}{a}$ **23.** $\dfrac{1}{x}$ **25.** $\dfrac{2a+3}{3a+4}$ **27.** $x < 1$

29. $x \geq -7$ **31.** $x < 6$ **33.** $x \leq 2$

CHAPTER 6 TEST,   P. 244

1. $\frac{5}{9}$ **2.** $\frac{2}{5}$ **3.** $\frac{5}{6}$ **4.** $\frac{2}{3}$ **5.** $\frac{2}{3}$ **6.** $\frac{7}{8}$ **7.** 3   **8.** $(x-7)(x-1)$

9. $\dfrac{x+4}{3x-4}$ **10.** $\dfrac{-3}{x-2}$ **11.** $\dfrac{2x+3}{(x+3)(x-3)}$ **12.** $\dfrac{-x(x+4)}{(x+1)(x-1)(x-2)}$

13. $x + 3 + \dfrac{-8}{x+1}$ **14.** $x^2 + 3x + 7 + \dfrac{24}{x-3}$ **15.** $\frac{11}{5}$ **16.** 1   **17.** 6   **18.** $\frac{4}{3}$ or $\frac{3}{4}$

19. 60 hours **20.** $\dfrac{x+4}{x+2}$

CHAPTER 7

PROBLEM SET 7.1, P. 248
1. 3 **3.** -3 **5.** Not a real number **7.** -12 **9.** 25 **11.** Not a real number
13. -8 **15.** -10 **17.** 35 **19.** 1 **21.** -2 **23.** -5 **25.** -1 **27.** -3
29. -2 **31.** $\frac{1}{3}$ **33.** $\frac{2}{3}$ **35.** $\frac{4}{5}$ **37.** $\frac{11}{7}$ **39.** x **41.** $3x$ **43.** xy
45. $a+b$ **47.** $7xy$ **49.** x **51.** $2x$ **53.** $3+4=7$ **55.** $\sqrt{25}=5$
57. $12+5=17$ **59.** $\sqrt{169}=13$ **61.** 9 **63.** 25 **65.** 225 **67.** 27 **69.** 4
71. 8 **73.** $x+3$ **75.** Both equal 5. **77.** 5 and 7 **79.** $2^2\cdot 5$ **81.** $3\cdot 5^2$
83. $2^2\cdot 3^2$ **85.** 3^3

PROBLEM SET 7.2, P. 254
1. $2\sqrt{2}$ **3.** $2\sqrt{3}$ **5.** $5x\sqrt{2}$ **7.** $3ab\sqrt{5}$ **9.** $4x^2\sqrt{2}$ **11.** $-4\sqrt{5}$ **13.** $-2\sqrt{7}$

15. $11\sqrt{2}$ **17.** $15\sqrt{5}$ **19.** $-105\sqrt{2}$ **21.** $15\sqrt{2x}$ **23.** $14x\sqrt{3y}$ **25.** $\dfrac{\sqrt{6}}{5}$

27. $\dfrac{\sqrt{3}}{7}$ **29.** $\frac{4}{5}$ **31.** $\frac{2}{3}$ **33.** $2x$ **35.** $3ab$ **37.** $\dfrac{5\sqrt{2}}{3}$ **39.** $\sqrt{3}$ **41.** $\dfrac{8\sqrt{2}}{7}$

43. $\dfrac{12\sqrt{2x}}{5}$ **45.** $\dfrac{3a\sqrt{6}}{5}$ **47.** $\dfrac{15\sqrt{2}}{2}$ **49.** $\dfrac{14y\sqrt{7}}{3}$ **51.** $5ab\sqrt{2}$ **53.** $-6x\sqrt{2y}$

55. $\dfrac{-8xy\sqrt{3y}}{5}$ **57.** x^2-25 **59.** $9x^2-16$ **61.** a^2-b^2 **63.** $49y^2-1$

PROBLEM SET 7.3, P. 260
1. $\dfrac{\sqrt{2}}{2}$ **3.** $\dfrac{\sqrt{3}}{3}$ **5.** $\dfrac{\sqrt{10}}{5}$ **7.** $\dfrac{2\sqrt{15}}{3}$ **9.** $\dfrac{\sqrt{30}}{2}$ **11.** $\dfrac{\sqrt{5}}{10}$ **13.** $2\sqrt{5}$

15. $2\sqrt{3}$ **17.** $\dfrac{-20\sqrt{6}}{3}$ **19.** $xy\sqrt{2}$ **21.** $\dfrac{4a^2\sqrt{5}}{5}$ **23.** $\dfrac{6a^2\sqrt{10a}}{5}$ **25.** $\dfrac{4x\sqrt{5}}{3}$

27. $\dfrac{18ab\sqrt{6}}{5}$ **29.** $-18x$ **31.** $\dfrac{\sqrt[3]{4}}{2}$ **33.** $\dfrac{\sqrt[3]{3}}{3}$ **35.** $\dfrac{\sqrt[3]{12}}{2}$ **37.** $\dfrac{5+\sqrt{3}}{-11}$

39. $2+\sqrt{5}$ **41.** $\dfrac{6-3\sqrt{6}}{2}$ **43.** $\dfrac{15-5\sqrt{2}}{7}$ **45.** $\dfrac{24+8\sqrt{2}}{7}$ **47.** $2\sqrt{5}+2\sqrt{2}$

49. $\dfrac{2\sqrt{2}-2\sqrt{5}}{3}$ **51.** Both equal 0.707. **53.** Both equal 3.464. **55.** $(3+7)x=10x$

57. $(15 + 8)x = 23x$ **59.** $(7 - 3 + 6)a = 10a$ **61.** $(2 - 9 + 50)y = 43y$

1. $7\sqrt{2}$ **3.** $2\sqrt{5}$ **5.** $7\sqrt{3}$ **7.** $5\sqrt{5}$ **9.** $13\sqrt{13}$ **11.** $6\sqrt{10}$ **13.** $6\sqrt{5}$
15. $4\sqrt{2}$ **17.** 0 **19.** $-30\sqrt{3}$ **21.** $-6\sqrt{3}$ **23.** $4\sqrt{3}$ **25.** $16\sqrt{2} + 25\sqrt{3}$
27. 0 **29.** $9\sqrt{2}$ **31.** $-3\sqrt{7}$ **33.** $35\sqrt{3}$ **35.** $-6\sqrt{2} - 6\sqrt{3}$ **37.** $2x\sqrt{x}$
39. $4a\sqrt{3}$ **41.** $15x\sqrt{2x}$ **43.** $13x\sqrt{3xy}$ **45.** $-b\sqrt{5a}$ **47.** $27x\sqrt{2x} - 8x\sqrt{3x}$
49. $23x\sqrt{2y}$ **51.** 3.968 is not equal to the decimal approximation of $\sqrt{8}$, which is 2.828.
53. $9\sqrt{3}$ **55.** $15x + 6y$ **57.** $x^2 - 3x - 10$ **59.** $9x^2 + 6xy + y^2$ **61.** $9x^2 - 16y^2$

PROBLEM SET 7.5,   P. 268

1. $\sqrt{6} - \sqrt{2}$ **3.** $2\sqrt{14} - 5\sqrt{7}$ **5.** $\sqrt{6} + 2$ **7.** $2\sqrt{6} + 3$ **9.** $6 - \sqrt{15}$
11. $2\sqrt{6} + 2\sqrt{15}$ **13.** $30 - 15\sqrt{6}$ **15.** $5 + 2\sqrt{6}$ **17.** $-1 + 2\sqrt{2}$ **19.** -31
21. $3 + 2\sqrt{2}$ **23.** $9 + 6\sqrt{2}$ **25.** $27 - 10\sqrt{2}$ **27.** $16 + 6\sqrt{7}$
29. $\sqrt{6} + \sqrt{2} + 3\sqrt{3} + 3$ **31.** $\sqrt{14} + \sqrt{2} + \sqrt{7} + 2$ **33.** $\sqrt{22} - 6\sqrt{2} + 6\sqrt{11} - 2$
35. $5\sqrt{6} - 7\sqrt{2} - 7\sqrt{3} + 15$ **37.** $\dfrac{\sqrt{15} + \sqrt{6}}{3}$ **39.** $\dfrac{5 - \sqrt{10}}{3}$ **41.** $-2\sqrt{6} - 5$
43. $\dfrac{5 - \sqrt{21}}{2}$ **45.** $5\sqrt{3} - 8$ **47.** $\dfrac{2\sqrt{15} - \sqrt{10} - 2\sqrt{6} + 2}{3}$ **49.** $19 - 7\sqrt{6}$
51. $6\sqrt{5}$ **53.** $14\sqrt{3}$ is missing.   **55.** 42 **57.** 4 **59.** $-6, 1$ **61.** $0, 3$

PROBLEM SET 7.6,   P. 272

1. 3 **3.** 44 **5.** $\varnothing$ **7.** $\varnothing$ **9.** 8 **11.** 4 **13.** $\varnothing$ **15.** $-\frac{2}{3}$ **17.** 25
19. 4 **21.** 7 **23.** 8 **25.** $\varnothing$ **27.** Possible solutions 1 and 6; only 6 checks
29. $-2, -1$ **31.** -4 **33.** Possible solutions 5 and 8; only 8 checks   **35.** $0, 3$
37. x^8 **39.** x^2 **41.** $\frac{1}{8}$

CHAPTER 7 TEST,   P. 275

1. 4 **2.** -6 **3.** $7, -7$ **4.** 3 **5.** -2 **6.** -3 **7.** $5\sqrt{3}$ **8.** $4\sqrt{2}$
9. $\dfrac{\sqrt{6}}{3}$ **10.** $\dfrac{\sqrt[3]{2}}{2}$ **11.** $15x\sqrt{2}$ **12.** $\dfrac{2xy\sqrt{15y}}{5}$ **13.** $2\sqrt{2} + 2$ **14.** $\sqrt{3} + \sqrt{2}$
15. $4\sqrt{3}$ **16.** $\sqrt{15} - 2\sqrt{3}$ **17.** $-1 + 2\sqrt{2}$ **18.** $26 + \sqrt{10}$ **19.** $8 - 2\sqrt{15}$
20. $-5 - 2\sqrt{6}$ **21.** $\dfrac{6 - \sqrt{42} + 2\sqrt{6} - 2\sqrt{7}}{2}$ **22.** -1 **23.** 12 **24.** $\varnothing$

25. Possible solutions 2 and 6; only 6 checks

CHAPTER 8

PROBLEM SET 8.1, P. 280

1. ± 3 **3.** ± 5 **5.** $\pm\sqrt{7}$ **7.** $\pm 2\sqrt{2}$ **9.** $\pm 4\sqrt{2}$ **11.** $\pm 5\sqrt{2}$ **13.** $\pm 4\sqrt{3}$

15. $0, -4$ **17.** $4, -6$ **19.** $5 \pm 5\sqrt{3}$ **21.** $-1 \pm 5\sqrt{2}$ **23.** $-8 \pm 2\sqrt{6}$

25. $2, -3$ **27.** $\frac{11}{4}, -\frac{1}{4}$ **29.** $2, -\frac{4}{3}$ **31.** $\frac{1}{3}, -1$ **33.** $\dfrac{1 \pm 3\sqrt{3}}{3}$ **35.** $\dfrac{-2 \pm 3\sqrt{5}}{3}$

37. $\dfrac{5 \pm 2\sqrt{2}}{2}$ **39.** $\dfrac{-1 \pm 6\sqrt{2}}{11}$ **43.** $(x+3)^2 = 16; -7, 1$ **45.** $x^2 + 6x + 9$

47. $x^2 - 10x + 25$ **49.** $x^2 + 14x + 49$ **51.** $(x-6)^2$ **53.** $(x+2)^2$ **55.** $(x-1)^2$

PROBLEM SET 8.2, P. 284

1. 9 **3.** 1 **5.** 16 **7.** 1 **9.** 64 **11.** $\frac{9}{4}$ **13.** $\frac{49}{4}$ **15.** $\frac{1}{4}$ **17.** $\frac{9}{16}$

19. $-6, 2$ **21.** $8, -2$ **23.** $-3, 1$ **25.** $0, 10$ **27.** $3, -5$ **29.** $-2 \pm \sqrt{7}$

31. $4, -1$ **33.** $8, -1$ **35.** $-1 \pm \sqrt{2}$ **37.** $-2, 1$ **39.** $-1, 3$ **41.** $\dfrac{1 \pm \sqrt{3}}{2}$

43. $\frac{1}{2}$ **45.** $\dfrac{9 \pm \sqrt{105}}{6}$ **47.** $-2 + 2.646 = 0.646; -2 - 2.646 = -4.646$

51. $(-2 + \sqrt{7}) + (-2 - \sqrt{7}) = -4$ **53.** 4 **55.** -24 **57.** $2\sqrt{10}$

PROBLEM SET 8.3, P. 289

1. $-1, -2$ **3.** $-2, -3$ **5.** -3 **7.** $-4, 3$ **9.** $-\frac{3}{2}, -1$ **11.** $\frac{2}{3}, -\frac{1}{2}$ **13.** 1

15. $\dfrac{5 \pm \sqrt{53}}{2}$ **17.** $\dfrac{5 \pm \sqrt{37}}{6}$ **19.** $\dfrac{1 \pm \sqrt{21}}{2}$ **21.** $\dfrac{-1 \pm \sqrt{57}}{4}$ **23.** $\frac{5}{2}, -1$

25. $\dfrac{-3 \pm \sqrt{65}}{4}$ **27.** $\dfrac{-2 \pm \sqrt{10}}{3}$ **29.** $\dfrac{1 \pm \sqrt{11}}{2}$ **31.** $0, \dfrac{-3 \pm \sqrt{41}}{4}$ **33.** $0, \frac{4}{3}$

35. $\dfrac{3 \pm \sqrt{21}}{6}$ **37.** $-2x + 5$ **39.** $6x^2 + 8x$ **41.** $\dfrac{3 - \sqrt{5}}{2}$

PROBLEM SET 8.4, P. 293

1. $3 + i$ **3.** $6 - 8i$ **5.** 11 **7.** $9 + i$ **9.** $-1 - i$ **11.** $11 - 2i$ **13.** $4 - 2i$
15. $8 - 7i$ **17.** $-1 + 4i$ **19.** $6 - 3i$ **21.** $14 + 16i$ **23.** $9 + 2i$ **25.** $11 - 7i$

27. 34 **29.** 5 **31.** $\dfrac{6 + 4i}{13}$ **33.** $\dfrac{-9 - 6i}{13}$ **35.** $\dfrac{-3 + 9i}{5}$ **37.** $\dfrac{3 + 4i}{5}$

39. $\dfrac{-6 + 13i}{15}$ **41.** $x^2 + 9$ **45.** $-2, 8$ **47.** $1, 5$ **49.** $-3 \pm 2\sqrt{3}$

PROBLEM SET 8.5, P. 296

1. $4i$ **3.** $7i$ **5.** $i\sqrt{6}$ **7.** $i\sqrt{11}$ **9.** $4i\sqrt{2}$ **11.** $5i\sqrt{2}$ **13.** $2i\sqrt{2}$

15. $4i\sqrt{3}$ **17.** $2, 3$ **19.** 2 **21.** $-4, \frac{3}{2}$ **23.** $2 \pm 2i$ **25.** $\dfrac{-1 \pm 3i}{2}$

27. $\dfrac{1 \pm i\sqrt{3}}{2}$ **29.** $\dfrac{-1 \pm i\sqrt{3}}{2}$ **31.** $2, 3$ **33.** $\dfrac{-1 \pm i\sqrt{2}}{3}$ **35.** $\frac{1}{2}, -2$

37. $\dfrac{1 \pm 3\sqrt{5}}{2}$ **39.** $\dfrac{-11 \pm \sqrt{33}}{4}$ **41.** Yes   **43.** $3 - 7i$

45.

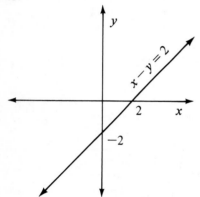

47.

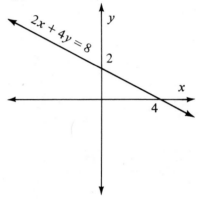

49.

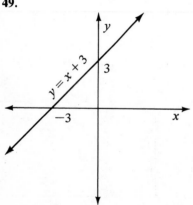

PROBLEM SET 8.6, P. 301

1.

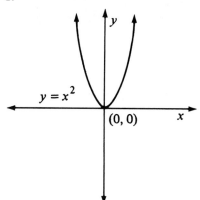

$y = x^2$ (0, 0)

3.

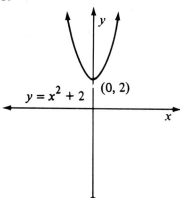

$y = x^2 + 2$ (0, 2)

5.

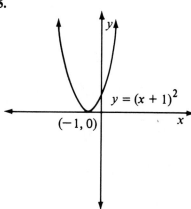

$y = (x + 1)^2$ (−1, 0)

7.

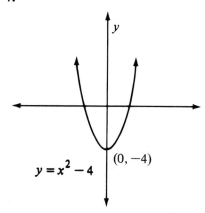

$y = x^2 - 4$ (0, −4)

9.

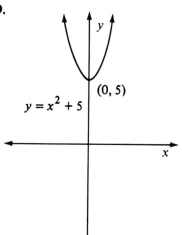

$y = x^2 + 5$ (0, 5)

11.

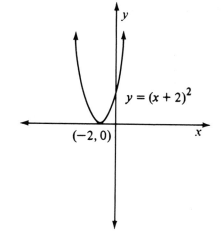

$y = (x + 2)^2$ (−2, 0)

13.

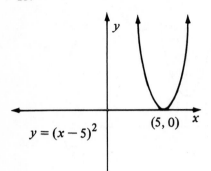

$y = (x - 5)^2$

$(5, 0)$

15.

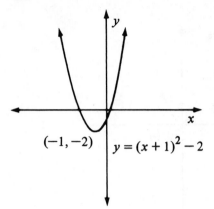

$(-1, -2)$ $y = (x + 1)^2 - 2$

17.

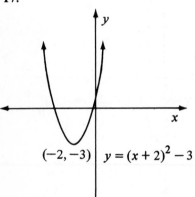

$(-2, -3)$ $y = (x + 2)^2 - 3$

19.

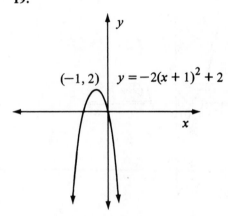

$(-1, 2)$ $y = -2(x + 1)^2 + 2$

21.

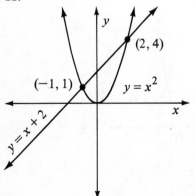

$(2, 4)$

$(-1, 1)$ $y = x^2$

$y = x + 2$

23.

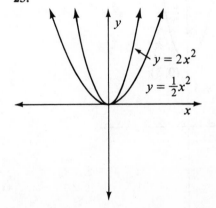

$y = 2x^2$

$y = \frac{1}{2}x^2$

25. $\dfrac{1}{x-3}$ **27.** $\dfrac{x+2}{x+4}$ **29.** $\dfrac{x+6}{x-6}$

CHAPTER 8 TEST,   P. 304

1. $\dfrac{7 \pm \sqrt{17}}{2}$ **2.** $3 \pm 2\sqrt{3}$ **3.** $\dfrac{5 \pm 5i\sqrt{3}}{2}$ **4.** $3 \pm \sqrt{15}$ **5.** $\dfrac{3 \pm i\sqrt{31}}{4}$

6. $\frac{1}{3}, -1$ **7.** $\dfrac{-1 \pm \sqrt{33}}{2}$ **8.** $-\frac{2}{3}$ **9.** $3i$ **10.** $11i$ **11.** $6i\sqrt{2}$ **12.** $3i\sqrt{2}$

13. $3 + 8i$ **14.** $-1 + 2i$ **15.** 5 **16.** $1 + 5i$ **17.** $\dfrac{-1 + 3i}{10}$ **18.** $\dfrac{3 + 4i}{5}$

19.

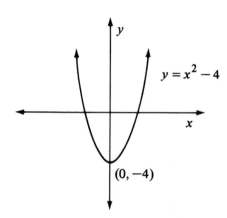

20.

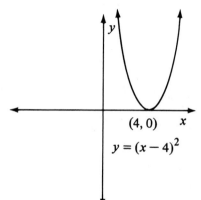

CHAPTER 9

PROBLEM SET 9.1,   P. 312

1. 2 **3.** 1 **5.** 2 **7.** 1 **9.** -1 **11.** $\frac{7}{8}$ **13.** $\frac{5}{2}$ **15.** 0 **17.** $\dfrac{b-d}{a-c}$

19. $\dfrac{1}{2b}$ **21.** $x = 3$ **23.** $y = 3$ **25.** $x = 2, \ y = -4$ **27.** $x = 2, y = -5$

29. $x = 2, y = -4$ **31.** $x = 5, y = -4$ **33.** $x = 4, y = -4$ **35.** $x = 3, y = -6$
37. $x = -\frac{1}{2}, y = 1$ **39.** $x = \frac{7}{4}, y = -7$ **41.** $x = -\frac{1}{2}, y = -1$

43.

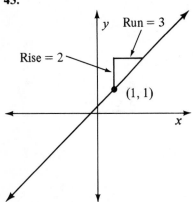

45.

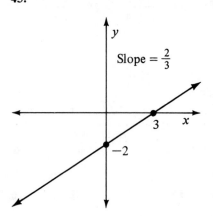

47. -11 **49.** 2   **51.** 3   **53.** $-2x$ **55.** $-3, 0, 2$

PROBLEM SET 9.2,   P. 316

1. $y = 5x - 13$ **3.** $y = 4x - 16$ **5.** $y = -2x + 9$ **7.** $y = x - 4$
9. $y = -2x - 15$ **11.** $y = \frac{1}{2}x + \frac{7}{2}$ **13.** $y = -\frac{4}{3}x + \frac{7}{3}$ **15.** $y = 5x - \frac{43}{3}$
17. $y = 2x + 1$ **19.** $y = -4x + 2$ **21.** $y = -x - 1$ **23.** $y = x + \frac{2}{3}$
25. $y = -\frac{1}{3}x - \frac{1}{2}$ **27.** $y = -\frac{3}{4}x - \frac{1}{4}$ **29.** $m = 2, b = 5$ **31.** $m = 5, b = -1$
33. $m = -3, b = -2$ **35.** $m = 5, b = -4$ **37.** $m = -2, b = 4$ **39.** $m = -\frac{2}{3}, b = 2$
41. $m = -\frac{2}{3}, b = \frac{5}{3}$ **43.** $y = 4x - 7$ **45.** $y = \frac{3}{2}x + 3$ **47.** -2 **49.** -4

51. $x < 4$

53. $x \geq -5$

55. $-1 \leq x \leq 2$

PROBLEM SET 9.3,   P. 319

1. $D = \{1, 3, 5\}, R = \{2, 4, 6\}$ **3.** $D = \{1, 3, 4\}, R = \{a, b, c\}$
5. $D = \{1, 2, 3, 4\}, R = \{1, 2, 3, 4\}$ **7.** $D = \{1, 3, 5, 7, 9\}, R = \{2, 8, 14, 20, 26\}$
9. $D = \{0, 2, 4\}, R = \{2, \frac{2}{3}, -\frac{2}{3}\}$ **11.** $D = \{3, 4, 5, 6, 7\}, R = \{2, 4, 6, 8, 10\}$ **13.** 2
15. -3 **17.** $-1, -2$ **19.** $-5, 1$ **21.** $3, -3$ **23.** $2, 3$ **25.** $-\frac{1}{2}, 3$
27. $0, 2, 4$ **29.** $-3, -5, -6, (a - 5)$ **31.** $-1, 1, -2, (2z - 3)$ **33.** $5, 10, 17, (a^2 + 1)$
35. $-1, -\frac{1}{2}, \frac{1}{3}, 2, \frac{3}{4}$ **37.** $15, 0, (a^2 + 2a), (a^2 - 2a)$ **39.** $0, 7, -28, (a^3 - 1)$ **41.** $(3, 1)$
43. Lines are parallel; $\emptyset$ **45.** $(3, 17)$

47.

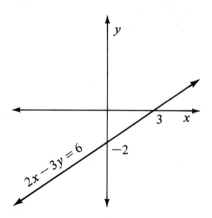

PROBLEM SET 9.4,   P. 324

1. 2   **3.** 4   **5.** 3   **7.** 5   **9.** 3   **11.** 9   **13.** 4   **15.** 25   **17.** 8
19. 64   **21.** 8   **23.** 4   **25.** -2 **27.** 5   **29.** 17   **31.** 6   **33.** x **35.** x^2

37. $a^{2/5}$ **39.** $3y^2$ **41.** $3a^2b$ **43.** x **45.** $\frac{1}{5}$ **47.** $\frac{1}{4}$ **49.** $\frac{1}{9}$ **51.** $\frac{5x^3}{y^4}$

53. $6x^4 + 6x^3 - 2x^2$ **55.** $10x^2 + 11x - 6$ **57.** $a^3 - 27$

PROBLEM SET 9.5,   P. 327

1. 3.75×10^3 **3.** 7.63×10^8 **5.** 2.5×10^1 **7.** 4×10^0 **9.** 3×10^{-3}
11. 4.8×10^{-2} **13.** 7×10^{-10} **15.** 4230   **17.** 5,800,000   **19.** 24   **21.** 6
23. 0.071   **25.** 0.00000411   **27.** 0.4   **29.** 6×10^8 **31.** 1.75×10^{-1}
33. 1.21×10^{-6} **35.** 9×10^{10} **37.** 1.5×10^2 **39.** 4.2×10^3 **41.** 3×10^{10}
43. 5×10^{-3} **45.** 6×10^{-2} **47.** 6×10^{-10} **49.** 2×10^6 **51.** 1×10^1
53. 4.2×10^{-6} **55.** 6×10^4 **57.** $2 \cdot 3^2 \cdot 5 \cdot 7$ **59.** $(x - 9)(x + 9)$

61. $(2x + 7)(x - 3)$ **63.** 3, 6   **65.** $\frac{-3}{2}, \frac{5}{4}$

PROBLEM SET 9.6,   P. 332

1. $\frac{4}{3}$ **3.** $\frac{4}{5}$ **5.** $\frac{8}{1}$ **7.** $\frac{3}{5}$ **9.** $\frac{8}{9}$ **11.** $\frac{5}{8}$ **13.** $\frac{2}{3}$ **15.** $\frac{12}{11}$ **17.** $\frac{22}{125}$ **19.** $\frac{1}{6}$
21. 1   **23.** 10   **25.** 40   **27.** $\frac{5}{4}$ **29.** $\frac{7}{3}$ **31.** $\frac{15}{11}$ **33.** $-4, 3$ **35.** $-4, 4$
37. $-4, 2$ **39.** $-1, 6$ **41.** 15 hits   **43.** 21 milliliters alcohol   **45.** 45.5 grams of fat

47. 14.7 inches   **49.** 343 miles   **51.** $\frac{x + 2}{x + 3}$ **53.** $\frac{2(x + 5)}{x - 4}$ **55.** $\frac{1}{x - 4}$

57. $\frac{x + 5}{x - 3}$

PROBLEM SET 9.7, P. 339

1. 8   **3.** 130   **5.** 5   **7.** -3 **9.** 3   **11.** 243   **13.** 2   **15.** $\frac{1}{2}$ **17.** 1
19. 5   **21.** 25   **23.** 9   **25.** 84 pounds   **27.** 367.5   **29.** \$235.50
31. 96 pounds   **33.** 12 amps   **35.** 7   **37.** $5\sqrt{2}$ **39.** $21\sqrt{3}$ **41.** $-4 - 3\sqrt{6}$
43. $4\sqrt{5} + 4\sqrt{3}$

PROBLEM SET 9.8, P. 343

1. $l = \dfrac{A}{w}$ **3.** $r = \dfrac{d}{t}$ **5.** $h = \dfrac{V}{lw}$ **7.** $R = \dfrac{V}{I}$ **9.** $P = \dfrac{nRT}{V}$ **11.** $R = \dfrac{PV}{nT}$

13. $a^2 = c^2 - b^2$ **15.** $a = p - b - c$ **17.** $l = \dfrac{p - 2w}{2}$ **19.** $C = (\frac{5}{9})(F - 32)$

21. $b = \dfrac{2A}{h} - B$ **23.** 25° Celsius   **25.** 24 square inches   **27.** $-3, 2$ **29.** $\dfrac{2 \pm \sqrt{10}}{3}$

31. $2 + 8i$ **33.** $23 - 2i$

35.

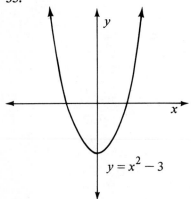

$y = x^2 - 3$

CHAPTER 9 TEST, P. 346

1.

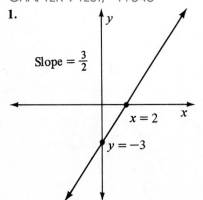

Slope $= \dfrac{3}{2}$

$x = 2$

$y = -3$

2.

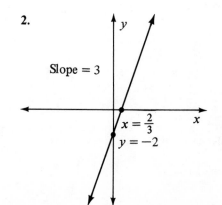

Slope $= 3$

$x = \dfrac{2}{3}$

$y = -2$

3.

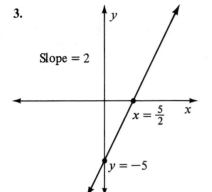

Slope = 2

$x = \frac{5}{2}$

$y = -5$

4.

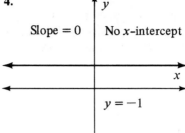

Slope = 0 | No x-intercept

$y = -1$

5. $y = 3x + 11$ **6.** $y = 4x + 8$ **7.** $y = -\frac{3}{5}x + \frac{14}{5}$ **8.** $y = 2x - 2$ **9.** (a) 26
(b) 11   (c) $3a^2 - 1$ (d) 2   (e) -23 (f) $5b + 2$ **10.** (a) 4   (b) 19   (c) $3^{-1/6}$
11. (a) 7.8×10^4 (b) 3.97×10^{-3} **12.** (a) 8×10^2 (b) 6×10^{12} **13.** (a) $\frac{5}{2}$ (b) $\frac{14}{3}$
14. 100   **15.** 2   **16.** $F = \frac{9}{5}C + 32$

Index

FORMULAS FROM GEOMETRY

RECTANGLE

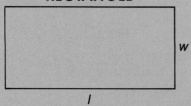

Area $= A = lw$
Perimeter $= P = 2l + 2w$

PARALLELOGRAM

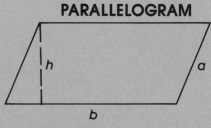

Area $= A = bh$
Perimeter $= P = 2a + 2b$

TRAPEZOID

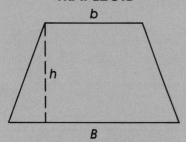

Area $= A = \frac{1}{2}(b + B)h$

CIRCLE

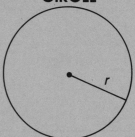

Area $= A = \pi r^2$
Circumference $= C = 2\pi r$

TRIANGLE

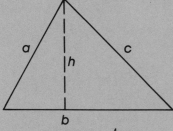

Area $= A = \frac{1}{2}bh$
Perimeter $= P = a + b + c$

RIGHT TRIANGLE

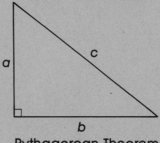

Pythagorean Theorem
$c^2 = a^2 + b^2$